IMPLEMENTING THE CIRCULAR ECONOMY FOR SUSTAINABLE DEVELOPMENT

IMPLEMENTING THE CIRCULAR ECONOMY FOR SUSTAINABLE DEVELOPMENT

HANS WIESMETH

TU Dresden, Faculty of Economics, Dresden, Germany

ELSEVIER

Elsevier
Radarweg 29, PO Box 211, 1000 AE Amsterdam, Netherlands
The Boulevard, Langford Lane, Kidlington, Oxford OX5 1GB, United Kingdom
50 Hampshire Street, 5th Floor, Cambridge, MA 02139, United States

Notices
Knowledge and best practice in this field are constantly changing. As new research and experience broaden our understanding, changes in research methods, professional practices, or medical treatment may become necessary.

Practitioners and researchers must always rely on their own experience and knowledge in evaluating and using any information, methods, compounds, or experiments described herein. In using such information or methods they should be mindful of their own safety and the safety of others, including parties for whom they have a professional responsibility.

To the fullest extent of the law, neither the Publisher nor the authors, contributors, or editors, assume any liability for any injury and/or damage to persons or property as a matter of products liability, negligence or otherwise, or from any use or operation of any methods, products, instructions, or ideas contained in the material herein.

Library of Congress Cataloging-in-Publication Data
A catalog record for this book is available from the Library of Congress

British Library Cataloguing-in-Publication Data
A catalogue record for this book is available from the British Library

ISBN: 978-0-12-821798-6

For information on all Elsevier publications
visit our website at https://www.elsevier.com/books-and-journals

Publisher: Candice Janco
Acquisitions Editor: Peter Llewellyn
Editorial Project Manager: Devlin Person
Production Project Manager: Paul Prasad Chandramohan
Cover Designer: Greg Harris

Typeset by SPi Global, India

Working together
to grow libraries in
developing countries

www.elsevier.com • www.bookaid.org

Contents

V

Implementing a circular economy

VI

Concluding remarks

1

Introduction

The circular economy respects and sustainably preserves the fundamental functions of the environment as supplier of natural resources, as recipient of all types of waste, and as direct provider of utility. Since countries differ in terms of natural resources, size, demographic, climatic and economic conditions, "the" circular economy is likely to vary from country to country – similar to the regular economy. In view of these differences, it is therefore advisable to speak of the implementation of "a" circular economy.

What distinguishes a circular economy from a regular market economy? How do we implement a circular economy? And, of course, what are appropriate instruments in this regard? These are just some of the questions, which are asked in this book and to which an answer is given.

1.1 Characterisation of the circular economy

This section briefly characterises the circular economy, the concept and its origins. The questions raised above will be covered in the following subsections.

1.1.1 What distinguishes a circular economy?

In a pure market economy, there are individual decisions on production and consumption, which depend on the local situation. Interestingly, under certain conditions, the "invisible hand", the famous metaphor of A. Smith, will lead individual decisions to an efficient, optimal result. One of the reasons for this optimality outcome is the personal interest of consumers and producers in combination with exclusivity: other agents can be excluded from the benefits of consumption or production of these commodities.

Unfortunately, this result, which explains part of the success of market economies, does not allow an immediate extension to environmental commodities. These environmental commodities, such as reductions of all kinds of air, soil, water and atmospheric pollutants, are certainly beneficial for all people. However, if someone reduces air pollution, for example, then all other economic agents will also benefit from such an environmentally friendly action, exclusivity is no longer available, and people will wait for others to take the first step towards protecting the environment.

Implementing the Circular Economy for Sustainable Development
https://doi.org/10.1016/B978-0-12-821798-6.00001-6

In addition, environmentally friendly designs, known as "designs for environment" (DfE), need not be in the commercial interest of producers. They, however, have the expertise for such design changes. Reusing certain commodities to extend their lifespan and thereby prevent waste can also reduce revenue and profit for businesses. With regard to the implementation of a circular economy, this means that not all producers are likely to be enthusiastic supporters. In general, mechanisms such as the Tragedy of the Commons and the Prisoners' Dilemma influence relevant decisions of consumers and producers, and information asymmetries and a lack of information characterise all kinds of decisions in such an environmental context. "Economy-guidance", using the market system for implementing a circular economy, thus faces various challenges.

For the implementation of a circular economy, it is therefore necessary to look at "societal path dependencies", such as a focus on profitable business activities. Of course, implementing a circular economy does not mean at all that we should "dispose" of the market economy. On the contrary, we should use the basic principles of a market economy such as decentralised decision-making whenever possible. The shortcomings of the market mechanism in the context of a circular economy must, however, be taken into account.

This sounds simple: take the market mechanism with some addendums or modifications, apply it to environmental commodities that are relevant to a circular economy, and that is it. Unfortunately, there are a number of other issues, which need to be respected.

1.1.2 How to implement a circular economy?

What is the best way to sustainably respect and preserve the fundamental functions of the environment? The book focuses on all aspects of waste management, on the implementation of the waste hierarchy, more precisely. As usual, waste prevention leads the waste hierarchy, followed in its simplest form by reuse and recycling.

Because there is usually limited knowledge on the assimilative capacity of the environment regarding waste, the priority goal should indeed be waste prevention. Moreover, the more waste we generate, the more stays in the environment. Plastic waste everywhere and micro-plastics in the food chain are proof of this. Also the reuse of commodities can prevent waste, and both goals together help in addition to save natural resources. Recycling of waste should be the final step with recovery of resources and energy. Nevertheless, recycling reduces the amount of waste to be landfilled and thus supports the assimilative capacity of the environment.

The sustainable implementation of the waste hierarchy helps to save natural resources and to maintain the assimilative capacity of the environment to receive waste. This includes greenhouse gases that appear to exceed the amount the atmosphere can absorb without rising global temperatures. "Recycling" of greenhouse gases in carbon sinks such as forests plays a certain role, but the focus is more on prevention.

The function of the environment as direct provider of utility should not be forgotten. However, to prevent waste, save resources and reduce landfilling can also keep intact an environment that is less disturbed by excessive and environmentally harmful mining or similarly problematic landfilling.

To sum up, the sustainable implementation of the waste hierarchy is of great importance for achieving the objectives of a circular economy in terms of the fundamental functions of the environment. It is, of course, possible to support circular economy activities through additional measures: establishing smart cities, meeting the Sustainable Development Goals, implementing the European Green Deal and others. Such activities can also help to raise awareness of the need for a circular economy.

Here, too, implementing the waste hierarchy seems to be simple: most people want to get rid of waste, they do not want to be bothered by it. Unfortunately, however, this attitude and its possible effects do not necessarily correspond to the goals of the waste hierarchy. Disposing of waste in a bin is not synonymous with the prevention of waste in the sense of the waste hierarchy. However, quite a few people regard waste, which is collected, perhaps even recycled or disposed of as residual waste in landfills as avoided or prevented waste. A societal path dependency leads to this assessment, which is hardly compatible with the objectives of a circular economy.

Therefore, respecting and preserving the waste hierarchy in a sustainable way is not an easy task. The following section highlights some aspects of appropriate tools for implementing the waste hierarchy and a circular economy.

1.1.3 What are the appropriate instruments?

In view of the above comments on the market system, additional mechanisms are required for implementing a circular economy. Since a circular economy with all its measures to implement the waste hierarchy appears to depend on appropriate technologies and their further development, some guidance through technologies seems to be adequate.

According to the Ellen MacArthur Foundation, one of the roots of the circular economy concept is in industrial ecology, which "aims at creating closed-loop processes in which waste serves as an input, thus eliminating the notion of an undesirable by-product". With its systemic point of view, "designing production processes in accordance with local ecological constraints whilst looking at their global impact from the outset, … ", industrial ecology could certainly provide support for the implementation of a circular economy, could thus provide "technology-guidance".

However, a more careful analysis shows that, again due to information asymmetries and a lack of information, technology-guidance can

challenge and produce results, which are not in line with a circular economy. Environmental technologies such as collection systems and recycling technologies, as well as collection and recycling targets, should be "economically reasonable". This property depends not only on technological aspects, but also on the local situation of an economy. Moreover, there are the rebound effects, which can significantly weaken the environmental impact of new technologies, and there are again societal path dependencies that seem to motivate, for example, the expansion of recycling waste at the expense of waste prevention.

Consequently, as neither economy-guidance nor technology-guidance alone can sufficiently support the introduction of a circular economy, a combination of policy tools seems necessary to guide relevant decisions in an appropriate way. In shaping these policies, preference should be given to decentralised decision-making whenever possible. This enables the use of individual knowledge and expertise, similar to a market economy. The technological framework must thereby be respected and the policy should incentivise DfEs or other pro-environmental technologies. Again due to a lack of knowledge, the policy tools must be appropriately linked. If, for example, producers are financially responsible for the collection and recycling of their products, then the incentives for a DfE increase with high collection rates. In this sense, the motivation of consumers to separate waste is linked to the motivation of producers for a DfE.

This holistic approach results in "Integrated Environmental Policies" (IEP), characterised by "constitutive elements" that relate the policies to important principles and features of the market mechanism. One of the principles, the locality principle, seems natural and easy to implement. However, the requirement of carefully taking into account the local situation when designing an IEP, poses immediate challenges: how to respect this principle, this constitutive element, in international contexts such as climate change mitigation or the reduction of pollution of rivers and seas with plastic waste?

Of course, the book presents examples of such IEPs for different types of waste. Due to different information requirements it is, however, necessary to draft different policies for the different waste streams, with the constitutive elements providing guidance.

The book deals in detail with all these questions and familiarises the reader with the concept of the circular economy, in its relation to the technological environment, but with a clear focus on the economic context. Moreover, the reader will find the tools needed to design an IEP, clues as how to establish the necessary links between the policy tools and how to reduce the possibilities of vested interests interfering with the objectives of the IEP.

The following section provides an outline of the book and gives an overview on the topics covered in relation to the general objective of implementing a circular economy.

1.2 The outline of the book

The book is structured into six parts, focusing on different aspects of the implementation of a circular economy. To strengthen the practical context, various case studies provide further insights from practice. Here, too, the focus is on the economic context:

- The refillable quota issue in Germany (Section 7.3)
- Sustainable use of the earth's biodiversity (Section 8.4)
- Promoting renewable energy sources in Germany (Section 11.3)
- E-commerce and circular economy (Section 13.3)
- Emission standards for vehicles (Section 15.4)
- Germany on the road to a circular economy (Section 19.4)

The following subsection introduces the parts of the book and briefly presents some of the aspects covered in the chapters.

1.2.1 Part I: The circular economy – Concept and facts

The first part of the book introduces the concept of the circular economy, both from an academic and a practical point of view. Some remarks on the relevance of societal path dependencies already point to challenges in the implementation of a circular economy at this stage.

Thereafter, we examine the perception of the circular economy in the extensive literature. Business models, a sustainable development, environmental innovations, and regenerative systems play a role in the literature. The perception of the circular economy in a practical context is of relevance because, as already indicated, practitioners had and continue to have an important influence on shaping the concept. The Ellen MacArthur Foundation with its mission "to accelerate the transition to a circular economy" has to be mentioned in this context, but also the Circular Economy Package of the European Union with its detailed recommendations for what needs to be done for the transition to a circular economy. As an example for a variety of other initiatives, the Russian TIARCENTER is presented as an independent think tank and advisory firm, which offers "strategic advice to corporations and government bodies on the sustainable development principles implementation".

The hierarchy of circular economy leaders and followers is the topic of the next chapter. Reasons for a country to early adopt circular economy strategies have to be found in the local situation regarding abundance of natural resources, perhaps also space for landfilling, or rather for hiding waste. Of importance is, however, also the level of environmental awareness, which is likely to depend on economic wellbeing. Germany and China are then portrayed as circular economy leaders, while Russia and Georgia seem to be, for different reasons, late-movers regarding a circular economy. The United States (U.S.) should be seen as "a country in between": there is potential, there are many local initiatives, but a coherent strategy on the federal level seems to be lacking. Strategic considerations for the positioning of a country can, of course, also play a role.

The final chapter of this part examines various environmental regulations in the context of a circular economy with regard to achieving its objectives. The discussion of environmental policies of the European Union (EU), Germany and others points to shortcomings, which need to be investigated in view of designing IEPs.

1.2.2 Part II: Integrating the economy and the environment

The second part of the book deals comprehensively with the economic context of a circular economy. The economic foundation introduces environmental commodities, but also the very important concepts of a "perceived scarcity" of these commodities, and of a "perceived feedback" from environmentally friendly actions. It then introduces the basic structures of the market mechanism, with a view to other allocation mechanisms, such as central planning.

The important aspect of the allocation of environmental commodities is addressed next with a focus on the mechanisms of the Tragedy of the Commons and the Prisoners' Dilemma. These two mechanisms have a decisive influence on many issues of relevance for implementing a circular economy.

Undoubtedly, behavioural economics, in particular behavioural environmental economics, has a lot to offer with respect to a transition to a circular economy. Indeed, it seems necessary to establish appropriate social norms, such as

the correct separation of waste, to give just one example. Of course, these attitudes can significantly support the development of a circular economy and need to be cultivated.

The final chapter in this part refers to the allocation problems in a circular economy, the concept of a sustainable development and waste management from an economic point of view and introduces the waste hierarchy.

1.2.3 Part III: The circular economy in a technological context

This part examines the technological framework of a circular economy. The first reference is, of course, to industrial ecology, in particular its perception in theory and practice. In this context, the markets for environmental technologies, especially the global markets, need to be considered. Given the aspect of technology-guidance, export promotions for specific technologies, which are themselves related to societal path dependencies with regard to international trade, can establish further technological path dependencies. The example of promoting e-mobility in various countries could belong to this category.

Information, rather its absence, and information asymmetries are considered in the next chapter. In particular the issue of the "economic reasonableness" of certain technologies and environmental standards requires some attention. The important fact is that this concept depends on the local situation, implying also that a technology, which is reasonable in this sense in one country need not be economically reasonable in another one. However, the practical application of this concept depends on quite extensive cost-benefit analyses. The chapter discusses also information asymmetries regarding a DfE and the consequences for implementing a circular economy.

Rebound effects characterise the reaction of consumers on new environmentally friendly technologies such as energy-efficient household appliances. The rebound effects are classified, including psychological rebound effects, and empirical results are presented, also in their relationship to societal and technological path dependencies.

Finally, the digital transformation is addressed as an ongoing process. The example of a smart city is discussed for the relation of the digital economy to a circular economy. The chapter refers also to the challenges the digital economy poses for a circular economy, in particular to various aspects accompanying the sharing economy and online shopping.

1.2.4 Part IV: Features of environmental policies

This part moves gradually to the environmental policies, which are needed for implementing the waste hierarchy and the circular economy. At first, the question of how to allocate environmental commodities is asked once more, thereby pointing to various policy tools, such as laissez-faire, technology-guidance, command-and-control policies, and market-oriented policies. These tools, which are briefly discussed in the first chapter, are considered as possible components of more complex policies. In fact, with some structural requirements, the already mentioned constitutive elements, these tools can be combined to holistic environmental policies. Special adaptations will then lead to the IEPs for the implementation of a circular economy in Part V.

Environmental standards replace the generally unknown efficient levels of the environmental commodities and are therefore of high relevance in all kinds of environmental policies. However, due to the locality principle, they represent a challenge in an international context. Moreover, the question of how and to what extent these environmental standards are to be raised is important for the reduction of pollution, but is at the same time difficult to answer due to a lack of information.

Market-oriented policy tools are discussed in more detail in the following chapter. These tools are important for implementing a circular economy because they are characterised by decentralised decision-making, enabling the use of individual information. In particular the pollution tax, together with the polluter pays principle and the role of avoidance possibilities are presented. In addition, markets for tradable emission certificates are of relevance for climate protection policies. There are also brief references to the Coase Theorem, to voluntary contributions in the environmental context and to flexible, information-based policies.

Thereafter, holistic policy approaches are considered. The EPR principle, the principle of extended producer responsibility, is of importance for these policies. This principle refers to the end of a product's life, which must already be taken into account when designing the product. After briefly reviewing the constitutive elements, the chapter discusses various collection systems and introduces then producer responsibility organisations (PRO), which are of relevance for implementing the EPR principle, for making various holistic policy approaches functional.

Because of its importance for the implementation of a circular economy, the last chapter of this part examines the economics of the waste hierarchy. After a review of the literature, the focus is on the priority goal of waste prevention, the "forgotten child". Of course, also the reuse of old and recycling of waste commodities are considered, including relations with societal and technological path dependencies.

1.2.5 Part V: Implementing a circular economy

The ground is now prepared for designing appropriate environmental policies, IEPs, for the implementation of a circular economy. The first chapter of this part briefly reviews relevant issues to note where we are on the road to the circular economy. The roots of the circular economy, the role of economics and environmental technologies including holistic policies are once more discussed.

IEPs are then proposed for various waste streams. For each case, there are references to currently existing practical policies, to their shortcomings regarding the objectives of a circular economy. Then, with the exception of the policy for mitigating climate change, appropriate collection systems based on laissez-faire, on technology-guidance, on refunds or a deposit system are presented. Subsequently, the implementation of the EPR principle is recommended through a system of independent PROs, compliance schemes in competition. For mitigating climate change, markets for tradable certificates are proposed for the specific environmental commodity "reduction of greenhouse gas emissions". The aim of all these constructions is always to reconcile the interests of consumers and producers with the objectives of a circular economy. Given the various possibilities for vested interests, this requires some consideration.

Policies for the following environmental areas are designed:

- Packaging waste
- Waste electrical and electronic equipment
- End-of-life vehicles
- Climate change mitigation
- Plastic waste
- Textile waste

With these examples, it should then be possible to develop IEPs for other areas of relevance for a circular economy and which are not covered in this book, such as food waste.

1.2.6 Part VI: Concluding remarks

There is just one chapter in this last part of the book: a summary on the circular economy, also with regard to the possible consequences of the Corona crisis. The summary reiterates the relevance of the waste hierarchy for implementing a circular economy, and reviews important characteristics of the IEPs designed in Part V. Afterwards, some challenges and opportunities arising from the Corona crisis are presented. On the one hand, the crisis offers the possibility of a systems change, which is also necessary for the transition to a circular economy. On the other hand, however, the economic recession accompanying this crisis seems to divert interest from environmental issues. It remains to be seen, whether, in the end, there will be a systems change, a reorientation of societal path dependencies with a greater focus on the implementation of a circular economy.

The circular economy – Concept and facts

This first part of the book introduces in Chapter 2 the circular economy as the natural economic system which respects the fundamental functions of the environment as supplier of natural resources, receiver of waste, and as direct provider of utility. This leads then to a generic definition of "the" circular economy, which leaves enough room for country-specific implementations of "a" circular economy – taking into account local framework conditions.

Important roots of the concept, both in the more theoretic, economic, but also in the more practical, applied context are investigated, pointing to roots in environmental economics on the one hand and in more technical fields such as industrial ecology on the other. In each case the focus is on observance of the waste hierarchy and sustainability with assigning technical and technological issues somewhat different roles in the two approaches. A further remark refers, already at this stage, to societal path dependencies, which may impede the implementation of a circular economy.

Chapter 3 considers different perceptions of a circular economy in literature and practice. The review reveals some shifts in the understanding of the concept, in particular regarding the waste hierarchy. Various contributions emphasise the role of appropriate business models for implementing a circular economy. Potential business opportunities are propagated on a large scale with the Circular Economy Package of the European Union, the European Green Deal or the mission of the Ellen MacArthur Foundation. Other perceptions of a circular economy focus more on innovative technologies or, more generally, on a sustainable development. The chapter then presents and discusses some practical approaches: of the Ellen MacArthur Foundation, the European Union with its Circular Economy Package, and the Russian TIARCENTER.

Chapter 4 addresses the fact that some countries adopt circular economy strategies earlier than others. What are reasons for this observation, generating circular economy leaders and followers? The investigation points to the substantial influence of availability of natural resources and land for dumping waste, but soft factors such as environmental awareness, to some extent depending on economic wealth, seem to play a role, too. Strategic behaviour regarding first-mover or late-mover advantages are of relevance in addition. The detailed situation of various countries: Germany and China as leaders, Russia and Georgia as followers, and the United States of America as a special case are investigated in more detail.

The last chapter of Part I analyses various environmental regulations with a focus on implementing a circular economy. The analysis shows that these regulations have deficiencies as their goals are not really achieved. Individuals obviously react on these regulations and the available technologies in not always completely predictable ways. Thus these case studies allow valuable insight into the interaction of individuals with environmental regulations and environmental technologies.

Summarising, this first part of the book lays the foundation for the further investigations. There is a formal definition of the circular economy, there is a review of various perceptions, there is an investigation of leaders and followers, and there are experiences with attempts to implement a circular economy. All these aspects need to be explored and exploited in the following parts and chapters, before the issue of implementing a circular economy can be fully addressed in Part V.

2

The circular economy – Understanding the concept

Understanding the circular economy means understanding the perceptions of the concept. The more theoretical, academic perception has one of its strongest roots in the sub-discipline of environmental economics, orientating therefore more on the behaviour of the economic agents, consumers and producers, in particular. The academic field of environmental economics emerged gradually in the 1960s, also as a response to the more and more noticeable discrepancies between the widely accepted social desideratum of a lasting economic growth, and a troubling and increasing environmental degradation, visible in many parts of the world: in developing and industrialised countries, in transition and emerging economies, in market and centrally planned economies – regardless of the political system. Disturbing reports, such as "Limits to Growth" by the Club of Rome (Meadows, Meadows, Randers, & Behrens, 1972), pointed in particular to the "complex of problems troubling men of all nations" (p. 9), likely constraining future living conditions, if nothing is done to address these problems.

The more technical, practice-oriented perceptions of the concept have roots in disciplines such as industrial ecology, focusing on material and energy flows in industrial systems, and are thus deeply grounded in science, technology and engineering. Environmental issues are often related to technological activities: waste and pollution are not accidents, but result from inappropriate designs. The Ellen MacArthur Foundation, established in 2010 "to accelerate the transition to a circular economy", refers to waste and pollution "as consequences of decisions made at the design stage, where around 80% of environmental impacts are determined". Thus it seems to be obvious that – according to this view – further advances in science and technology should play a major role in implementing a circular economy, which is based on the principles of "designing out waste and pollution, keeping products and materials in use, and regenerating natural systems".

The growing concern with air, soil and water pollution, and persistent environmental degradation in general, has kept motivating scientists from all areas, including economists and engineers, to reconsider and preserve the fundamental role the environment has in the economies of both the developed and the developing world. It is in this situation that the concept of the circular economy emerged.

The more or less simultaneous appearance of both theoretical and practical perceptions reveals the highly interdisciplinary character of the circular economy: economics, management, science, technology and engineering play a crucial role, in particular regarding the implementation of a circular economy. Whereas scientists and engineers provide insights and technologies, economists and managers have the explicit task to design and implement appropriate environmental policies, so-called "circular economy (CE) policies", which motivate consumers and producers to adequately support the goals of a circular economy. It will turn out that the interaction between humans and technology is not always without challenges for implementing a circular economy – pointing again to some discrepancies between the various perceptions of a circular economy.

This chapter introduces first a generic academic concept of the circular economy, and discusses thereafter aspects of prominent practical conceptualisations. Various publications, such as Heshmati (2015) and Antikainen, Lazarevic, and Seppälä (2018), provide further detailed remarks on the history of "circularity" in economics and especially in environmental economics. Ghisellini, Cialani, and Ulgiati (2016) present a survey on the origins of the circular economy (see, in particular, Fig. 1, p. 13), whereas Murray, Skene, and Haynes (2017) turn to the origins of the circular economy term itself (see p. 371), and Hartley, van Santen, and Kirchherr (2020) review briefly some definitions. On the other hand, Winans, Kendall, and Deng (2017) consider the history of the more practice-oriented concepts of the circular economy, and identify challenges and research gaps. The last section of this chapter points to societal path dependencies and their role in the context of implementing a circular economy.

The following section, based on Pearce and Turner (1989), presents the relevant interactions between the economy and the environment, thereby suggesting that a closed, circular structure is the native structure of an economic system, not the traditional open and linear structure. This leads, then, to the fundamental characteristics of "the" circular economy, providing the basic guidelines in this book for implementing "a" circular economy.

2.1 The academic concept of a circular economy

Pearce and Turner (1989) introduce and investigate the concept of the circular economy in their textbook, devoting a complete chapter to explaining "the fundamental ways in which consideration of environmental matters affect our economic thinking" (p. 29). They discuss the obvious interactions between the environment and the economy: the environment as a direct source of utility, as a supplier of natural resources for production, and as receiver of waste, which could not be assimilated or recycled by nature itself. Andersen (2007) additionally emphasises the function of the environment as a life-support system for the biosphere as a separate fourth item.

Ignoring these vital tasks of the environment with the possible consequences of a depletion of natural resources and the further more or less uncontrolled landfilling of waste, leaves us with the well-known paradigm of the linear economy (see Fig. 1.2 in Ellen MacArthur Foundation, 2017), which is, or rather was, based on the implicit assumption that the natural system, the environment, will continue to provide appropriate and necessary resources, and that it has the unlimited capacity to "digest" or assimilate all waste products – leading to the air, water and soil pollution accompanying the process of industrialisation in the developed world and the emerging economies, and spilling over to the developing countries by means of trade relations, tourism and their own desire and need for economic growth.

However, noticeable depletions of certain resources and the more and more recognisable environmental pollution with its adverse effects on the bio-system and even on the economic system, reveal the obvious capacity constraints of the environment to serve as a supplier of resources and as a repository of waste products, to assimilate waste and turn it back into harmless products. The focus must therefore shift to the excessive exploitation of resources and the generation of amounts of waste, which impair the interactions of the environment with the economy, thereby also harming the functioning of the economic system: "everything is an input into everything else" (Pearce & Turner, 1989, p. 37). This is then their motivation for referring to a closed and circular picture of an economy, "the circular economy", as the natural economic system. As the situation of the economy and thus also the situation of the environment result to a large extent from decisions of economic agents, it is the behaviour of these economic agents, mainly producers and consumers, that needs to be adjusted – certainly with adequate support from science and technology.

Pearce and Turner (1989) stress the relevance of the laws of thermodynamics and the materials balance model to justify this academic picture of a circular economic system (p. 40). They refer in particular to the Second Law of Thermodynamics to point out that not all waste can be recycled. Taking additionally into account the empirical observation that it is almost impossible or at least very costly to collect all waste leads quite naturally to the necessity to prevent waste as the priority task of the waste hierarchy (see Chapter 18), beyond that preventing waste also helps to save resources.

Without going into more details of these laws of physics (see, for example, Ayres, 1998), the consequences of these considerations are clear: for long-run economic growth, for "sustaining

an economy" (p. 41), it is of utmost importance to prevent waste and limit the uncontrolled disposal of waste in order to not endanger the assimilative capacity of the environment. In addition to that, it is necessary to observe and sustain the stocks of exhaustible resources and preserve the environment itself as a direct provider of utility, the other vital functions of the environment. Consequently, Pearce and Turner (1989) point to the importance of establishing conditions "for the compatibility of economies and their environments" for a sustainable development.

These basic interactions are indicated in Fig. 2.1, redrawn from Pearce and Turner (1989): resources are taken from the environment and used for the production of consumption commodities. At each stage of this process waste arises, which can be partially recycled to replenish resources, or ends up in the environment with its limited assimilative capacity. The circular economy does not exceed this assimilative capacity and not deplete natural resources. Pearce and Turner (1989) develop this figure further into more complex diagrams of the circular economy, pointing also to possible consequences of exceeding the limited assimilative capacity of the environment regarding waste

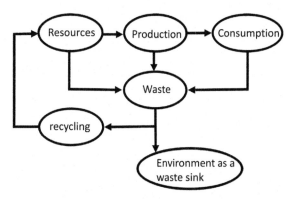

FIG. 2.1   A basic model of a circular system. Protecting the environment and saving resources through recycling. *Source: Own drawing after Fig. 2.2 in Pearce, D. W., & Turner, R. K. (1989). Economics of natural resources and the environment. Johns Hopkins University Press.*

(see Figs. 2.3 and 2.4). These consequences, not indicated in Fig. 2.1, include the potential negative feedback of excessive waste and excessive depletion of resources on production and consumption possibilities – on economic activities in general.

Thus, according to Pearce and Turner (1989), the academic concept of the circular economy points to the inherently circular structure of an economic system. It is, therefore, from a formal, theoretical point of view, quite natural to adopt the following generic concept of "the" circular economy:

Definition 2.1 In a generic sense, the circular economy fully respects the interdependencies between the environment and the economy and preserves the fundamental functions of the environment in a sustainable way.

In principle, implementing a circular economy in this generic sense means that economic activities should respect this particular structure, these interdependencies with the environment. The dominant issues are sustainability and all aspects of handling waste. Regarding the latter, waste prevention should lead the hierarchy, and by means of recovery and recycling, landfilling of waste should be reduced to a minimum. Sustainability should refer to saving and preserving resources, but also to observing the assimilative capacity of the natural environment.

To accomplish these tasks is, however, anything but simple: the interactions between the economy and the environment depend on the special situation in a particular country or region. Abundance of certain natural resources or ample availability of land for dumping waste on the one hand, or a high environmental awareness on the other, may lead to different perceptions regarding sustainability and suitable features of the circular economy to be considered for implementation. It is therefore, as already indicated, advisable to speak of the implementation of "a" circular economy in this case, referring to such

a specific conceptualisation. As these interdependencies may vary from region to region, this generic definition allows for many different interpretations and just as many different proposals for implementing a circular economy.

Already at this point it is necessary to draw the attention to increasingly interconnected economies: intensifying trade relations and tourism tend to internationalise initially local or regional environmental issues with immediate consequences for the degrees of freedom regarding the choice of circular economy strategies. Moreover, there are the inherently global issues such as climate change, which require a global approach in a circular economy context. As will be seen, these different perceptions regarding the concept and the implementation of a circular economy may and do create problems in a highly interconnected world.

These aspects are critical for all practical matters and will be further discussed (see Chapter 17, for example). The following section considers more technical, practice-oriented concepts of the circular economy.

2.2 Practice-oriented concepts of a circular economy

The Ellen MacArthur Foundation points to various "schools of thought", which have contributed "to refining and developing the circular economy concept". Among them, following the characterisation of the Ellen MacArthur Foundation, are the "cradle to cradle" concept with its special design philosophy, perceiving nature's biological metabolism as a model for a technical metabolism, and the "Performance Economy" with its closed loop approach to production processes with the goals of product-life extension, long-life goods, reconditioning activities, and waste prevention, and its focus on selling services, rather than products, in view of a sharing economy. "Biomimicry" is based on nature's best ideas, and imitates these ideas to solve human problems, and, finally, "industrial ecology" is the "study of material and energy flows through industrial systems". Fig. 1.4 in Ellen MacArthur Foundation (2017) characterises material, energy and resources flows in a circular economy.

These thoughts materialised and led to applications already in the late 1970s, before Pearce and Turner (1989) introduced their concept of the circular economy. This time lag might help to explain the dominant role science, technology and engineering play in many aspects of implementing a circular economy. Not surprisingly, this entails some interesting and far-reaching proposals for implementing a circular economy.

In the context of industrial ecology, for example, Lifset and Graedel (2002) point to the role of companies as a "locus of technological expertise". They are therefore "an important agent for accomplishing environmental goals" (see p. 8), who should thus have an intrinsic motivation for accomplishing environmental goals. This implies a more active, more cooperative role for business in an environmental context as a policy-maker rather than a policy-taker in "historic" command-and-control schemes. This throws then a new light on environmental policies, brings in more than a touch of behavioural economics, and poses interesting new questions regarding the design of policies, which ascribe companies a more active role and help to turn them into policy-makers by further developing their intrinsic motivation.

For Lifset and Graedel (2002) industrial ecology "focuses on product design and manufacturing processes" with "firms as agents for environmental improvement because they possess the technological expertise that is critical to the successful execution of environmentally informed design of products and processes" (see p. 3). There is thus a systems perspective, and eco-design, or a design for environment (DfE), is considered a "conspicuous element of industrial ecology" (see p. 7). Industrial ecologists rather incorporate environmental considerations into product and process design ex ante and thus seek to avoid environmental impacts (see p. 7). There remains, however, the important question for implementing a circular economy: how can firms be motivated to make optimal use of their technological expertise exactly to this regard?

This strong focus on technologies is also of relevance for the concept of a circular economy promoted by the Ellen MacArthur Foundation. By changing the mindset waste should be viewed as a "design flaw": with appropriate designs and new materials and technologies it should be possible to ensure that waste and pollution are not created in the first place. Moreover, according to this view, products should and could be designed such that they can be reused, repaired and remanufactured.

Regarding this last position, there is once more the immediate question about the implementation of a circular economy. If this technological approach to a circular economy is considered the heart of a circular economy, then not only this question arises, but also the more general question regarding the role of economics at all. As indicated earlier, Lifset and Graedel (2002) refer to the doubts in industrial ecology circles regarding the role of environmental policies, in particular of traditional command-and-control policies.

There are a few approaches which try to bridge this gap between more human- and more technology-centred perceptions of a circular economy: according to Choudhary (2012), industrial ecology "explores the idea that industrial activities should not be considered in isolation from the natural world but rather as a part of the natural system" (see p. 2). "Industrial" thereby refers to all human activities with tourism, housing, medical services, transportation, agriculture and others among them, and "ecology" means the science of ecosystems (see p. 2). Industrial ecology, as well as other technical schools of thought, is therefore a branch of systems science and systems thinking (see p. 3).

Systems thinking is also inherent in the concept of the circular economy (see Definition 2.1), and that all human activities should be considered as part of the natural system is a decisive feature of a circular economy, too. Also Hanumante, Shastri, and Hoadley (2019) recommend a "balanced approach for the adoption of the circular economy", and Schröder, Lemille, and Desmond (2020) address this gap by relating the circular economy to human development.

So far these considerations of more technical, practice-oriented approaches to a circular economy. The next section addresses some general aspects of implementing a circular economy – with a view on societal path dependencies, which are of some relevance in this context. These remarks will be reconsidered at a later stage (see Chapter 12).

2.3 Remarks on societal path dependencies

Adopting circular economy strategies or moving towards a circular economy requires, according to general consent, systems thinking and the willingness to accept changes, fundamental changes perhaps. Societal path dependencies can stand in the way of necessary changes and may thus prevent or at least slow down the implementation of a circular economy – not a particularly exciting perspective.

A short remark in this context – further remarks will follow – refers to economic growth, a globally accepted goal of economic policy, a societal path dependency so to say: the "limits to growth", addressed by the Club of Rome, posed by the exhaustible resources and the limited assimilative capacity of the environment, fit into an interesting series of mostly pessimistic views on the prospects of long-term economic growth, which were, equally interesting, most often related to environmental issues, in particular to the limited supply of agricultural land.

Adam Smith (1723–1790), Thomas Malthus (1766–1834), David Ricardo (1772–1823) and John Stuart Mill (1806–1873) are among the prominent philosophers and classical economists, who considered economic growth a temporary phase due to the limits posed by the natural system. Karl Marx (1818–1883), with the background of the classical labour theory of value, had similar pessimistic views on the long-run economic development (see, for example, Pearce & Turner, 1989, Ch. 1, or Shanahan, 2018).

This leaves us with some important questions: will it, by means of a transition to a circular economy, be possible to overcome the constraints of the assimilative capacity of the environment? Will it be possible to get around and compensate for the limited supply of exhaustible resources? After all, the predictions of the classical economists and of Karl Marx did not come true: a continuous stream of new scientific knowledge, and technological and social innovations helped to accommodate a growing global population – at least in most regions and at least to some extent. Economic growth as a societal path dependency has not really been challenged so far. But will it be possible to "convince" societies for the systems changes necessary for a circular economy – even if this implies less economic growth? Will it be possible to overcome relevant societal path dependencies?

After the publication of Pearce and Turner (1989), there was some scepticism, in particular regarding the sustainability goal. For example, Bennett (1991), in his review of the book points to the "sustainability requirement", which involves the maintenance of the currently available stock of natural capital – at least in principle. Any reductions to this stock "would need to be compensated for by additions to that stock achieved by other projects …", and "most economists will see it as a prescription for the creation of a poorer society, less able to afford environmental protection" (Bennett, 1991, p.

228), thus a clear hint to societal path dependencies.

Segerson (1991), while "sympathetic to the arguments in favour of sustainability", draws the attention to the scarce information on "what sustainability would mean in terms of either theory or practical implementation" (p. 273). In fact, the question of how sustainability goals, or, more generally, the sustainable development goals (SDG), should or could be implemented, is still of utmost importance and under rigorous discussion in the context of implementing a circular economy (see the platform "Science for Sustainable Development"). Needless to say, this is again an issue related to societal path dependencies.

These aspects are obviously crucial for the successful implementation of a circular economy, and it is indeed necessary to motivate all stakeholders to provide their valuable support for the transition to a circular economy. Probably in order to encourage participation, to dissipate fear about possibly higher costs due to new environmental regulations, to stay with familiar framework conditions to some extent, many proponents of a transition to a circular economy keep pointing to profitable business models and a resulting higher competitiveness of business companies in the context of implementing a circular economy (see, for example, EU, 2015, or Kalmykova, Sadagopan, & Rosado, 2018). Of course, such a positive outlook, if it comes true, could considerably accelerate the implementation of a circular economy, at least in decentralised market economies. Some important, also critical aspects of business models in this context will be addressed later (see Section 3.1 and Chapter 18).

Thus not only different perceptions of the concept of a circular economy, but also the imprecise nature of the concept of a sustainable development point to a multitude of possible and viable procedures regarding the implementation of a circular economy – whichever it might be. And, in addition, societal path dependencies might be in the way of a fast transition.

2.4 The circular economy— Understanding the concept

Summarising this chapter, the circular economy, as proposed and introduced in Pearce and Turner (1989), emphasises the inherent circular structure of an economic system due to the necessary and – due to the laws of physics – indispensable interactions between the economy and the environment. There is a focus on sustainability regarding resources, and there is a special focus on all aspects of waste in order to restore and retain the assimilative capacity of the environment in a sustainable way, a focus, which is also shared by proponents of more technical, practice-oriented concepts of a circular economy. Without sufficient attention given to these interactions, rebounds from the environment might endanger the functioning of the economy at large. At least there is the risk that more and more resources are required to compensate for the environmental degradations.

As an immediate consequence, the challenges of implementing a circular economy in the sense of the detailed perceptions of the concept (see Chapter 3) consist in keeping these interactions in good order for sustaining the economy. Therefore, all measures taken to implement a circular economy have to be evaluated with a view on this background and on the increasing importance sustainability gained with the Earth Summit in Rio de Janeiro in 1992.

After these remarks on the academic and practical roots of the circular economy, the next chapter turns to its detailed perceptions in the recent literature and in practice. It is thereby important to understand to what extent a particular perception is still in agreement with the generic definition of the circular economy provided in Definition 2.1. The concept proposed, in principle, by Pearce and Turner (1989) is thereby considered a reference point.

References

Andersen, M. S. (2007). An introductory note on the environmental economics of the circular economy. *Sustainability Science*, 2, 133. https://doi.org/10.1007/s11625-006-0013-6.

Antikainen, R., Lazarevic, D., & Seppälä, J. (2018). Circular economy: Origins and future orientations. In H. Lehmann (Ed.), *Earth and environmental science*. Cham: Springer. https://doi.org/10.1007/978-3-319-50079-9_7.

Ayres, R. U. (1998). Eco-thermodynamics: Economics and the second law. *Ecological Economics*, 26(2), 189–209. https://doi.org/10.1016/S0921-8009(97)00101-8.

Bennett, J. W. (1991). Pearce, D. W., and R. K. Turner. Economics of Natural Resources and the Environment. Baltimore MD: Johns Hopkins University Press, 1990, 378 pp., $42.50, $19.50 paper. *American Journal of Agricultural Economics*, 73(1), 227–228. https://doi.org/10.2307/1242904.

Choudhary, C. (2012). Industrial ecology: Concepts, system view and approaches. *International Journal of Engineering Research & Technology*, 1(9) Retrieved from https://www.ijert.org/research/industrial-ecology-concepts-system-view-and-approaches-IJERTV1IS9396.pdf.

EU. (2015). *Closing the loop—An EU action plan for the circular economy*. Retrieved from https://eur-lex.europa.eu/resource.html?uri=cellar:8a8ef5e8-99a0-11e5-b3b7-01aa75ed71a1.0012.02/DOC_1&format=PDF.

Ghisellini, P., Cialani, C., & Ulgiati, S. (2016). A review on circular economy: The expected transition to a balanced interplay of environmental and economic systems. *Towards Post Fossil Carbon Societies: Regenerative and Preventative Eco-Industrial Development*, 114, 11–32. https://doi.org/10.1016/j.jclepro.2015.09.007.

Hanumante, N. C., Shastri, Y., & Hoadley, A. (2019). Assessment of circular economy for global sustainability using an integrated model. *Resources, Conservation and Recycling*, 151, 104460. https://doi.org/10.1016/j.resconrec.2019.104460.

Hartley, K., van Santen, R., & Kirchherr, J. (2020). Policies for transitioning towards a circular economy: Expectations from the European Union (EU). *Resources, Conservation and Recycling*, 155, 104634. https://doi.org/10.1016/j.resconrec.2019.104634.

Heshmati, A. (2015). *A review of the circular economy and its implementation. IZA Discussion Papers, (9611)*. Retrieved from http://ftp.iza.org/dp9611.pdf.

Kalmykova, Y., Sadagopan, M., & Rosado, L. (2018). Circular economy—From review of theories and practices to development of implementation tools. *Sustainable Resource Management and the Circular Economy*, 135, 190–201. https://doi.org/10.1016/j.resconrec.2017.10.034.

Lifset, R., & Graedel, T. E. (2002). Industrial ecology: Goals and definitions. In R. U. Ayres, & L. W. Ayres (Eds.), *A handbook of industrial ecology*. Cheltenham: Edward Elgar Publishing. https://doi.org/10.4337/9781843765479.00009.

Meadows, D. H., Meadows, D. L., Randers, J., & Behrens, W. W., III (1972). *The limits to growth: A report of the Club of Rome's project on the predicament of mankind*. A Potomac Associates Book. Retrieved from http://www.donellameadows.org/wp-content/userfiles/Limits-to-Growth-digital-scan-version.pdf.

Murray, A., Skene, K., & Haynes, K. (2017). The circular economy: An interdisciplinary exploration of the concept and application in a global context. *Journal of Business Ethics*, 140, 369. https://doi.org/10.1007/s10551-015-2693-2.

Pearce, D. W., & Turner, R. K. (1989). *Economics of natural resources and the environment*. Johns Hopkins University Press.

Schröder, P., Lemille, A., & Desmond, P. (2020). Making the circular economy work for human development. *Resources, Conservation and Recycling*, 156, 104686. https://doi.org/10.1016/j.resconrec.2020.104686.

Segerson, K. (1991). Reviewed Work: Economics of natural resources and the environment by D. W. Pearce, R. K. Turner. *Land Economics*, 67(2), 272–276. https://doi.org/10.2307/3146419.

Shanahan, M. (2018). Can economics assist the transition to a circular economy? In *Unmaking waste in production and consumption: Towards the circular economy* (pp. 35–48). Emerald Publishing Limited. https://doi.org/10.1108/978-1-78714-619-820181004.

Winans, K., Kendall, A., & Deng, H. (2017). The history and current applications of the circular economy concept. *Renewable and Sustainable Energy Reviews*, 68, 825–833. https://doi.org/10.1016/j.rser.2016.09.123.

The circular economy in literature and practice

The generic definition of the circular economy in Section 2.1 presents two major issues of utmost importance for implementing a circular economy: sustainability and all aspects of handling waste, in particular preventing waste and reducing landfilling. Sustainability means also saving and preserving resources, and paying attention to the assimilative capacity of the natural environment. Moreover, first evidence of societal path dependencies interfering with the implementation of a circular economy point to possible challenges, perhaps complications.

How and to what extent do detailed perceptions of the circular economy take these issues into account? Regarding these aspects, this chapter focuses first on the literature in general, but also, not less importantly, on perceptions in more practical contexts. The literature has grown considerably since the publication of Pearce and Turner (1989), and this review attempts to relate various categories of perceptions of "a" circular economy to the generic concept of "the" circular economy introduced in Pearce and Turner (1989), and substantiated in Definition 2.1.

The course of this review reveals some interesting shifts in the understanding of a circular economy (Fig. 3.1). This is due to the fact that most of these definitions give already some more or less concrete recommendations for a transition to a circular economy, thus allowing some insight into the respective understanding of "the" circular economy. The resulting deviations do play a role in efforts to implement these concepts and in indicators measuring the degree of implementation, and are therefore of relevance for practical considerations. The varying interpretations, in particular of the waste hierarchy and aspects thereof, deserve special attention: they represent current and adopted practices of circular economy strategies with, in view of the goals of the circular economy, possibly significant deviations from the generic concept.

The review of practical usages of the concept of a circular economy refers to various institutions and organisations, which promote their view of a circular economy – regionally or even globally. Among these institutions and organisations are the already introduced Ellen MacArthur Foundation, a charity in the UK, with the mission to accelerate the

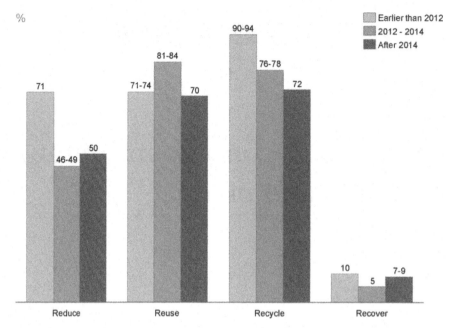

FIG. 3.1   Perceptions of a circular economy. Development of circular economy definitions over time. © *From Kirchherr, J., Reike, D., & Hekkert, M. (2017). Conceptualising the circular economy: An analysis of 114 definitions. Resources, Conservation and Recycling, 127, 221–232. https://doi.org/10.1016/j.resconrec.2017.09.005 226.*

transition to a circular economy, but also the European Union (EU) with its circular economy package, and, exemplarily, the TIARCENTER, an independent think tank and advisory firm in Russia with the objective to promote circular economy strategies in Russia, in particular in the area of waste management. Also for these applications it is necessary to investigate the understanding of the core concept of the circular economy, which is contained in these recommendations.

The following sections consider the fundamental concepts of sustainability and the waste hierarchy in order to explore the economic background of the multitude of different perceptions and approaches regarding a circular economy. The next sections discuss "appearances" of the concept of a circular economy in the literature, including some practical applications. There are also some links to related concepts such as the "smart city" and the "green economy". The economic background is briefly addressed in a

separate section. Thereafter, as already indicated, we turn in more detail to the applications of the concept of a circular economy by the Ellen MacArthur Foundation, the European Union, and the TIARCENTER. Finally, conclusions from the literature review point to tasks necessary for implementing a circular economy.

3.1 The perception of the circular economy in the literature

A recent count of definitions of a circular economy by Kirchherr, Reike, and Hekkert (2017) resulted in 114 different approaches to this concept, published in peer-reviewed journals, in policy papers and reports – Fig. 3.1 indicates the shift of the focus regarding the waste hierarchy over time. Reike, Vermeulen, and Witjes (2018) found – by means of a Scopus search on the term – an increase of 50% in academic publications in a recent period of five years. Similarly, Prieto-Sandoval, Jaca, and Ormazabal (2018) selected hundreds of articles on the circular economy and related concepts from the Web of Science database with numbers tending to increase in the recent years (see p. 607). Finally, Murray, Skene, and Haynes (2017) investigate, among other issues, applications of the concept in practice and policy, and Schröder, Lemille, and Desmond (2020) combine the circular economy "with the approach for Human Development", an attempt to bridge the gap to the more technically-oriented perceptions.

Undoubtedly, the concept has gained momentum in the last years, both in theory and practice, both among scholars and practitioners, leaving a variety of questions: what are the main differences between these various approaches, what are the commonalities? To what extent are they related to the concept of Pearce and Turner (1989) and, therefore, to the generic Definition 2.1? And another important issue refers to the implementation: needless to say, different concepts of a circular economy may require different approaches regarding the implementation.

According to Kirchherr et al. (2017), a circular economy may mean many different things to different people, in particular to critics of the concept. They find that many definitions refer to the 3Rs: reduce, reuse, recycle, sometimes neglecting "reduce", sometimes entirely focusing on "recycling", often without emphasising the necessity of a systemic shift. Only three of the 114 definitions mentioned above include the 3R framework, the waste hierarchy, a systemic perspective and social equality, referring to a sustainable development (see p. 228). With regards to the implementation of a circular economy, they point to missing business models and the unclear role of various stakeholder groups as enablers of the circular economy (see p. 228). This role of business models, this interplay of the market system with the implementation of a circular economy, is indeed an interesting point, which needs further consideration (see Chapter 18).

Similarly, Reike et al. (2018) refer to the large differences, which manifest themselves globally with regard to the circular economy, but they point also to the widely shared potential ascribed to the circular economy to break the linear model of an economy (see p. 246). Interestingly, Korhonen, Honkasalo, and Seppälä (2018) argue that the concept of a circular economy and its practice "have almost exclusively been developed and led by practitioners ... ", with the consequence that the scientific research content remains largely unexplored (see p. 37). Consequently Prieto-Sandoval et al. (2018) mention the increasing importance of research topics related to the circular economy (see p. 608).

In the following subsections a few definitions which are, to some extent, representative for others will be considered more carefully – under a separate heading. It is interesting to note that all definitions reviewed introduce the concept of a circular economy already with details and recommendations regarding its implementation. In view of the analysis in Section 2.1, this is, however, not without problems, as there should be different possibilities for implementing "a" circular economy for approaching "the" circular economy. Thus, the generic definition, based on Pearce and Turner (1989), circumvents this issue by explicitly allowing for different implementations depending on the concrete framework conditions prevailing in the participating countries.

These subsections address also certain aspects of relevance regarding the implementation of the concept of a circular economy in consideration. In this sense the heading also provides a specific recommendation for the implementation of the concept.

3.1.1 Relevance of business models

Kirchherr et al. (2017) propose the following definition of a circular economy, without, however, completely ruling out other concepts (see p.

224): "A circular economy describes an economic system that is based on business models which replace the 'end-of-life´ concept with reducing, alternatively reusing, recycling and recovering materials in production/distribution and consumption processes, thus operating at the micro level (products, companies, consumers), meso level (eco-industrial parks) and macro level (city, region, nation and beyond), with the aim to accomplish sustainable development, which implies creating environmental quality, economic prosperity and social equity, to the benefit of current and future generations."

How and to what extent does this definition correspond to the generic Definition 2.1 referring to the concept of an inherently circular economy focusing on the fundamental functions of the environment? Obviously, with their concept the authors address both the waste hierarchy and the issue of a sustainable development – also of relevance in Pearce and Turner (1989). They refer to the operations at all levels of the economy, which is – in comparison to Pearce and Turner (1989) – certainly a more detailed picture of a circular economy pointing to the necessity of integrating all stakeholders for the implementation: at the micro level companies are focused on eco-innovation because of a positive impact on their prestige and associated reduction of costs, the meso level refers to companies, which will benefit from the cleaner natural environment, whereas the macro level is more oriented towards the development of eco-cities or eco-provinces.

Interestingly they, but also other authors introducing the concept of a circular economy, relate their concept to appropriate business models required for the implementation. There are, in particular, various practical approaches to a circular economy, which support this view, among them the Ellen MacArthur Foundation and the EU with its action plan for the circular economy. Business model are mentioned in the context of increasing recycling activities, relying on Extended Producer Responsibility (EPR), on environmental innovations in general,

and a Design for Environment (DfE) in particular, although it is not always straightforward to understand them as viable business models, profitable under most circumstances and framework conditions. In this context, Antikainen, Lazarevic, and Seppälä (2018) argue that the "circular economy is expected to bring multiple benefits to the environment and the economy, but only a few examples have demonstrated the circular's economy potential economic benefit for industrial actors" (see p. 115).

Although there is no unique definition, smart cities are often brought in relation to a sustainable development (Martin, Evans, & Karvonen, 2018). As information and communication technologies (ICT) in the context of the digital transformation are of importance both for a circular economy and for a smart city, establishing smart cities seems to be a profitable business model – not too far from implementing a circular economy (see Section 13.1 for more details).

Similarly, the "green economy" and "green growth" have gained momentum in recent years – Merino-Saum, Clement, Wyss, and Baldi (2020) found about a 140 definitions of these concepts. They have "a multidimensional notion, whose focus is on the potential trade-offs and synergies between economic and environmental dimensions (without ignoring social issues)", and are thus also close to sustainability.

The United Nations Environment Programme (UNEP) defines the green economy "as low carbon, resource efficient and socially inclusive. In a green economy, growth in employment and income are driven by public and private investment into such economic activities, infrastructure and assets that allow reduced carbon emissions and pollution, enhanced energy and resource efficiency, and prevention of the loss of biodiversity and ecosystem services."

Similarly, in the context of its Circular Economy Package, a greener economy means for the EU new growth and job opportunities. "Eco-design, eco-innovation, waste prevention and the reuse of raw materials can bring net savings for EU businesses of up to EUR 600 billion" (Fig. 3.2), if the EU can be turned "into a resource-efficient, green, and competitive low-carbon economy". Similarly, the European Green Deal "is about improving the well-being of people". It is considered the "new growth strategy" – cutting "emissions while creating jobs".

In conclusion, there are substantially differing opinions in the literature on the possibility of large-scale profitable business models supporting the implementation of a circular economy – obviously mirroring practical experiences. Indeed, it seems that references to and examples of viable business models are also meant to further attract business to the world of the circular economy, which is certainly important. One is then, however, tempted to ask: if there are profitable business models, why this is not yet happening at a larger scale. Or, put the other way around, what needs to be done to make business better "aware" of the opportunities in the context

of the circular economy? This is, in fact, an important and crucial issue, and it needs to be discussed more thoroughly (see Part II). Environmental innovations, discussed in the next subsection, pose an important area in this regard.

3.1.2 Environmental innovations

Prieto-Sandoval et al. (2018) also mention the growing importance of the concept of the circular economy for attaining a sustainable development with its supposed and expected positive impacts on economic prosperity, on environmental quality and social equity. Many definitions reveal linkages to sustainability. Among the dominant determinants of a circular economy, which they find in their literature review, they mention the waste hierarchy, both as a conceptual basis for a circular economy and a guiding principle for implementing a circular economy.

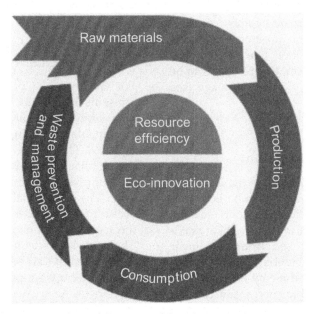

FIG. 3.2 **Green growth and circular economy.** Green growth based on eco-innovation, resource efficiency, waste prevention and reuse of raw' materials – the vision of the EU. © *From EC – DG Environment. Green growth and circular economy. Retrieved from https://ec.europa.eu/environment/green-growth/index_en.htm (Original work published 2019).*

They then propose a "cohesive and inclusive concept" of the circular economy: "The circular economy is an economic system that represents a change of paradigm in the way that human society is interrelated with nature and aims to prevent the depletion of resources, close energy and materials loops, and facilitate sustainable development through its implementation at the micro (enterprises and consumers), meso (economic agents integrated in symbiosis) and macro (city, regions and governments) levels. Attaining this circular model requires cyclical and regenerative environmental innovations in the way society legislates, produces and consumes" (Prieto-Sandoval et al., 2018).

Again, a brief analysis of this concept shows the close relationship to Definition 2.1 and to the concept of Pearce and Turner (1989). There is, however, beyond the reference to the operations at all levels of the economy, a clear remark on the relevance of environmental innovations, addressing again the implementation of a circular economy. Moreover, a change of paradigm is mentioned as a prerequisite for the implementation of a circular economy. This is more general than postulating appropriate business models, but leaves still the open question, what this new paradigm should look like and how it should be achieved? Again, this issue has to be addressed more carefully at a later stage (see Part II).

3.1.3 Sustainable development

Korhonen et al. (2018) propose the following slightly different definition: "Circular economy is an economy constructed from societal production-consumption systems that maximises the service produced from the linear nature-society-nature material and energy throughput flow. This is done by using cyclical materials flows, renewable energy sources and cascading-type energy flows. Successful circular economy contributes to all the three dimensions of sustainable development. Circular economy limits the throughput flow to a level that nature tolerates and utilises ecosystem cycles in economic cycles by respecting their natural reproduction rates" (see p. 39).

Similar to the concept of Kirchherr et al. (2017), this definition, although more technically oriented, orientates on a sustainable development – in fact, the circular economy is meant for sustainable development. The functions of the environment, emphasised in Pearce and Turner (1989) and Definition 2.1 are somewhat hidden behind the "linear nature-society-nature material and energy throughput flow".

It also remains somewhat unclear, what is meant by "maximising the service produced from the linear nature-society-nature material and energy throughput flow", because this can obviously have many dimensions: consumer welfare, volume of any kind of products and services, quality of these products and services and, of course, environmental issues – just to name a few. Observe that aspects of an implementation play some role in this definition, too.

3.1.4 Integration of economic activity and environmental wellbeing

For Murray et al. (2017) the circular economy "represents the most recent attempt to conceptualise the integration of economic activity and environmental wellbeing in a sustainable way" (see p. 369). They argue that operationalisations of the circular economy in business and policy place "emphasis on the redesign of processes and cycling of materials", thereby limiting its ethical dimension, mainly through the absence of the social dimension (see p. 369, 376).

Their "revised" definition of a circular economy considers it "an economic model wherein planning, resourcing, procurement, production and reprocessing are designed and managed, as both process and output, to maximise ecosystem functioning and human well-being" (see p. 377).

Again, there is the idea of "maximising ecosystem functioning and human wellbeing", without, however, making precise what is meant. Although this definition refers to the functions of the environment as proposed in Pearce and Turner (1989) and to some aspects of sustainability and the implementation, it is not straightforward to implement it, to make it operational.

3.1.5 Regenerative systems

Returning to a sustainable development in its relation to a circular economy, Geissdoerfer, Savaget, Bocken, and Hultink (2017) refer to the ambiguity blurring the similarities and differences between the two concepts already addressed in Pearce and Turner (1989). By means of an extensive literature review they aim at providing conceptual clarity. They define the circular economy "as a regenerative system in which resource input and waste, emission, and energy leakage are minimised by slowing, closing, and narrowing material and energy loops. This can be achieved through long-lasting design, maintenance, repair, reuse, remanufacturing, refurbishing, and recycling" (see p. 759).

Moreover, considering sustainability as "the balanced and systemic integration of intra and intergenerational economic, social, and environmental performance" (see p. 759), Geissdoerfer et al. (2017) then identify similarities and differences between sustainability and the circular economy on the basis of a literature review. They particularly point to the non-economic aspects inherent in both concepts and to the necessity of a system change (see Table 2, p. 764). Regarding the differences they stress, among others, the stronger focus on economic and environmental issues in the context of the circular economy. Thus, also due to a different history, the concept of sustainability seems to be broader, although less clearly defined in comparison to the circular economy (see Table 3, p. 765).

Geissdoerfer et al. (2017) with their definition and their proposal of a "regenerative system" clearly refer to the generic concept of the circular economy in Pearce and Turner (1989). They do, however, consider more carefully the role of sustainability and provide some rules for the implementation.

One is thus left with a variety of concepts for the circular economy: the waste hierarchy in different versions, sustainability concepts, regenerative systems, references to business models, the necessity of environmental innovations, and perhaps some more, which are in one way or the other related to the generic definition, to each other, and with many references to the literature. Moreover, most of these definitions provide or, rather, prescribe some concrete ways for the implementation. This, to a large extent, explains the multitude of different definitions of a circular economy.

Once again, different approaches to a circular economy are justified in view of the different environmental situations of the various countries and regions, although the perceptions presented above reveal large variances and do not always fully comply with the generic concept of Section 2.1.

After this survey and review of recent definitions and perceptions of the principles of a circular economy in the literature, the following section tries to provide a structure to this confusing variety of different, but nevertheless related concepts. In particular, the literature linking a circular economy more directly to economics and environmental economics is addressed. So far, mainly technologies, innovations and business models have helped to bridge the gap between a circular economy and the economic system.

3.2 Economy and circular economy in the literature

Although the concept of "a" or "the" circular economy has important roots in economics (Pearce & Turner, 1989), the concept itself has been used and further developed in recent years mostly by practitioners, i.e., policy makers, businesses, consultants, foundations and others (Korhonen et al., 2018, p. 37). Nevertheless, there are various publications elaborating on the links between a circular economy and economics in general, and environmental economics in particular.

There is, for example, Andersen (2007), who mainly refers to Pearce and Turner (1989) and provides an introduction to environmental economics, thereby indicating its potential to develop a circular economy. A circular economy "will turn negative external effects into positive ones by connecting waste streams to possible beneficiaries" (see p. 137). For this it is necessary to apply appropriate methods to account for external costs. Externality estimates for air pollution in various countries are given (see Table 1, p. 138).

Shanahan (2018) asks the fundamental question, whether economics can assist the transition to a circular economy, and highlights the important role economics can play in this context. It is thereby necessary to overcome market imperfections, such that "the price signals that are used in markets be an accurate reflection of the true costs and benefits of resource consumption and material production" (see p. 39), in view of externalities. Moreover, Shanahan (2018) points to the necessity of a "transformation in human thinking", and the role institutions can play in this context (see p. 41f). In summary, according to this view "a combination of economic instruments, institutions and ideological change" is necessary for a transition to a circular economy (see p. 43).

Ghisellini, Cialani, and Ulgiati (2016) refer to a distinction between a circular economy and "mainstream" neoclassical economics, with the latter failing "to provide analytical tools that take into account the limited and exhaustible nature of natural resources" (see p. 16). The circular economy "operates around the neoclassical economy framework" and seems to threaten some of its key pillars, if, for example, it proposes a rethinking of ownership in favour of models where products are only leased to consumers, who then become users of a service (see p. 17).

This is, indeed, one of the more difficult questions regarding the implementation of a circular economy: the strict observance of the waste hierarchy, which is anything else than simple or straightforward.

Milios (2018) identifies three policy options, "lacking specific attention in the current policy landscape of the EU" for advancing to a circular economy: policies for reuse, repair and manufacturing; public procurement for resource efficiency; strengthening secondary resource markets. A mix of these policies is then recommended for an approach towards a circular economy (see p. 868, 872). The background is that the relevant EU policy landscape focuses on the promotion of the waste hierarchy. However, according to Milios (2018), the "majority of product-related policies fail to incorporate any material resource efficiency clauses in a meaningful way". Consequently, these additional policies are expected to have "a significant potential for promoting higher resource efficiency throughout the life cycle of a product" (see p. 874).

Although Milios (2018), pointing to deficiencies of current attempts to develop a circular economy, is recommending environmental policies to support the development of a circular economy, those policies need to be thoroughly embedded into economics. After all, these policies are meant to change environmentally and economically relevant behaviour of consumers and producers, an important aspect of economics.

Finally, Ermolaeva (2019) collects opinions of experts on problems of modernising the waste management sector in Russia. One of the experts, an eco-sociologist, attributes the problem of waste to "the principle of capitalist production", which "does not take into account the last stage of the production process, namely the need for waste disposal" (see p. 62). In view of the environmental degradations in the former socialist and communist countries, this is indeed a surprising statement. Beyond some remarks on Russia's abundance with natural resources, most other statements of those asked, whether correct or not, lack any deeper economic content.

The following subsection introduces and discusses the practice-oriented circular economy concepts of the Ellen MacArthur Foundation, the European Union, and the Russian TIARCENTER.

3.3 The perception of the circular economy in a practical context

As already indicated, practitioners played and continue to play an important role in shaping the concept of a circular economy. Thus it is not too surprising that their concepts of a circular economy are very practically oriented with a focus on inputs from science, industry and engineering, but also on business models. As already indicated, the business models often refer to EPR and a DfE, immediately proposing the question, how to introduce EPR and how to motivate business companies for a DfE, given that they have the knowledge necessary for a DfE regarding their products?

Moreover, this emphasis on business models, which is also prevalent in quite a few other definitions, refers implicitly to market economies, at least to some important features of a market economy. But although markets and the generation of profits are of relevance for the economic systems of many countries, the rules for the operation of business companies are nevertheless differing and may lead to decisions, which are not necessarily in accordance with those of a "classical" market economy, especially regarding efficiency of the resulting allocations (see Chapter 6). This should at least be taken into account when

"business models" become relevant in practical perceptions of a circular economy.

These aspects and others need to be observed in the review of the following applications. Moreover, there is again a strong focus on the implementation of a circular economy, which represents a particular strategic recommendation, a guideline. The question, however, is to what extent this guideline restricts the options of the countries, which need to respect their particular situation, their concrete framework conditions, for a successful transition to a circular economy.

3.3.1 The Ellen MacArthur Foundation

The general mission of the well-known Ellen MacArthur Foundation is "to accelerate the transition to a circular economy". With the support of global business partners and philanthropic partners, the foundation is active in many regions "as a global thought leader, establishing the circular economy on the agenda of decision makers across business, government, and academia". It also considers a sustainable development an important dimension of a circular economy: "Circular economy is an industrial system that is restorative or regenerative by intention and design. It replaces the 'end-of-life' concept with restoration, shifts towards the use of renewable energy, eliminates the use of toxic chemicals, which impair reuse, and aims for the elimination of waste through the superior design of materials, products, systems, and, within this, business models" (Ellen MacArthur Foundation, 2013).

According to the Foundation, a circular economy aims to prevent waste, as "products are designed and optimised for a cycle of disassembly and reuse". Circularity introduces a strict differentiation between consumable and durable components of a product, with consumables largely made of biological ingredients, and durables made of technical ingredients, like metals and plastics, designed for reuse.

In addition, the energy required should be renewable (Ellen MacArthur Foundation, 2017, Fig. 2, p. 4).

The systemic shift required for the implementation of a circular economy should, again according to the Foundation, replace the concept of a consumer with that of a user, implying a new contract between businesses and their customers based on product performance. Durable products are leased or shared, and there are incentives in place to ensure the return and thereafter the reuse of the products or its components (Ellen MacArthur Foundation, 2013, p. 7).

The circular economy is, thus, meant to replace the existing model of a linear economy (Ellen MacArthur Foundation, 2017, Fig. 1.2, p. 12), which is characterised as a "throughput economy", a take, make and dispose economy based on the use of fossil fuels. This linear economy has obviously been very successful for many decades in terms of economic growth as measured by GDP per capita. This is in contrast to subsistence or rural economies that till today prevail in some parts of the world. The success of the linear economy is, of course, a consequence of the framework conditions determining economic systems in earlier times: "there was plenty to take and plenty of room to dispose". This situation seems to change in view of global warming and other local and global environmental issues increasingly affecting current economies on a large scale (Ellen MacArthur Foundation, 2017, p. 13).

According to the Foundation, economies will benefit, in particular, from substantial material savings and the long-term resilience of the economy. Companies can gain from reduced costs and new business opportunities, for example, in reverse cycle services (collection, sorting, funding and financing new business models). Consumers will also profit from reduced total ownership costs. The main differences between a linear and a circular economy are discussed and illustrated in (Ellen MacArthur Foundation, 2013, p. 22).

In view of Pearce and Turner (1989), the core functions of the environment are indirectly addressed, although there is a clear commitment to prevent waste through appropriately designed products. There is also a strong reference to the role of technologies ("industrial system") and of business models in developing a circular economy, and the Foundation provides many examples and case studies in support of the viability of such business models.

Nevertheless, the question remains, how to implement such a circular economy associated with a system change? The Foundation refers to business models such as EPR and DfE: how can they become role models for economic systems? This, as already mentioned in the context of other definitions of a circular economy, needs to be investigated more carefully (see Part II). The implicit recommendations that, for example, consumers will profit from "reduced ownership cost" etc., depend, of course, on the concrete situation in a country or region, and can, probably, not be generalised in a straightforward way.

As the transition to a circular economy requires specific strategies, the rather concrete recommendations provided by the Foundation with their focus on technologies and the industrial system might not be appropriate for each country. In particular developing countries with their lower wages have different possibilities regarding the implementation of a circular economy. On the other hand, these cases prove that there are business opportunities associated with implementing a circular economy – they only need to be discovered and developed. The question remains: can this process be accelerated?

3.3.2 The circular economy package of the European Union

In its press release from 4 March 2019, the European Commission declares that all 54 actions under the Circular Economy Action Plan have now been delivered or are being implemented.

In the corresponding report (EU, 2019) the European Commission emphasises that the Circular Economy Action Plan "has helped to put the EU back on a path of job creation", although "the competitive advantage" it brings to EU business needs yet to be secured. In 2016, jobs in sectors of relevance to the circular economy increased by 6% compared to 2012. Moreover, again in 2016 "circular activities such as repair, reuse or recycling generated almost €147 billion in value added while standing for around €17.5 billion worth of investments" (see p. 1).

These are impressive numbers although one must not forget that the GDP of the EU amounted to 18.8 trillion US-Dollar in 2018, reducing the financial volume of circular activities to less than 1% of all economic activities within the EU. Nevertheless, the EU can demonstrate a variety of achievements on its way towards a circular economy.

In its action plan (EU, 2015), the EU argues that in a circular economy "the value of products, materials and resources is maintained in the economy as long as possible, and the generation of waste is minimised". This is considered to be "an essential contribution to the EU's efforts to develop a sustainable, low carbon, resource efficient and competitive economy" (see p. 2).

As already indicated above, possible gains, in particular economic gains, from a circular economy play an important role in the EU's promotion of the concept: "The circular economy will boost the EU's competitiveness by protecting businesses against scarcity of resources and volatile prices, helping to create new business opportunities and innovative ways of producing and consuming. … At the same time, it will save energy and help avoid the irreversible damages caused by using up resources at a rate that exceeds the earth's capacity to renew them in terms of climate and biodiversity, air, soil and water pollution. … Action on the circular economy therefore ties in closely with key EU priorities, including jobs and growth, the investment agenda, climate and energy, the social agenda and industrial innovation, and with efforts on sustainable development" (EU, 2015, p. 2). This, of course, explains the focus of the press release cited earlier on additional employment in circular activities in the EU.

The EU, thus, considers business but also consumers as key in driving this development towards a circular economy. And besides local, regional and national authorities, it also assumes a fundamental role in supporting this transition. The aim thereby is to provide the right regulatory framework for the development of a circular economy. Appropriate measures should promote economic incentives and improve EPR schemes and commitments on DfE. Moreover, targeted actions in areas such as plastics, food waste, construction, critical raw materials, industrial and mining waste, consumption, public procurement, fertilisers and water reuse are or will get funding under the EU's Horizon 2020 research programme.

According to the EU, circular economy strategies are supposed to start at the very beginning of a product's life: both the design and production processes have important impacts on resource use and waste generation throughout a product's life. By means of an improved labelling system for the energy performance of household appliances, for example, the EU wants to direct consumer demand to the most efficient products. A product's lifetime can be extended through reuse and repair, thereby reducing waste, supported through other initiatives to reduce waste. The waste hierarchy plays a central role in waste management and aims at encouraging the options that deliver the best environmental outcome (EU, 2015, p. 4ff).

So far the circular economy concept of the EU and the proposals for its implementation, mostly based on framework conditions established through an appropriate legislation. The existing legislative framework is, however, still incomplete and is not yet functioning completely satisfactorily. Moreover, various recommendations or prescriptions, the EU Strategy for Plastics in a Circular Economy, for example, and also the links to the energy and climate policy of the EU require a critical analysis. Other sectors with high environmental impact and potential for circularity such as information and communication technologies, electronics, mobility, the built environment, mining, furniture, food and drinks or textiles, are not yet included in the action plan.

Similar to the concept of the Ellen MacArthur Foundation, also the concept of the EU is accompanied by detailed recommendations on what needs to be done in view of a transition to a circular economy. Given the large differences among the EU member states regarding the energy policies and aspects of the waste management policies, this will pose a challenge. The question thereby is whether the member states are ready to adopt a more homogeneous policy with respect to a circular economy. If not, how can this transition be achieved in this case?

These and other aspects related to the Circular Economy Package of the EU will be investigated more carefully in later chapters.

3.3.3 The Russian TIARCENTER

The TIARCENTER, which will be presented here as an exemplary local initiative, presents itself as an independent think tank and advisory firm, which provides "strategic advice to corporations and government bodies on the sustainable development principles implementation". Its mission is to support the transformation of the traditional linear economy into a circular model – with a focus on Russia, of course. The current projects refer to, among others, energy efficiency, urban mobility, stakeholder engagement, and the circular economy, considered to be an ongoing study to promote the principles of a circular economy.

TIARCENTER characterises the circular economy as follows: "The circular economy, an alternative to the traditional linear economic model, enables economic growth via more effective use of available resources, collaborative and repeat consumption of manufactured goods, waste recycling, and producing goods from recycled resources. Transition to the principles of a green economy will mean meeting the targets set out in a series of government papers, in particular the presidential decree on national objectives and strategic challenges in the Russian Federation's development up to 2024, with regard to increasing efficiency in manufacturing and consumer waste management, reducing air pollution, and introducing environmental regulation systems based on the best available technologies (BAT)".

The focus is undoubtedly again on potential economic effects of a circular economy, probably of importance for motivating Russia as a "follower" to adopt certain features of a circular economy. The original goals of a circular economy as discussed in Pearce and Turner (1989) – supporting the fundamental functions of the environment – vanish in the background. Beyond recycling of waste, the waste hierarchy does not play much of a role, and sustainability, although a topic at the TIARCENTER, seems not to be linked with a circular economy. Moreover, the strong reliance on presidential decrees, which are, without any doubts, important, might lead to disappointment. That's at least the experience with this kind of command-and-control policies from other countries (see Section 14.1).

Beyond that also the TIARCENTER provides concrete guidelines for the transition to a circular economy focusing on Russia, where they can expect to meet a homogeneous audience. Whether this audience is willing to receive the messages of the TIARCENTER remains to be seen. Interestingly, in addition to the link to a circular economy, there is again a link to the green economy (see Section 3.1.1). The reference to the eco-system services of a green economy provides a bridge to the concept of Pearce and Turner (1989).

So far this collection of articles, which refer, in one way or the other, to the economic background of the circular economy. The following section provides some conclusions on this literature review.

3.4 Conclusions from the literature review

This review of the literature on the background of the circular economy in economics in general and in environmental economics in particular demonstrates that deep economic principles are not really embodied in the theoretical conceptualisations and the practical applications of a circular economy. On the small basis of the literature reviewed and the other accessible literature the situation is as follows:

There is a variety of publications, which establish a link between (environmental) economics and a circular economy. These articles are, however, most often rather introductory and do not overly address the obvious complexity of a circular economy in its relationship to the economic system.

Then there are publications, which refer in a more detailed way to the issues of externalities, market failures etc., which are quite regularly associated with environmental commodities of relevance in a circular economy. In these cases, the potential of economics to deal with these issues, is often not further taken into account. In some few cases, the solution capacity of a market-based approach towards a circular economy is outrightly denied. This might be a hint to quite a few environmental issues related to market systems (see Chapter 7).

And, finally, there are those articles, which discuss the necessity of appropriate environmental policies, without, however, embedding these policies into a sound economic framework. The cases considered in Chapter 5 will show that this is, indeed, an issue, which needs to be taken care of: the design of appropriate integrated environmental policies (see Part V).

Thus, it seems to be true that the circular economy, despite its deep economic roots, has been further developed mostly by non-economists: by researchers in industrial ecology, by engineers developing environmental technologies, and by policy makers. It is indeed, surprising that the economists left this important arena more or less completely to practitioners. If the transition to a circular economy requires a new thinking, as many authors claim, then economics has the tools to understand and study the behaviour of individuals regarding the allocation of scarce resources and commodities, including natural resources and environmental services, and it has the tools to influence and change the mindset of the people in a society regarding the requirements of a circular economy. This does, however, not mean that it is always easy and straightforward to handle these issues in an adequate way.

In order to prepare the careful investigation of the behaviour of stakeholders the following chapter addresses the decisions of governments to adopt circular economy strategies sooner or later, thus introducing "circular economy leaders and followers". What are possible reasons for these observations, why do governments decide in this way or another?

References

Andersen, M. S. (2007). An introductory note on the environmental economics of the circular economy. *Sustainability Science*, 2, 133–140. https://doi.org/10.1007/s11625-006-0013-6.

Antikainen, R., Lazarevic, D., & Seppälä, J. (2018). *Circular economy: Origins and future orientations.* Cham: Springer International Publishing. https://doi.org/10.1007/978-3-319-50079-9_7.

Ellen MacArthur Foundation. (2013). *Towards the circular economy: Economic and business rationale for an accelerated transition.* Retrieved from https://www.ellenmacarthur foundation.org/assets/downloads/publications/Ellen-MacArthur-Foundation-Towards-the-Circular-Economy-vol.1.pdf.

Ellen MacArthur Foundation. (2017). *Circular economy and curriculum development in higher education.* Retrieved from https://www.ellenmacarthurfoundation.org/assets/downloads/EMF_HE-Curriculum-Brochure-03.10.17.pdf.

Ermolaeva, Y. V. (2019). Problems of modernization of the waste management sector in Russia: Expert opinions. *Revista Tecnologia e Societa, 15*(35), 56. https://doi.org/10.3895/rts.v15n35.8502.

EU. (2015). *Closing the loop—An EU action plan for the circular economy.* Retrieved from https://eur-lex.europa.eu/resource.html?uri=cellar:8a8ef5e8-99a0-11e5-b3b7-01aa75ed71a1.0012.02/DOC_1&format=PDF.

EU. (2019). *Report from the Commission to the European Parliament, the Council, the European Economic and Social Committee and the Committee of the Regions on the Implementation of the Circular Economy Action Plan.* Retrieved from https://eur-lex.europa.eu/legal-content/EN/TXT/PDF/?uri=CELEX:52019DC0190&from=EN.

Geissdoerfer, M., Savaget, P., Bocken, N. M. P., & Hultink, E. J. (2017). The Circular Economy—A new sustainability paradigm? *Journal of Cleaner Production, 143*, 757–768. https://doi.org/10.1016/j.jclepro.2016.048.

Ghisellini, P., Cialani, C., & Ulgiati, S. (2016). A review on circular economy: The expected transition to a balanced interplay of environmental and economic systems. *Towards Post Fossil Carbon Societies: Regenerative and Preventative Eco-Industrial Development, 114*, 11–32. https://doi.org/10.1016/j.jclepro.2015.09.007.

Kirchherr, J., Reike, D., & Hekkert, M. (2017). Conceptualizing the circular economy: An analysis of 114 definitions. *Resources, Conservation and Recycling, 127*, 221–232. https://doi.org/10.1016/j.resconrec.2017.09.005.

Korhonen, J., Honkasalo, A., & Seppälä, J. (2018). Circular Economy: The concept and its limitations. *Ecological Economics, 143*, 37–46. https://doi.org/10.1016/j.ecolecon.2017.06.041.

Martin, C. J., Evans, J., & Karvonen, A. (2018). Smart and sustainable? Five tensions in the visions and practices of the smart-sustainable city in Europe and North America.

Technological Forecasting and Social Change, 133, 269–278. https://doi.org/10.1016/j.techfore.2018.01.005.

Merino-Saum, A., Clement, J., Wyss, R., & Baldi, M. G. (2020). Unpacking the Green Economy concept: A quantitative analysis of 140 definitions. *Journal of Cleaner Production, 242*, 118339. https://doi.org/10.1016/j.jclepro.2019.118339.

Milios, L. (2018). Advancing to a Circular Economy: Three essential ingredients for a comprehensive policy mix. *Sustainability Science, 13*(3), 861–878. https://doi.org/10.1007/s11625-017-0502-9.

Murray, A., Skene, K., & Haynes, K. (2017). The circular economy: An interdisciplinary exploration of the concept and application in a global context. *Journal of Business Ethics, 140*(3), 369–380. https://doi.org/10.1007/s10551-015-2693-2.

Pearce, D. W., & Turner, R. K. (1989). *Economics of natural resources and the environment.* Johns Hopkins University Press.

Prieto-Sandoval, V., Jaca, C., & Ormazabal, M. (2018). Towards a consensus on the circular economy. *Journal of Cleaner Production, 179*, 605–615. https://doi.org/10.1016/j.jclepro.2017.12.224.

Reike, D., Vermeulen, W. J. V., & Witjes, S. (2018). The circular economy: New or refurbished as CE 3.0?—Exploring controversies in the conceptualization of the circular economy through a focus on history and resource value retention options. *Sustainable Resource Management and the Circular Economy, 135*, 246–264. https://doi.org/10.1016/j.resconrec.2017.08.027.

Schröder, P., Lemille, A., & Desmond, P. (2020). Making the circular economy work for human development. *Resources, Conservation and Recycling, 156*, 104686. https://doi.org/10.1016/j.resconrec.2020.104686.

Shanahan, M. (2018). Can economics assist the transition to a circular economy? In *Unmaking waste in production and consumption: Towards the circular economy* (pp. 35–48). Emerald Publishing Limited. https://doi.org/10.1108/978-1-78714-619-820181004.

Circular economy – A hierarchy of leaders and followers

The following sections investigate the reception of the concept of a circular economy in different countries. There are countries, which adopted circular economy strategies rather early, and there are many more, which are currently struggling with their more or less first efforts implementing features of a circular economy.

There is consequently a hierarchy of early adopters of circular economy strategies (CE leaders) and late-comers (CE followers), with a few other countries somewhere in between. As there are obvious differences among CE leaders and CE followers, it is interesting to understand, why some countries prefer to turn towards a circular economy earlier than others.

Another interesting question in this context is: to what extent can or do CE leaders motivate other countries for circular economy strategies? The organisations characterised in Section 3.3 and others can and do spread news about successful projects in various countries. Nevertheless, the special situations of the CE followers, requiring possibly different approaches, should thereby be taken into account.

There is yet another issue of interest in this context: the possibility that countries are behaving strategically. This implies that some countries could decide to act as "aggressive" CE leaders in certain areas, which are of importance for the development of a circular economy – mitigation of climate change, for example. There could be various reasons for adopting such a strategy, among them expected returns from developing, producing and selling innovative technologies to other countries, in particular to CE followers. In these cases, CE leaders could profit from a first-mover advantage. On the other hand, a country might as well opt for the status of a CE follower, with the possibility to profit from the efforts of other countries, likely from the CE leaders in the area under consideration. It could as well be possible that a country could profit just from waiting, from letting other countries go ahead, thereby "protecting" its own industry. In such a case there could be a late-mover or even last-mover advantage.

This kind of strategic behaviour plays a role in other societal constellations with countries pursuing a common goal, such as joining forces

---

in a union or an alliance, or implementing a circular economy. A formal analysis is possible, but will not be attempted for the cases considered here. (Weber, Weber, & Wiesmeth, 2020) investigate strategic behaviour regarding contributions to NATO in a formal context along these lines.

Instead of looking for and investigating economic reasons for strategic actions, characteristics of some CE Leaders and CE followers will be considered. It seems quite obvious to first take into account the availability of natural resources. Abundant resources do not necessarily provide strong signals to save resources through preventing waste and appropriate recycling activities. Moreover, if a country has enough space to landfill and thereby dump and "hide" all kinds of waste, then again the pressure for a serious waste management need not be very high.

The level of "environmental awareness" in a country is another factor. This seems to be a rather complicated construct, which has some roots in behavioural economics (see Chapter 8) and which cannot be handled straightforwardly. In principle, it means that people are aware of environmental issues, of environmental pollution and degradation, locally and/or globally, and take therefore measures to protect the environment, or start thinking about and campaigning for such measures. Thus, a high level of environmental awareness in a country or region does not necessarily mean that everything is fine regarding the environment in this country, but the situation may be on the verge to change.

How to get a general idea on the level of environmental awareness in a country? Khakimova, Lösch, Wende, Wiesmeth, and Okhrin (2019), and Lösch, Okhrin, and Wiesmeth (2019) investigate environmental awareness, respectively awareness of climate change in Russian regions. They introduce environmental awareness as a latent, not directly observable variable, which expresses itself through various indicators, such as specific queries in search engines, and which is dependent on certain causal variables, among

them economic wealth given by the gross regional product per capita. Weber and Wiesmeth (2018) provide a formal approach for the role of awareness for mitigating climate change in various OECD countries.

Thus, for a first approach environmental awareness could be related to economic wealth in a country: in the sense of an Environmental Kuznets Curve (EKC), one could assume that a higher gross domestic product (GDP) per capita might eventually raise environmental awareness. There are some studies supporting this view, but also various others, which are, in most cases, not directly against this view, but less supportive (for issues to and related to the EKC see, for example, Dinda, 2004; Grossman & Krueger, 1995; Huang, Lee, & Wu, 2008; Stern, 2004). This is therefore only a first approach, which has to be investigated more carefully at a later stage.

There are, thus, the following preliminary results: in order to categorise CE leaders and CE followers, and in order to relate their decisions regarding a circular economy, the following characteristics of a country should be taken into account:

- Endowment with natural resources
- Density of the population, respectively land available for landfilling
- Level of environmental awareness
- Motivation for strategic behaviour

As mentioned, there is only a rough indicator regarding the level of environmental awareness, namely the level of economic wealth given by GDP per capita. The possibly strategic dimension of the decisions is not yet respected.

4.1 Circular economy hierarchy: Leaders

In view of these remarks, this section addresses exemplarily two countries, Germany and China, which among various others (Austria, Japan, the Netherlands, Sweden, the UK …) are early adopters of strategies leading to a circular economy. The analysis of these two countries reveals the role of the potential drivers for early developing a circular economy. Heshmati (2015), Antikainen, Lazarevic, and Seppälä (2018), and Reike, Vermeulen, and Witjes (2018), among others, provide detailed reviews of the strategies of those two and other countries.

The following questions – some of them will be addressed again in later chapters – are thereby of interest: what are particular reasons of these countries for turning early towards a circular economy? Can the expected role of the above drivers be confirmed? Are there any strategic differences between Germany (a highly decentralised market economy) and China (with features of a centrally planned economy)? What are the specific measures taken in these countries to prepare the path towards a circular economy?

4.1.1 Germany's closed substance cycle waste management

Germany is one of the forerunners with respect to adopting the basic principles of a circular economy, such as reducing and recycling waste in order to limit landfilling and prevent negative externalities of waste on the economy. Some of the reasons for these early attempts in support of the assimilative capacity of the environment can easily be found. On the one hand, there are the high growth rates of the economy in the 1960s and thereafter, accompanied by all kinds and increasing amounts of waste. On the other hand, there is the rather high density of population, which increased further in the 1960s, leaving not many appropriate sites for dumping waste in unregulated and uncontrolled ways without risking the pollution of the soil and the groundwater. Schnurer (2002) points to these framework conditions and provides more details and additional information on the development of the German waste legislation on its way to the current Circular Economy Act (Germany, 2012), which is presented in more details in Nelles, Grünes, and Morscheck (2016), for example.

There was, in the very beginning, the Waste Disposal Act of 1972 with the goal to regulate dump sites, in particular, to stop uncontrolled landfilling. Waste incineration was then considered a means to extract electrical and thermal energy from waste, not unwelcome after the energy crisis of 1973. Gradually increasing resistance in the population against further incineration plants and new landfill sites, brought the Waste Avoidance and Management Act of 1986, which established the basic waste hierarchy in Germany: waste avoidance and recycling were given precedence over waste disposal. Thus, it is fair to say that the feedback from generating too much waste clearly revealed potential negative consequences of overextending the assimilative capacity of the environment. This confirms the picture of the circular economy proposed by Pearce and Turner (1989), although the adoption of the Waste Disposal Act in 1972 preceded the generic economic concept of the circular economy.

Further environmental regulations regarding waste and the implementation of the waste hierarchy followed, among them the Closed Substance Cycle Waste Management Act of 1994. This Act was also inspired by the Earth Summit in Rio de Janeiro in 1992 with its focus on a sustainable development. Reasons for these regulations, which mainly aimed at keeping waste in the cycle of the economic activities, can again be found in the characteristics of the paradigm of a circular economy: to reduce the negative effects of waste overextending the assimilative capacity of the environment, and use recycled resources as a substitute for exhaustible resources. The energy crisis of 1973 has clearly demonstrated that Germany, lacking natural resources of crucial importance for a highly industrialised country, is especially vulnerable regarding this aspect.

This Act, whose German name includes "Kreislaufwirtschaft", meaning "circular economy", has the purpose "to promote closed substance cycle waste management (Kreislaufwirtschaft), in order to conserve natural resources and to ensure environmentally safe disposal of waste" (Germany, 1994, Art. 1), obviously genuine tasks of the circular economy in the sense of Pearce and Turner (1989).

The Circular Economy Act, superseding the Closed Substance Cycle Waste Management Act, entered into force in 2012, and is meant to tighten the regulations of the preceding legal acts. In particular, the waste hierarchy is now extended and ranks waste management measures as follows: prevention; preparation for recycling; recycling, other types of recovery, particularly use for energy recovery; disposal. This Act thus has the purpose "to promote circular economy in order to conserve natural resources and to ensure the protection of human health and the environment in the generation and management of waste" (Germany, 2012, Section 1).

With this Act Germany finally arrived also formally in the sphere of the circular economy, after having adopted its main principles already some 25 years earlier. Again, the driving forces were the presumably high environmental awareness, but especially the already visible or at least expected negative impacts of too much waste, threatening the vital functions of the environment in a densely populated country with further consequences for the economic system. In addition, the lack of vital natural resources inspired a multitude of recycling activities and a desire to be a leader regarding this kind of activities. According to the German Federal Ministry for the Environment, Nature Conservation and Nuclear Safety (BMU) "high-end recycling", recycling of batteries from e-vehicles, for example, reduces the dependence on import of precious resources, but could also raise the competitiveness of German companies.

There is thus without any doubt a strategic component in the German approach towards a circular economy. This becomes even more apparent, when we investigate the German activities in the context of renewable energies

and mitigation of climate change (see Section 11.3).

Regarding the relationship to Definition 2.1 and to the principles of the circular economy proposed by Pearce and Turner (1989), there is a clear focus on waste management in general and on the waste hierarchy in particular, beyond all activities related to the mitigation of climate change. At first glance, sustainability seems to play a smaller role, but there is also the German National Sustainable Development Strategy of 2016, which is linked with the circular economy strategies.

The situation regarding the goals and the implementation of further features of a circular economy in Germany will be investigated in various contexts in later chapters. The next subsection turns to China, a country, which has started rather early with the adoption of principles of a circular economy.

4.1.2 China's eco-industrial development

There is the following situation with natural resources in China: on the one hand, endowment with natural resources in some areas is low and not all resources are developed, thus preventing the "resource curse" in the context of eco-efficiency, in contrast to the situation in other areas (Wang & Chen, 2020). On the other hand, according to Song, Zhu, Wang, and Wang (2019) it is one of the top priorities of the Chinese government's oversight "to address the conflicts between economic growth and resource consumption and between economic development and ecological damage".

Thus, the reasons that brought Germany onto the path towards a circular economy, are to some extent also of relevance for China: a decades-long rapid economic growth in the course of China's transformation from a centrally planned to a more market-based economy has led to serious depletions of natural resources, such as deforestation, water shortages and loss of biodiversity, and to increasing environmental pollution, in particular air pollution and waste generation. All this and a still growing population in combination with a rapid urbanisation motivated, perhaps even forced the government to serious actions to prevent further environmental degradation. The Chinese government has long since become aware of this highly problematic development, which certainly raised environmental awareness and even environmental concern, at least in the many urban agglomeration areas.

With scholarly discussions starting already in the late 1990s, the circular economy in China emerged out of this problematic situation as a new development strategy, which was formally adopted by the government in 2002 (see, for example, Heshmati, 2015; Su, Heshmati, Geng, & Yu, 2013; Yong, 2007; Yuan, Bi, & Moriguichi, 2006). The main goals of this development strategy are to face the environmental challenges mentioned above, to reduce the gaps between resource requirements and available supply, and to prove higher environmental standards – gaining increasing importance in international trade (Heshmati, 2015, p. 5f). Thus already at this stage there is some strategic behaviour with the aim to – by means of circular economy activities – favour international business activities, which are of great importance for the Chinese economy.

There are a number of laws and regulations, which are meant to support the development of a circular economy in China. The Cleaner Production Promotion Law was enacted in 2003, and the Circular Economy Promotion Law in 2009, which considers the circular economy as "the general term for the activities of decrement, recycling and resource recovery in production, circulation and consumption" (China, 2008, Art. 1). Moreover, an increasing number of environmental NGOs shall help to change attitudes and expectations in the society, necessary for a successful implementation of this new development strategy, which is predominantly a top-down strategy initiated by the government.

Implementation efforts refer to promoting an eco-industrial development, extending to the various levels of the economic system: the micro-level (producers, consumers), the meso-level (eco-parks, eco-agriculture, …), and the macro-level (regional circular industry, …). As will be analysed later in more details, this "integration" of relevant stakeholders into the efforts to establish a circular economy is of importance for designing incentive compatible environmental policies (see Section 17.3).

To what extent are the Chinese circular economy strategies in agreement with the generic definition of a circular economy and the proposal of Pearce and Turner (1989)? There is, of course, a clear commitment to recycling and resource recovery, but there is less talk on waste reduction, and there is an emphasis on the eco-industrial development, meaning a focus on technological changes and innovations, a focus on the feasibility of the required and necessary changes in society in general. The role of economics remains, however, in the background of this top-down approach.

There is a high political commitment in China to promote the circular economy in order to improve resource efficiency and to reduce environmental pollution. There are, however, indications, that this high political commitment is sometimes confronted with poor enforcement of the environmental regulations. Once again, Heshmati (2015), Su et al. (2013), Yong (2007), Yuan et al. (2006) and Antikainen et al. (2018), among others, provide more details on successes and failures of the circular economy strategy in China.

Again these brief remarks on the reception of the circular economy in China will be supplemented by further remarks and investigations in later chapters. For now, the main driving forces of the circular economy strategy in China seem to be – similar to the situation in Germany – the clearly observable and noticeable limitations of the environment to assimilate all waste resulting from the economic development, in particular in the major urban agglomeration areas, and the consequences of the depletion of natural resources, aggravated by the further increasing urbanisation resulting in a rather high and still increasing density of population in vast regions in China. In addition to that, high levels of soil, water and air pollution in combination with rising economic welfare raise environmental awareness, thus supporting a pressure for changes.

4.1.3 Circular economy leaders: A critical view

In summary, the countries, which adopted the circular economy early on – independent of their economic and/or political system, typically experienced or expected negative effects of a too high burden of waste on their economies, in addition to a threatening depletion of their exhaustible resources. The concrete level of a "too high burden" seems to depend, however, on the level of economic wealth, thus supporting, to some extent, the view that environmental awareness, triggering pressures for changes in environmental policies, is related to economic well-being.

Moreover, there are differences between Germany and China regarding their paths towards a circular economy. For example, when the German Circular Economy Act defines: "Circular economy within this Act shall be the prevention and recovery of waste" (Germany, 2012, Section 3(19)), whereas the Chinese Circular Economy Promotion Law declares: "The promotion of circular economy is an important strategy for the national economic and social development, … " (China, 2008, Art. 3), then the significance of the circular economy in the much more centrally organised Chinese economy becomes obvious. This extends to the measures taken and, in particular, to the management of the systems, with "goal-responsibility systems" and many planning requirements on all levels of the government in China (China, 2008, Art. 6–9). Of course, there are also "waste management plans" and "waste prevention programmes" with public participation in Germany (Germany, 2012, Sections 31–33), but they appear in the later sections of the Act and play therefore a less dominant role. However, as will be seen later, legal regulations alone are not always sufficient for a successful implementation of a circular economy, neither in Germany, nor in China (see Chapter 5).

Thus it seems that the early emergence of the circular economy in quite a few of the early adopting countries is substantially linked to the painful experience of wasting scarce resources and overly polluting the environment. This results in quite a natural classification into "leaders" and "followers" with respect to adopting the circular economy. The two countries use their positions as CE leaders also in a strategic sense: they export environmental technologies to support their economic development.

For a more or less global implementation of the circular economy, it is important that other countries follow the example of the leaders. For obvious reasons, this is of particular interest in the context of reducing greenhouse gas emissions, where this leader-follower situation can already be observed. Germany, for example, has been pivotal in the foundation of the International Renewable Energy Agency (IRENA), which also provides information on renewable energy technologies and innovations worldwide, and promotes "the further development of technologies to generate electricity from renewable sources", according to the Renewable Sources Act of Germany (see Germany, 2017, Section 1), also with the goal of exporting these technologies.

Establishing a serious leader-follower relationship in this context of the circular economy is not straightforward. With a share of approximately 2% of the global greenhouse gas emissions, Germany is too small to substantially mitigate climate change. It can, however, develop and provide appropriate technologies and try to motivate other countries to follow suit.

In order to have a somewhat contrasting picture, the situation and in particular the motivation of late adopters of circular economy strategies, of "followers", will be explored in the following subsection.

4.2 Circular economy hierarchy: Followers

This section addresses Russia and Georgia, two countries, which are, again among many others, late adopters of circular economy strategies. Similar to the situation with the analysis of the leaders, the discussion of these two countries representing "followers" provides some insight regarding their interest in developing a circular economy rather lately. As the focus in the literature seems to be typically more on success stories with respect to a circular economy, there are not many publications covering the efforts of these late adopters, although this could provide some interesting insights. Nevertheless, Reike et al. (2018) sketch the situation in a variety of countries, both leaders and followers regarding the adoption of a circular economy.

The following questions, some of them to be addressed again in later chapters, are thereby of interest: what are particular reasons in these countries for turning rather late towards a circular economy? Are there any strategic differences between Russia (an economy abundant with natural resources) and Georgia (an economy in transition with no noteworthy natural resources, but associated with the European Union (EU))? What are the specific measures taken in these countries? Can or do these countries learn or profit from the experiences of the leaders? Are there any indications of a strategic positioning of these countries?

4.2.1 Russia's first steps towards a circular economy

Russia is different in many ways from the countries discussed so far with respect to their attitudes and strategies towards a circular economy. First of all, Russia is abundant with developed natural resources, in particular oil and gas. Large-scale exports of oil and gas allow the import of all kinds of goods and consumption commodities. In terms of GDP per capita (PPP, current international US-Dollar (USD); data taken from the World Bank), Russia is with 27,100 USD in 2018 richer than China with 18,200 USD. However, there are 148 people per square km land area in China, with many huge and densely populated urban centres, but only nine people per square km land area in Russia with only few larger urban centres (again, data for 2018 taken from the World Bank).

Consequently, if we keep in mind the motives of the CE leaders discussed earlier, the pressures in Russia to adopt a development strategy for a circular economy are probably lower than in China. Regarding the few large urban centres in Russia there seems to be enough empty land for landfilling waste. Although issues regarding improper handling of waste have appeared and keep appearing more often in recent years, environmental awareness seems to be still on a lower level, not yet putting too much pressure on policy makers, at least for the time being.

Thus, beyond various recycling activities, which were probably of more economic relevance during the time of the Soviet Union in order to save valuable resources, and attempts to improve solid municipal waste management, there are right now not many aspects pointing to the widespread introduction of a circular economy in Russia. However, there are efforts to learn and gain from international experience, an indicator for some strategic positioning (see Ermolaeva, 2018, 2019; Korobova et al., 2014; Larionov & Ecorem, 2012 for further details regarding waste management in Russia).

Current approaches focus clearly on the improvement of the existing waste management systems: assess the systems, engage business companies for collecting, sorting and recycling of waste, and prepare for a transition to some features of a circular economy. Plastinina, Teslyuk, Dukmasova, and Pikalova (2019) analyse this situation, also the factors hindering respectively fostering the development of a circular economy. A particular reason for the still quite low level of circular economy activities in Russia seems to be the insufficient involvement of the government so far. Waste recovery and recycling standards are substantially lower than in the countries of the EU. Moreover, the financial situation of companies handling and processing waste is often not satisfactory. This points again to still insufficient and incompletely implemented or controlled environmental regulations (Ermolaeva, 2019, p. 71f; Plastinina et al., 2019, p. 13f).

Thus according to understandable reasons, Russia is still at the beginning of introducing principles of a circular economy, mainly in waste management, which is one of the more urgent environmental issues. It should therefore be considered a "follower", and the "leaders" should try to support the activities of the followers in view of Boulding's conception of planet earth as a spaceship (Boulding, 1966). In fact, this is already happening: in addition to a variety of other activities, the International Finance Corporation (IFC), an organisation of the World Bank Group, in partnership with the Free State of Saxony in Germany, the Netherlands, and Sweden, investigated the waste-recycling sector in Russia, offering recommendations for improvement (see Korobova et al., 2014; Larionov & Ecorem, 2012).

Other aspects regarding the development of a circular economy in Russia will be considered in various contexts in later chapters. The next subsection turns to Georgia, a transition country, which has also started with the adoption of principles of a circular economy – for reasons different from those of Russia, however.

4.2.2 Georgia's association with the European Union

In terms of economic development, Georgia with a GDP per capita of 11,400 USD (PPP, current international USD; data for 2018 taken from the World Bank) is trailing both China and Russia. The density of population is with 65 people per square km also rather low – however, a large part of the country is covered by high mountains and therefore uninhabitable. There are not many natural resources, although hydro power stations provide "green energy", at least in summer. Regarding the economy, Georgia is still on the way to transform its economy from part of a centrally planned economy with quite specific economic tasks within the Soviet Union to a diversified economic system based on the principles of a free market economy in the context of the association with the EU.

There are various issues with dump sites, which take away valuable land and threaten to pollute the ground water. Moreover, air pollution in Tbilisi is severe and people gradually start to complain about it. The natural beauty of the country is for sure one of its economic assets, which has to be preserved as one of the vital functions of the environment. The development in recent years regarding deforestation and air pollution in particular, but also waste management, should provide enough incentives for a change. Support for a change is now coming through the association of the country with the EU.

Consequently, the legal background of current and future activities of Georgia in the environmental arena in general and in waste management in particular is and will remain dominated by the "Association Agreement between the European Union and Georgia", which was signed in June 2014, and which entered into force on July 1, 2016. This agreement refers to tending to environmental issues in a manifold of ways, including waste management (EU, 2014, Ch. 3), and many measures have to be implemented by 2020 or 2021. Thus, the directives of the EU in all areas of waste management are and will be of relevance for Georgia, too. And being an important topic in the EU, this will also extend to the implementation of a circular economy, which, so far, is not directly visible in the legal regulations of the country.

The Law on Environmental Protection, the basic environmental legislation of Georgia (Georgia, 1996), refers to a "stable development", which in principle means a sustainable development (Art. 4 (k), (l)). Moreover, this law mentions the polluter-pays principle (Art. 5 (e)), waste prevention and recycling (Art. 5 (g), (i)).

The Waste Management Code (Georgia, 2014) establishes "a legal framework to implement measures that will facilitate waste prevention and its increased reuse as well as environmentally safe treatment of waste" (Art. 1), thus clearly pointing to the waste hierarchy, which is detailed again in Art. 4. In Art. 9, Extended Producer Responsibility (EPR) is mentioned to address issues such as product design and others.

As both the waste hierarchy and EPR are of relevance for a circular economy, there are thus some important first steps towards a circular economy, without explicitly addressing this issue. Despite the fact that Georgia is cutting back on plastic waste by discussing a ban of plastic bags, the environmental regulations are lacking compliance from part of the population, probably resulting from a low level of environmental awareness in combination with an insufficient knowledge and information on environmental issues.

Although still far away from a somewhat functioning circular economy, the situation in Georgia is interesting in this chapter on the emergence of circular economy strategies for a variety of reasons.

First of all, Georgia is not abundant with any valuable natural resources beyond the beauty of its landscape, which is, however, threatened by environmental pollution and deforestation. As tourism in Georgia depends to a large extent on an intact and beautiful environment, an increasing environmental devastation is not without risk for the economy of the country: the assimilative capacity of the environment, especially in the mountainous areas, is not large enough to handle a significant load of environmental pollution. Depletion of the exhaustible natural resources becomes visible, threatening the further economic development.

Nevertheless, as indicated above the pressure in the population to reduce pollution, to clean up the environment, is still within narrow bounds. This is, on the one hand, surprising in view of various environmental degradations and the limited space for landfilling. On the other hand, there is not yet sufficient pressure from the population for changes, although things start to move, supported by and accelerated through the association with the EU.

The problematic aspect seems to be the low level of environmental awareness. Grossman and Krueger (1995) found that countries with a GDP per capita above 8000 USD (in 1985 USD) tend to become increasingly aware of environmental pollution (air, soil and water pollution). It is, therefore, possible that this "environmental awareness" is, due to a still comparatively low level of economic wealth, not yet sufficiently developed in Georgia. After all, according to the Bureau of Labor Statistics prices in USD are 133.37% higher in 2018 than in 1985, and the Georgian GDP per capita of 11,400 USD in 2018 amounts to only 4884.95 USD in 1985 USD, far below the threshold value indicated by Grossman and Krueger (1995). This is also true for Georgian GDP per capita in current USD, which amounts to 4344.60 USD in 2018 (data taken from the World Bank). Thus there is the chance that environmental awareness increases with increasing economic wealth according to the EKC – gradually improving the situation.

However, Georgia is in a particular situation due to the association with the EU – the various obligations regarding the adoptions of the environmental policies might also help to acquire an intrinsic motivation and raise environmental awareness more quickly. This is an issue of behavioural economics and needs further discussion (see Chapter 8). Moreover, the fact that Georgia asked for this association might be considered a strategic move, also regarding environmental issues. It could help the country to differentiate itself from neighbouring countries with potentially positive effects for further economic and social development.

4.2.3 Circular economy followers: A critical view

These brief remarks on the situation in Russia and Georgia in combination with the preceding remarks on the situation in Germany and China show that certain characteristics of the countries seem to further or delay the development of a circular economy and the adoption of adequate strategies. For the case of Russia: abundance in natural resources, and sufficiently many and appropriate spaces for dumping waste of all kinds are certainly among them. In addition to that a still comparatively low level of economic wealth can play a role: a low value of GDP per capita might affect environmental awareness, giving priority to economic issues without paying too much attention to the continuing degradation of the environment, although the situation starts to change.

For the case of Georgia: there is, beyond the natural beauty of the country, neither abundance of natural resources, nor abundant space for landfilling. Obviously, a low level of environmental awareness, probably also due to a low value of GDP per capita, has kept the country till recently from major efforts to reduce environmental pollution. The association with the EU might change this situation.

These observations point again to the question raised earlier: how can the leaders motivate these countries, the followers, for circular economy strategies? An additional question is closely related to this one: what is the interest of the leaders to convince the followers to adopt their example? Especially this second question will have to be reconsidered in the context of implementing a circular economy.

The examples of Russia and Georgia point to possible strategies of the leaders to convince and motivate other countries to prepare and develop paths towards a circular economy. There is first of all the offer to support the followers by sharing their experiences. In an economic context, this implies a reduction of costs of the circular economy strategies for the followers. These experiences could also include scientific findings on the health risks associated with environmental pollutions and degradations, thereby raising voices in those countries for turning towards a circular economy. All this is already happening, not only in Russia and Georgia, but in many countries all over the world. Especially the Ellen MacArthur Foundation works with all kind of stakeholders in various countries to "accelerate the transition to a circular economy".

Expectations and promises on economic effects, employment opportunities associated with an implementation of a circular economy play a role in promoting circular economy strategies – by governments, but also by charities and other NGOs (see Section 3.3). The important issue is, whether these promises can be fulfilled. The problem thereby is less that these opportunities do not exist. The problem is, rather, how to modify the existing economy such that producers and consumers grasp these opportunities offered by the prospect of a circular economy. And one must not forget that not all countries can offer conditions amenable to certain business models in a circular economy. These issues refer to the implementation of a circular economy and need to be reconsidered in Part V.

Why should leaders provide support to other followers? Trade relations with export of environmental technologies to followers, perhaps at preferred terms, could be one of the reasons, which also helps to accelerate the adoption of a circular economy strategy (see Section 10.2). Another reason could be the intrinsic nature of the environmental commodities: for the case of global warming, for example, it is necessary that many countries, including the fast growing emerging economies, participate in efforts to mitigate climate change. Therefore adequate support from the CE followers could provide the missing motivation.

Georgia gives an example for another path towards a circular economy: the association of Georgia with the EU not only postulates to strengthen further "democracy and market economy" (EU, 2014, Preamble), but also the commitment "to respecting the principles of sustainable development, to protecting the environment and mitigating climate change … and meeting environmental needs, including cross-border cooperation and implementation of multilateral international agreements" (EU, 2014, Preamble). Consequently, integrating a country into some international organisation offering certain benefits for the country can be used as a vehicle to promote a circular economy.

There is, as already indicated, the possibility and recommendation to investigate this leader-follower relationship by means of a formal, game-theoretic model. There are some observations in Germany, China, Russia, but also in Georgia, which point in this direction. This is of particular interest for mitigating climate change with the focus on the joint provision of the public commodity "reduction of global warming". With such a model it would be possible to analyse more carefully the interdependencies between the leaders and the followers regarding the provision of this public good. Moreover, such an analysis could be useful for studying different measures and policy tools to affect the behaviour of followers, for example. (Weber et al. (2020)) investigate this leader-follower paradigm for burden sharing in a military alliance with a hierarchical membership structure (NATO).

So far, this first analysis of leaders and followers with respect to adopting strategies of a circular economy. On the bottom line the above observations make clear: there must be some "pressure" on a country, a pressure resulting from environmental pollutions and degradations, from actual or imminent depletions of exhaustible resources, in order to implement at least waste management strategies. Interestingly, these more or less external linkages between framework conditions and intrinsic motivation to act accordingly pertain also to the internal operations of a circular economy. This principle is of utmost importance for environmental economics in general, and the design of environmental policies for circular economics in particular. It will be thoroughly discussed in Part IV of the book and applied in Part V.

The following section discusses the positioning of the United States (U.S.) regarding circular economy strategies. This country is rather peculiar, as it is neither a fully convincing CE leader, nor an outspoken CE follower.

4.3 A country in between: The United States of America

According to Goblon (2017), the U.S. has not been a trendsetter regarding the adoption of a circular economy. The country is abundant with natural resources, and has lots of empty spaces. Thus, landfilling waste does not seem to pose too much of a problem, and the possibility of exceeding the assimilative capacity of the environment has not really been visible so far. On the other hand, there are numerous activities, related to a circular economy: reducing greenhouse gas emissions, segregation of waste, deposit system for drinks packaging, for example, and a multitude of recycling activities, sometimes existing as long as 30 or 40 years.

The U.S. is a rich country with a high level of GDP per capita and a presumably high level of environmental awareness. The Ellen Mac Arthur Foundation took this as an opportunity and launched a chapter of its Circular Economy 100 (CE100) programme in the U.S. in 2016. This programme includes corporations, universities, city and government authorities, and "provides a pre-competitive space to learn, share knowledge, and build new collaborative approaches", regarding the circular economy.

Among the motivations for in particular global business companies to participate in this network and/or to adopt a circular economy strategy is the aspect of remaining competitive in a highly interdependent world: environmental technologies developed in the course of circular economy strategies could be of relevance in trade relations.

Regarding environmental legislation, there are so far no regulations explicitly postulating the development towards a circular economy (see also Ghisellini, Cialani, & Ulgiati, 2016, 4.2). There are, of course, many regulations regarding air, soil and water pollution with the famous Clean Air Act of 1970, serving as a role model for similar legislation in various other countries, among them Germany with its Federal Immission Control Act of 1972. Moreover, the U.S. Environmental Protection Agency (EPA) informs about and promotes the concept of "Greener Living" with various aspects of a sustainable development. In particular, the waste hierarchy is addressed with a variety of hints how to reduce and reuse, and about recycling possibilities and, of course, in the last decades, various organisations and companies have already been implementing circular economy activities, without referring to them as such. Moreover, greenhouse gas emissions decreased in the last years including 2019 (see IEA, 2019, Fig. 14A), although the U.S. announced on June 1, 2017, its intention to withdraw from the Paris Agreement.

In summary, the U.S. clearly has a great potential for further developing the circular economy. In comparison to other countries, however, there seems to be insufficient public support for these many bottom-up activities, and the pressure from within the society seems also to be limited. This refers especially to the much-debated issue of the degree of anthropogenic causes of global warming. High environmental awareness, and the abundance of natural resources and ample possibilities of

landfilling seem to neutralise each other. According to the EPA, as long as it is ensured that landfills can be "built in suitable geological areas away from faults, wetlands, flood plains or other restricted areas", there is obviously no need for a more sophisticated waste management. Moreover, it might be that more locally operating business companies are afraid of losing their competitive edge in international trade relations in view of potentially higher costs associated with circular economy activities.

Thus there remains the hope that initiatives of the Ellen MacArthur Foundation and other prominent institutions will help, together with the pressure from trade relations, to further promote circular economy strategies in the U.S. The country for sure has the capacity, the knowledge and the means to become a global leader, also in this regard.

The next section summarises the most important facts on the CE Hierarchy – the existence of CE leaders and CE followers.

4.4 Determinants of the circular economy hierarchy

Relative abundance of natural resources, availability of land for landfilling waste, and the level of environmental awareness were considered drivers or preventers of adopting circular economy activities. Strategic considerations can, however, also play an important role in positioning a country – sometimes regarding the positioning, sometimes making better use of this positioning.

Regarding the five countries briefly and exemplarily investigated in the last sections, Table 4.1 indicates the relevance of the characteristics for the countries.

Table 4.1 shows that availability of natural resources and land for dumping waste are of importance for early adopting circular economy strategies, at least for seriously taking care of waste management. However, the example of

TABLE 4.1 Characteristics of CE leaders and CE followers.

	Germany (leader)	China (leader)	Russia (follower)	Georgia (follower)	United States (leader/follower)
Natural Resources	Not Really	Yes, but not all of them developed	Yes	Not Really	Yes
Space for Landfilling	Not Really	Yes, but not in the vicinity of the cities	Yes	Not Really	Yes
Environmental Awareness	Supposedly High	Probably Growing	Probably Growing	Supposedly Low	Supposedly High
Strategic Behaviour	Export of Technologies	Export of Technologies	Not Yet Visible	Getting Support	Export of Technologies

Factors affecting the adoption of circular economy strategies.

China points to additional aspects: environmental degradation, in particular air, soil and water pollution, can "convince" governments to switch to more environmentally friendly, perhaps even circular economy strategies. The fact that abundant natural resources are not yet completely developed, and the imminent scarcity of some resources are in favour of such a move.

For the case of Georgia, the association of the country with the EU might prove quite helpful. Of course, the country is still struggling with its first moves towards a circular economy, but these efforts will soon gain velocity. Russia is again different: recent demonstrations against further huge dump site outside the major cities point to some resistance in the population and the necessity to organise waste management in a more environmentally friendly way. Russia probably wants to gain and can gain from the experience and technologies of other countries.

Finally, the U.S. with serious efforts to protect the environment, many of them on the level of the states, and the significant export of environmental technologies, reveals a huge potential combined with strategic behaviour. In fact the export of environmental technologies is one of the key drivers in most CE leaders, one of the successful business models, so to say (see Section 10.2). Regarding circular economy strategies, the U.S. are currently lagging behind on the federal level – for political reasons.

Availability of natural resources and landfills, perhaps in combination with strategic thinking, are certainly of importance in this context. One should, however, not forget environmental awareness as one of the decisive factors.

References

Antikainen, R., Lazarevic, D., & Seppälä, J. (2018). *Circular economy: Origins and future orientations*: (pp. 115–129). Cham: Springer International Publishing. https://doi.org/10.1007/978-3-319-50079-9_7.

Boulding, K. E. (1966). *The economics of the coming spaceship earth. In H. Jarrett (Ed.), Resources for the future.* Johns Hopkins University Press Retrieved from http://arachnid.biosci.utexas.edu/courses/THOC/Readings/Boulding_SpaceshipEarth.pdf.

China. (2008). *Circular economy promotion law of the People's Republic of China.* Retrieved from http://www.fdi.gov.cn/1800000121_39_597_0_7.html.

Dinda, S. (2004). Environmental Kuznets curve hypothesis: A survey. *Ecological Economics, 49*(4), 431–455. https://doi.org/10.1016/j.ecolecon.2004.02.011.

Ermolaeva, Y. V. (2018). Problems of institutionalization of waste management in Russia. *Revista Amazonia Investiga, 7*(12), 261. Retrieved from https://www.researchgate.net/publication/330366366_Problems_of_institutionalization_of_waste_management_in_RussiaAmazonia_Investiga_Vol_7_No_12_2018.

Ermolaeva, Y. V. (2019). Problems of modernization of the waste management sector in Russia: Expert opinions. *Revista Tecnologia e Societade, 15*(35), 56. https://doi.org/10.3895/rts.v15n35.8502.

EU. (2014). Association agreement between the European Union and Georgia. *Official Journal of the European Union, L 261*, 57. Retrieved from https://eeas.europa.eu/sites/eeas/files/association_agreement.pdf.

Georgia. (1996). *Law of environmental protection. Retrieved from* https://www.matsne.gov.ge/ka/document/download/33340/19/en/pdf.

Georgia. (2014). *Law of Georgia: Waste management code.* Retrieved from http://environment.cenn.org/app/uploads/2016/06/Waste-Management-Code_FINAL_2015.pdf.

Germany. (1994). *Closed substance cycle waste management act.* Retrieved from https://germanlawarchive.iuscomp.org/?p=303.

Germany. (2012). *Act to promote circular economy and safeguard the environmentally compatible management of waste.* Retrieved from https://www.bmu.de/fileadmin/Daten_BMU/Download_PDF/Abfallwirtschaft/kreislaufwirtschaftsgesetz_en_bf.pdf.

Germany. (2017). *Renewable Sources Act (EEG 2017).* Retrieved from https://www.bmwi.de/Redaktion/EN/Downloads/renewable-energy-sources-act-2017.pdf%3F__blob%3DpublicationFile%26v%3D3.

Ghisellini, P., Cialani, C., & Ulgiati, S. (2016). A review on circular economy: The expected transition to a balanced interplay of environmental and economic systems. *Towards Post Fossil Carbon Societies: Regenerative and Preventative Eco-Industrial Development, 114*, 11–32. https://doi.org/10.1016/j.jclepro.2015.09.007.

Goblon, A. (2017). *Circular economy growth in North America.* Retrieved from http://circularconstruction.eu/2017/06/28/circular-economy-growth-in-north-america/.

Grossman, G. M., & Krueger, A. B. (1995). Economic growth and the environment*. *The Quarterly Journal of Economics, 110*(2), 353–377. https://doi.org/10.2307/2118443.

Heshmati, A. (2015). *A review of the circular economy and its implementation. IZA Discussion Papers, 9611.* Retrieved from http://ftp.iza.org/dp9611.pdf.

Huang, W. M., Lee, G. W. M., & Wu, C. C. (2008). GHG emissions, GDP growth and the Kyoto protocol: A revisit of environmental Kuznets curve hypothesis. *Energy Policy, 36*(1), 239–247. https://doi.org/10.1016/j.enpol.2007.08.035.

IEA. (2019). *International Energy Agency. CO2 emissions from fuel combustion—Highlights.* Retrieved from https://webstore.iea.org/download/direct/2521?fileName=CO2_Emissions_from_Fuel_Combustion_2019_Highlights.pdf.

Khakimova, D., Lösch, S., Wende, D., Wiesmeth, H., & Okhrin, O. (2019). Index of environmental awareness through the MIMIC approach. *Papers in Regional Science, 98*(3), 1419–1441. https://doi.org/10.1111/pirs.12420.

Korobova, N., Larionov, A., Michelsen, J. D., Pulyayev, M., Ivanovskyy, S., Turilova, K., & Kuznetsova, M. (2014). *Waste in Russia: Garbage or valuable resource.* Washington, DC: IFC World Bank Group. Retrieved from http://documents.worldbank.org/curated/en/702251549554831489/pdf/Waste-in-Russia-Garbage-or-Valuable-Resource.pdf.

Larionov, A., & Ecorem, N. V. (2012). *Municipal solid waste management: Opportunities for Russia. Summary of key findings.* Washington, DC: IFC, World Bank Group. Retrieved from https://www.ifc.org/wps/wcm/connect/topics_ext_content/ifc_external_corporate_site/sustainability-at-ifc/publications/publications_report_russia-solidwaste.

Lösch, S., Okhrin, O., & Wiesmeth, H. (2019). Awareness of climate change: Differences among Russian regions. *Area Development and Policy, 4*(3), 284–307. https://doi.org/10.1080/23792949.2018.1514982.

Nelles, M., Grünes, J., & Morscheck, G. (2016). Waste management in Germany—Development to a sustainable circular economy? *Waste Management for Resource Utilisation, 35*, 6–14. https://doi.org/10.1016/j.proenv.2016.07.001.

Pearce, D. W., & Turner, R. K. (1989). *Economics of natural resources and the environment.* Johns Hopkins University Press.

Plastinina, I., Teslyuk, L., Dukmasova, N., & Pikalova, E. (2019). Implementation of circular economy principles in regional solid municipal waste management: The case of Sverdlovskaya Oblast (Russian Federation). *Resources.* https://doi.org/10.3390/resources8020090.

Reike, D., Vermeulen, W. J. V., & Witjes, S. (2018). The circular economy: New or refurbished as CE 3.0?—Exploring controversies in the conceptualization of the circular economy through a focus on history and resource value retention options. *Sustainable Resource Management and the Circular Economy, 135*, 246–264. https://doi.org/10.1016/j.resconrec.2017.08.027.

Schnurer, H. (2002). *German waste legislation and sustainable development. Guest lecture at the International Institute for Advanced Studies in Kyoto, Japan.* Retrieved from https://www.bmu.de/fileadmin/bmu-import/files/pdfs/allgemein/application/pdf/entwicklung_abfallrecht_uk.pdf.

Song, M., Zhu, S., Wang, J., & Wang, S. (2019). China's natural resources balance sheet from the perspective of government oversight: Based on the analysis of governance and accounting attributes. *Journal of Environmental Management, 248*, 109232. https://doi.org/10.1016/j.jenvman.2019.07.003.

Stern, D. I. (2004). The rise and fall of the environmental Kuznets curve. *World Development, 32*(8), 1419–1439. https://doi.org/10.1016/j.worlddev.2004.03.004.

Su, B., Heshmati, A., Geng, Y., & Yu, X. (2013). A review of the circular economy in China: Moving from rhetoric to implementation. *Journal of Cleaner Production, 42*, 215–227. https://doi.org/10.1016/j.jclepro.2012.11.020.

Wang, Y., & Chen, X. (2020). Natural resource endowment and ecological efficiency in China: Revisiting resource curse in the context of ecological efficiency. *Resources Policy, 66*, 101610. https://doi.org/10.1016/j.resourpol.2020.101610.

Weber, S., & Wiesmeth, H. (2018). Environmental awareness: The case of climate change. *RUJEC, 4*(4), 328–345. https://doi.org/10.3897/j.ruje.4.33619.

Weber, S., Weber, Y., & Wiesmeth, H. (2020). Hierarchy of Membership and Burden Sharing in a Military Alliance. *Defence and Peace Economics.* https://doi.org/10.1080/10242694.2020.1782584.

Yong, R. (2007). The circular economy in China. *Journal of Material Cycles and Waste Management, 9*(2), 121–129. https://doi.org/10.1007/s10163-007-0183-z.

Yuan, Z., Bi, J., & Moriguichi, Y. (2006). The circular economy: A new development strategy in China. *Journal of Industrial Ecology, 10*(1–2), 4–8. https://doi.org/10.1162/108819806775545321.

Environmental regulations with a view on the circular economy

In order to complete the review of the conceptualisation of the circular economy in the literature and in practical applications, this chapter presents current environmental laws, directives and ordinances from different countries aiming at certain aspects of a circular economy. Naturally, this review is a rather incomplete, but nevertheless exemplary survey. It will provide insight into recent efforts implementing certain features of a circular economy. Experiences with these regulations serve as kind of a testbed for investigating the reaction of economic agents on certain policy tools and the interaction with technical tools and technological issues.

5.1 Exemplary regulations for a circular economy

The regulations refer to the waste hierarchy, keeping in mind that it is one of the prominent goals of circular economy strategies to preserve or to restore the assimilative capacity of the environment regarding waste of all kinds. In this sense, policies with the goal to mitigate climate change through reducing greenhouse gas emissions are included as well.

Unfortunately, the following examples will show that pursuing and achieving the various environmental goals formulated in these regulations seems to be difficult and anything else than straightforward.

5.1.1 The waste directive of the European Union

This directive (EU, 2008) of the European Union (EU), based on the waste hierarchy (Art. 4), does not yet directly refer to the transition to a circular economy – the corresponding action plan of the EU (2015) appeared only later. It rather refers to Extended Producer Responsibility (EPR) to strengthen prevention, reuse, recycling and other recovery of waste (Art. 8) with prevention being "the first priority of waste management, and that reuse and material recycling should be preferred to energy recovery from waste, … " (Preamble 7). The costs of these activities in waste management shall be borne by the original waste producer, or the current or previous waste holders – in accordance with the polluter-pays principle (Art. 14), which plays an important and recurring role in most of these environmental regulations.

Moreover, the member states of the EU shall ensure that their authorities establish waste management plans analysing the current waste management situation, and the measures to be taken for preparing for reuse, recycling, recovery and disposal of waste (Art. 28). In addition, the directive postulates a waste prevention programme with clearly identified waste prevention measures (Art. 29). Of course, the Commission of the EU has to be informed regularly of the implementation of this directive (Art. 37).

Annex IV of the directive presents a list of examples of waste prevention measures. This list includes measures that can affect framework conditions related to the generation of waste, the design and production and distribution phase, but also the consumption and use phase. However, there are so far no quota regarding the prevention or reduction of waste, although there are clear references to recovery and recycling targets (see Art. 10, 11, for example).

As this basic directive and others pertaining to waste have to be adopted by the member states, all regulations regarding waste in the EU refer also to this directive and therefore to the waste hierarchy in particular. It remains to be repeated that the EU always emphasises waste prevention with the highest priority.

Regarding reduction of municipal waste, the directive has, so far, not been overly successful: although municipal waste decreased between 2005 and 2017 from 515 kg per capita to 486 kg per capita, there are cyclical fluctuations resulting in increases through the most recent years. Germany and Denmark are among the countries with especially strong increments in this period. Interestingly, during the financial crisis waste generation decreased, likely due to reduced consumption and production activities. Fig. 5.1 shows some details regarding municipal solid waste generated and treated in the EU. There has been a reduction of landfilling with increasing shares of waste incinerated or materially recycled, while the total amount of waste has been more or less constant. On a per capita level, municipal waste could thus not be prevented in

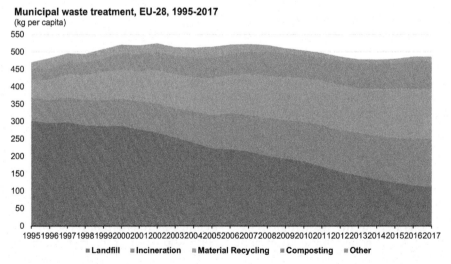

FIG. 5.1   **Municipal waste treatment.** EU-28 from 1995 to 2017 (kg per capita). *Source: Eurostat (online data code: env_wasmun). © From EUROSTAT. Municipal waste statistics. Retrieved February 13, 2020 from https://ec.europa.eu/eurostat/statisticsexplained/index.php?title=File:Municipal_waste_treatment,_EU-28,_1995-2018_(kg_per_capita).png (Original work published 2019).*

the EU-28, there was, however, a decoupling of municipal waste generation and economic growth.

Effects of economic developments on environmental issues are not unknown. They are, for example, also visible in the long-run development of global greenhouse gas emissions. Regarding the EU, greenhouse gas emissions have decreased since 1990 by more than 20%. But major upswings and downswings in the trend are most often connected with economic recoveries and downswings (see Fig. 5.5).

One possible explanation for this little or no progress in reducing waste or emissions across the EU (and elsewhere) and their relationship with major economic effects could be that practitioners are worried about negative economic consequences of serious waste prevention measures. Thus in a union with strong trade relations the Prisoners' Dilemma (see Section 7.1) might affect decisions with the consequence that major attempts to prevent waste or emissions are first left to other countries. There are, of course, countries, Germany among them, which understand that more and additional efforts are needed in order to reduce waste. This attitude at least points to some intrinsic motivations which need to be "nurtured" to turn waste management in the EU into a success story. The example of Germany shows that this is not easy to achieve.

Looking briefly at packaging waste with the EU Packaging Directive (EU, 2018) providing the basic regulations, the German Environment Agency (UBA) points to a total of 18.16 million tons of packaging waste generated in Germany in 2016, an increase of 0.05% over 2015, and of 19% over 2000. Plastic packaging increased by 74% between 2000 and 2016 (UBA, 2017a). These numbers imply 220.5 kg per capita packaging waste in Germany in 2016, compared to the 167.3 kg per capita consumption in the EU in 2015. 70% of the total packaging waste was recycled, with most of the remainder used for the production of energy ("thermal recovery"),

not completely in agreement with the regulations of the EU Waste Directive.

The recycling quota varies depending on the packaging: it is relatively high for glass (85.5%), paper/cardboard (88.7%), aluminium (87.9%) and steel (92.1%). Plastics (49.7%) and wood (26%) still have a lot of potential. Plastic packaging in particular – because of the diversity of the materials concerned – is difficult to sort and recycle. Nevertheless, the recycling of plastic packaging in 2016 was 0.9% higher than in the previous year – higher for the first time than the rate for energy production (UBA, 2017b).

It is also interesting to note that export of packaging waste amounted to 10.9% in 2016, all of which was reportedly destined for recycling. Also, 10.6% of plastic waste was exported, with no imports of same (see again UBA, 2017a). China was one of the major importers of plastic waste from Germany, until China banned these imports in 2018. Nevertheless, plastic waste from Germany continues to show up in various developing countries, not all of it recycled in a proper way, according to the regulation regarding the export of plastic (see Chapter 23 on plastics in a circular economy).

One of the conclusions of the UBA regarding prevention of packaging waste is that "waste prevention still remains just that – a principle for which no actual law has yet been enacted" (UBA, 2017b). It is, however, questionable, whether an additional law would help in this regard. Maybe looking at the incentives provided by the current legislation to motivate prevention of waste turns out to be more promising. There seems to be a need to better integrate all those individuals, who generate waste, into the policies. How this could be done, will be explored in Part V.

In conclusion, although a decoupling of waste generation and economic growth has been achieved, at least in the EU, both the Waste Directive and the Packaging Directive of the EU have not been overly successful in reducing waste with Germany assuming a rather low rank in the EU – unfortunately not only regarding packaging waste (UBA, 2017a). Interestingly, it is still the state of the economy, which seems to play an important role regarding these developments. Chapter 20 addresses further issues and introduces policy tools for handling waste, in particular packaging waste.

There are similarly interesting observations regarding the generation, collection and recycling of waste electric and electronic equipment (WEEE), which is gaining increasing importance all over the world, in particular again in the developing countries.

5.1.2 The directive of the European Union on waste electric and electronic equipment

This directive (EU, 2012), amending the original version of 1996, supplements the waste management legislation in an important field. The market for electronic equipment increases rapidly, innovation cycles get shorter and the replacement of equipment accelerates, making electric and electronic equipment a fast-growing source of waste. The subject of this directive is therefore to lay down measures "to protect the environment and human health by preventing or reducing the adverse impacts of the generation and management of waste from WEEE …" (Art. 1).

Important measures refer to the product design, facilitating reuse, dismantling and recovery of WEEE (Art. 4), the separate collection with collection rates (Art. 7) to minimise the disposal of WEEE in household waste (Art. 5), and proper treatment for recovery of WEEE using best available techniques (Art. 8), recovery targets (Art. 11). Shipments of WEEE to countries outside the EU are allowed, if they are in compliance with other regulations concerning the transboundary movement of waste (Art. 10). The financing of these activities has to come from the producers (Art. 12). Again, there are

requirements for the registration of the producers, for information and reporting on the quantities of equipment placed on the markets and the collected, reused, recycled and recovered WEEE (Art. 16).

From 2019, there is a required collection rate of 65% of WEEE, given the quantity (in terms of weight) sold on average in the last three years. In 2015, Germany reached 42.5%, still below the rate of 45% required for 2016. In absolute terms, these collection numbers range from 1.6 kg per capita in Romania to 16.5 kg per capita in Sweden, with a share of approximately 55% referring to large household appliances. Fig. 5.2 provides some information on WEEE collected pointing to a slight increase in more recent years, probably resulting from a similar increase in equipment put on the market.

Whether it is meaningful to use absolute or relative collection rates, remains to be discussed. There is, however, another issue, which deserves a closer look, and which will be investigated more carefully in Chapter 21: it is the often semi-legal or even illegal export of WEEE to developing countries. In 2008 some 155.000 tons of WEEE, declared as reusable, were exported from Germany, also to countries such as Nigeria, Ghana, India or South-Africa. In these countries, the old equipment is "recycled", often under conditions hazardous to health and environment (Sander & Schilling, 2010; UBA, 2010). Although this practice has likely slowed down in recent years due to stricter regulations and stricter enforcement at least in the EU, it seems to continue to endanger the health of people and to pollute the environment in developing countries. This could help to explain the still comparatively low collection rates of WEEE of less than 50% for most EU member states and less than 40% for quite a few of them. Fig. 5.3 shows that these numbers are still some distance away from the 65% collection target set for 2019.

Again, a closer look at the existing legislations can help to at least reduce these questionable WEEE exports, which are certainly not tolerable from the point of view of a sustainable

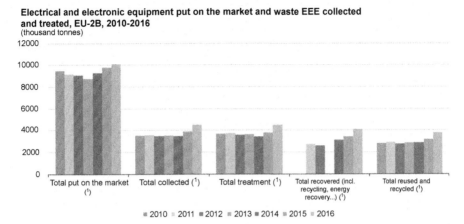

Electrical and electronic equipment put on the market and waste EEE collected and treated, EU-2B, 2010-2016
(thousand tonnes)

■ 2010 ■ 2011 ■ 2012 ■ 2013 ■ 2014 ■ 2015 ■ 2016

FIG. 5.2 **(Waste) Electrical and electronic equipment in the EU-28.** Equipment put on the market and WEEE collected. © *From EUROSTAT. Waste statistics—Electrical and electronic equipment. Retrieved February 13, 2020 from https://ec.europa.eu/ eurostat/statistics-explained/index.php?title=File:Figure1_Electrical_and_electronic_equipment_put_on_the_market_and_waste_EEE_ collected_and_treated,_EU-28,_2010–2016_(thousand_tonnes).png (Original work published 2019).*

development in a circular economy. It also remains to some extent unclear, whether manufacturers of electronic equipment are really interested in a Design for Environment (DfE), which might increase costs with perhaps uncertain return from demand. In this context, it has to be noted that it is the manufacturers, who have the necessary knowledge regarding a DfE, not the public authorities. Manufacturers might therefore want to wait with a DfE update till other manufacturers make the first move.

The main question regarding WEEE is again how to better integrate all the consumers, manufacturers and importers of electric and electronic equipment into the relevant policies to motivate them to comply with the regulations? How to stimulate a DfE, which seems to be particularly important for electronic equipment with its many different and also valuable components and materials, which, in addition, pose a serious health risk, if recycled in an improper way? These aspects will be further analysed and discussed in Chapter 21.

These conclusions apply mutatis mutandis also to car manufacturers and the fate of scrap cars, so-called end-of-life vehicles (ELV).

5.1.3 End-of-life vehicles legislation in Germany

The German end-of-life vehicle legislation is based on the corresponding directive of the EU (2011). The core regulations refer to the take-back requirement of scrap cars through the manufacturers at no cost to the owner. Moreover, the manufacturers and the recycling companies, acting on their behalf, have to observe the environmental standards – the given targets referring to material recovery and recycling, and are, of course, also obliged to observe health standards and other regulations regarding the export of old, but reusable cars. Again there is a reference to a DfE in order to reduce future waste by appropriately designing cars.

In 2017, the situation in Germany was as follows (Germany, 2019): of the 2.98 million cars, which were deregistered, only some 510,000 were carefully recycled in Germany – in compliance with the regulations. In particular, more than 98% of all metals could be recovered. 1.99–2.14 million used cars were exported to other member states of the EU, and 180,000–280,000 used cars were exported to non-EU countries (mainly to CIS members and some countries in Africa).

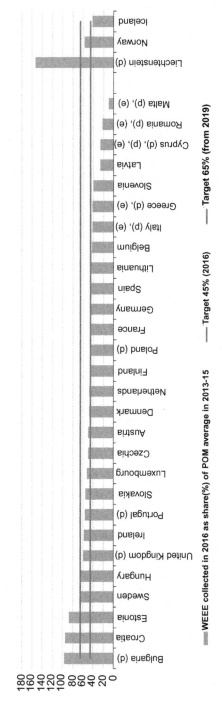

Rate of total collection of waste electrical and electronic equipment in 2016 in relation to the average weight of EEE put on the market in the three preceding years (2013-2015) (%)

Note: Ranked on 'Share of WEEE collected...' data.
(d) definition differs, see metadata
(e) estimated
(p) provisional

■ WEEE collected in 2016 as share(%) of POM average in 2013-15 ── Target 45% (2016) ── Target 65% (from 2019)

■ eurostat

FIG. 5.3 **Collection of WEEE in 2016.** Rate of collection of WEEE in relation to EEE put on the market. *Source: Eurostat (online data code: env_waselee). © From EUROSTAT. Waste statistics—Electrical and electronic equipment. Retrieved February 13, 2020 from https://ec.europa.eu/eurostat/statistics-explained/index.php?title=File:Figure_5_Rate_of_total_collection_for_WEEE_in_2016_in_relation_to_the_average_weight_of_EEE_POM_2013-2015_percentage.png (Original work published 2019).*

Thus, the statistical results of Germany, pointing to a recycling rate of 97%, has to be taken with a grain of salt, as it does not include the destiny of the exported cars, in particular of those, exported to countries outside the EU.

Even if most of the cars can be reused and are reused, this statistic does not provide much information on the fate of these cars, once they are outside Germany. Of particular relevance is that exports directly from Germany or from Germany to members state of the EU and then to non-EU countries can mean that appropriate maintenance of these cars is not guaranteed – with potentially aggravating consequences for air pollution and other environmental concerns. A similar consideration applies to recycling of these cars. The question is, how to modify this practice without preventing car drivers from buying used German cars? After all, reusing cars, which are still in a good condition, is for sure a matter of a circular economy.

For example, Germany exported some 10,800 used cars to Georgia in 2016 (Germany, 2019, p. 34), and Tbilisi is one of the cities with a rather high air pollution, not only, but also related to transport activities.

The aspect of a DfE, raised for the case of electrical and electronic equipment, also refers to the car manufacturers. The smaller the number of old vehicles that have to be recycled in Germany (or the EU), the lower the pressure for a possibly costly DfE, which reduces the total expenditures for recycling (Gerrard & Kandlikar, 2007).

Fig. 5.4 shows some more details: the numbers for Germany, for example, are with approximately 5 dismantled cars per 1000 inhabitants far below the EU-28 average which is above 10 car. This indicates once more that many cars produced in Germany are not dismantled in Germany. This seems to be different for other car manufacturing countries such as France with significantly higher dismantling numbers.

Consequently, a lot needs to be done regarding a DfE for electronic equipment and cars, and appropriate recycling activities in this context.

Chapter 21 discusses details of an EPR policy in this field. Also, the issue of mitigating climate change, as well of relevance for the transition to a circular economy, requires further attention: the transport sector is one of the major emitters of greenhouse gases, and aspects of an ELV policy are therefore related to efforts mitigating climate change.

5.1.4 Regulations to mitigate climate change

The 2030 Climate & Energy Framework of the EU adopted in 2014 with the targets for renewables and energy efficiency revised upwards in 2018, is a set of binding legal regulations to ensure the EU meets its climate and energy targets for the year 2030 (EU, 2014). The key targets include an at least 40% cut in greenhouse gas emissions from 1990 levels, a share of at least 32% of EU energy from renewables and an at least 32.5% improvement in energy efficiency.

To meet these targets the EU is employing its emission trading system (ETS) for cutting greenhouse gas emissions by 43% (compared to 2005). There are, in addition, national targets covering sectors not in the ETS, accounting for 55% of total EU emissions, such as housing, agriculture, waste and transport (excluding aviation) – the targets differ according to national wealth, but should cut emissions in total by 30%. There are also binding national targets for raising the share of renewable energies in their energy consumption by 2020 – again varying across countries, to reflect their relevant differences regarding the economy, the geography and other relevant characteristics.

Among the benefits, the EU counts an increasing energy security as well as advancing green growth and rendering the EU more competitive.

In Germany the share of renewable energies in electricity consumption increased from 6.3% to 36% between 2000 and 2017, the share in final energy consumption increased from 6.2% in

End-of-life vehicles, 2008 and 2016

(number of dismantled cars per 1000 inhabitants)

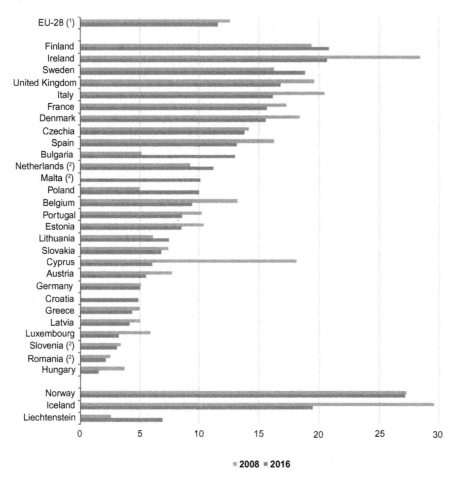

■ **2008** ■ **2016**

(¹) Eurostat estimates for 2008 and 2016.
(²) Eurostat estimates for 2016.

FIG. 5.4   End-of-life vehicles in Europe 2008 and 2016. Number of dismantled cars per 1000 inhabitants. *Source: Eurostat (online data code: env_waselvt). © From EUROSTAT. End-of-life vehicles statistics. Retrieved February 13, 2020 from https://ec.europa. eu/eurostat/statistics-explained/index.php?title=File:End-of-life_vehicles_2008_and_2016_number_of_dismantled_cars_per_1000_ inhabitants.png (Original work published 2019).*

2004 to 15.9% in 2017 (UBA, 2018). Nevertheless, Germany (and quite a few other EU member states) will likely miss their 2020 targets for reducing greenhouse gas emissions. Among the reasons is the too slow progress providing the grid infrastructure necessary to transport electricity from renewable sources.

Also on a global level energy-related CO_2 emissions continue to rise according to the International Energy Agency (IEA, 2019, p. 3). All attempts of the United Nations Framework Convention on Climate Change (UNFCCC) and other global endeavours did not really succeed in curbing greenhouse gas emissions. As already observed in earlier time periods, only economic downswings reduced global emissions for some time. With economic recovery, emissions picked up again (IEA, 2019, p. 24).

According to the IEA, these emissions rose again by 1.7% in 2018 over 2017 to a "historic high" of 33.1 Gt CO_2 of global energy-related CO_2 emissions, with the power sector contributing nearly two-thirds of emissions growth. China, India, and the United States (U.S.) accounted for 85% of the net increase in emissions, while emissions declined for Germany, Japan, Mexico, France and the United Kingdom. This is clearly in contrast with the sharp reduction needed to meet the goals of the Paris Agreement on climate change (IEA, 2019, p. 3).

However, one has to respect that China's emissions grew by just 2.5% in 2018, despite of

an economic growth of 6.6%. Interestingly, India's per-capita emissions remain still at some 40% of the global average, although emissions grew by 4.8%. Across Europe emissions fell by 1.3%, obviously not sufficient to compensate for the expected further increases in a variety of countries. Fig. 5.5 shows decreasing greenhouse gas emissions in the EU with economic recessions and crises leaving their temporary footprints. On a global level, however, emissions from fuel combustion keep rising – after three years of stability – "with robust economic growth and the slowdown in renewables penetration more than offsetting some improvement in energy productivity" (see IEA, 2019, p. 9, 22, Fig. 1).

This result seems again to be a consequence of economic necessities, aggravated through wide differences in awareness of climate change (see, for example, Lösch, Okhrin, & Wiesmeth, 2019 for the case of Russian regions). Without a powerful supranational organisation

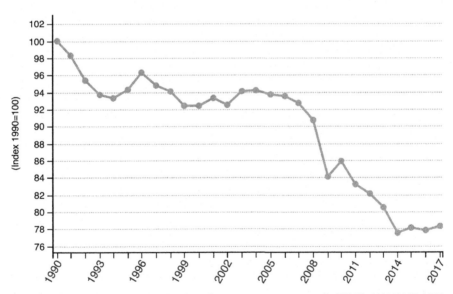

FIG. 5.5 **Greenhouse gas emissions in the EU.** Greenhouse gas emission trends, EU-28, 1990–2017, 1990=100. © *From EUROSTAT. Greenhouse gas emission statistics—Emission inventories. Retrieved February 13, 2020 from https://ec.europa.eu/ eurostat/statistics-explained/index.php?title=Greenhouse_gas_emission_statistics_-_emission_inventories (Original work published 2019).*

necessary changes to avoid a too rapid increase of the average global temperature might be difficult to achieve. Thus, an effective global climate change policy will probably remain a challenging task for the years to come. Chapter 22 details and discusses relevant features and aspects of such a policy.

For this reason quite a few countries started to prepare themselves for some likely effects of global warming such as floods or draughts, periods of excessive cold or excessive heat. Adapting to climate change seems to take over and partially displace the role of mitigating climate change. Whether this is a signal of surrendering to the facts, remains to be seen.

5.1.5 China's circular economy promotion law

The environmental challenges associated with and accompanying the industrialisation in China and its rapid economic growth led to the adoption of the circular economy as a new development strategy in 2002. The Circular Economy Promotion Law of 2009 (China, 2008) provides the main framework for environmental regulations on the national level, with various action plans with further details for specific sectors (McDowall et al., 2017). Although there is a broad engagement of the government, Zhu, Fan, Shi, and Shi (2019) argue that the current policy framework "is concerned more with the means rather than the ends of the circular economy, and relies too much on direct subsidies and other financial incentives" (see Abstract). In particular, they point to a missing "sustainable economy scale" (p. 111).

There certainly are visible improvements regarding various kinds of environmental pollution, although it is difficult to find comprehensive data. Lin and Wang (2018) refer to the "great importance" the government has given to soil protection, and other environmental issues (p. 55), and Silver, Reddington, Arnold, and Spracklen (2018) refer to positive, but also

negative changes regarding ambient air pollution in recent years. On the other hand, greenhouse gas emissions have doubled between 2005 and 2014 and continue to grow, although on a slower pace and with a small decrease between 2014 and 2016, and China's per capita emissions are also growing, with 6.6 tCO_2 in 2016 gradually approaching those of EU member states with 6.8 tCO_2 (see IEA, 2019, p. 47, 69).

Yuan, Bi, and Moriguichi (2006) explain China's three-layer approach to the implementation of a circular economy, based on cleaner production, industrial ecology, and ecological modernisation (p. 6). The centralised and government-controlled procedure certainly allows direct and immediate contacts with polluting enterprises. However, whether, companies will be "encouraged to design more environmentally friendly products ..." (p. 6), remains to be seen. After all this requires the direct support from the companies – also in China they typically have the knowledge required for a DfE.

McDowall et al. (2017) refer to the differences of China's approach to a circular economy in comparison to Europe's approach, "rooted in different industrial structures and different governance systems" (p. 7), and they also highlight relative advantages of the two systems (p. 8). It seems that a successful and efficient transition to a circular economy needs to be based, depending on the context, both on more centralised, but also on more decentralised strategies. This issue will be addressed again in the chapters of Part V.

In summary, some of the profound effects of China's circular economy strategies on all dimensions of the environment are currently and will further be overlaid by consequences on the environment of the still strong economic growth and the further increasing population. Furthermore, the "Social Credit System" that China will soon introduce, will probably also collect and evaluate data on the environmental performance and emissions of business companies. The resulting scores are likely to affect the activities of these companies in one way or the other. It will be interesting to observe this development.

In the end, the question, whether a centralised, top-down approach towards a circular economy is of advantage against the more decentralised attempts in other countries, will remain open for some time in the future. After all, the transition to a circular economy needs reliable guidance from the governments, but without the continuous support and the motivation of the "stakeholders" – in particular consumers and producers, the further transition to a circular economy will remain difficult. For sure, the relevant local conditions in a country have to be respected for a successful implementation of a circular economy. The chapters of Part V will look more carefully into this issue.

5.1.6 California's Bottle Bill

Enacted in 1987, "California's Bottle Bill", the "California Beverage Container and Litter Reduction Act", is considered one of the outstanding recycling and pollution prevention programmes in the U.S. As an initiative of California's Department of Resources, Recycling and Recovery (CalRecycle) its vision is "to inspire and challenge Californians to achieve the highest waste reduction, recycling and reuse goals in the nation". Its goal is to achieve an 80% recycling rate for all beverage containers sold in California. In addition, the programme is proud of many jobs created in the recycling industry.

The programme is funded by beverage distributors through redemption payments on each beverage container sold in the state. Redemption payment revenues are deposited in the California Beverage Container Recycling Fund (Fund). Payments are made out of the Fund to consumers in the form of California Refund Value (CRV) when they return empty beverage containers to certified recycling centres. The redemption payments paid by distributor and

the CRV paid to consumers are 0.05 US-Dollar (USD) per container less than 24 oz. Recycling centres are subsidised from the Fund in order to close the gap between recycling costs and the value of recyclables. Thus, there is kind of a public-private partnership regarding the recycling of the returned drinks packaging.

Recently, however, container recycling rates have fallen below 80% for the first time in a couple of years. Moreover, many recycling centres closed or, rather, had to close due to increasing operating costs resulting from higher wages, also the minimum wage, and requirements for health insurance. Additionally, commodity prices, in particular of aluminium and PET plastics, declined substantially. The payments and adjustment of the payments from the Fund were largely considered insufficient to close this gap between rising costs and decreasing revenue.

The question that is related with this and other observations in such contexts is, of course, the cooperation between private (recycling) companies and the public agencies. No doubt, private business is needed to carry out all the required tasks of collecting, sorting and recycling waste drinks containers. But how should this cooperation be structured in order to avoid these calamities?

Consequently, the transition to a circular economy is also associated with challenges regarding suitable and feasible private-public partnerships in various areas in waste management. Thereby the interests of various stakeholders meet in interesting ways: public agencies (reduce littering and environmental pollution, stimulate a DfE), distributors of drinks (earn profits, increase market share), producers of packaging (earn profits), private waste management companies (generate profits). The issue is that not all of these goals are compatible: waste reduction through innovations in drinks packaging, for example, is not necessarily compatible with higher profits in the packaging industry and the recycling industry. Thus, a DfE for drinks packaging may have negative effects for packaging producers and recyclers, in addition to higher costs for distributors of drinks.

The following section reviews these findings from current regulations in the context of the circular economy.

5.2 Conclusions from current policies

This short review of a selection of current policies related to characteristics and features of a circular economy reveals various challenges, which are associated with its implementation. There are, of course, positive effects on the environment associated with these policies, although it proves to be hardly possible to achieve the goals at large. What are reasons for these not really satisfactory results? What needs to be done in order to obtain better results?

A closer look at the above cases shows that in most cases a reduction of the specific waste could not be achieved so far, there are exports of waste and, thus, "exports" of the problem, also reducing the need for a DfE. The following observations require therefore a more careful investigation:

- the member states of the EU did not succeed to significantly reduce waste, in particular packaging waste;
- the environmental issues associated with WEEE, such as semi-legal exports or the requirement of a DfE, could not really be solved; this is also a problem of declaring old electronic equipment reusable;
- used cars still contribute to air pollution in many cities outside the EU, in particular in Africa and in Eastern Europe; there is, consequently, also a need for more carefully considering conditions for reusing old cars;
- efforts to mitigate climate change have, so far, not led to a decrease of global greenhouse gas emissions; only wealthy industrialised countries succeeded in reducing emissions;
- issues of the structure and the management of policies requiring the cooperation between

public agencies and private companies need a clarification in view of achieving the policy goals;
- and the questions of the advantages or the disadvantages of a top-down vs. a bottom-up approach for strategies regarding the transition to a circular economy have to be discussed.

One of the main reasons for these and other observations has to do with the fact that individual producers and consumers do not sufficiently contribute towards the goals of the various policies. It is, of course, not possible, and in most countries politically also not feasible to fully control compliance with the regulations, or to motivate these individuals to "collectively" follow the regulations. The question that remains is, whether these observations are a consequence of inadequate policies, of an insufficient environmental awareness, or rather of both?

Another question in this context refers to the role of innovative technologies for accelerating the development of a circular economy. This becomes immediately clear with respect to new technologies for recycling waste products. Moreover, there are indeed various publications investigating the relationship between the organisation of waste management and required technologies (Cossu & Masi, 2013), and the effects of different collection systems on collection rates of municipal solid waste (Gallardo, Bovea, Colomer, Prades, & Carlos, 2010). Thus there is an issue, which also requires a further and more detailed analysis: what is the role of technologies in general and of innovative technologies in particular for a transition to a circular economy.

Together with attempts to raise environmental awareness, this issue touches aspects of behavioural economics, which, due to the necessity of an appropriate integration of all stakeholders, is of utmost importance for implementing a circular economy. For example, consumers and producers have to be motivated to reduce waste, producers need more incentives for a DfE, especially regarding cars and electronic equipment, and many countries need to be convinced that participating in efforts to mitigate climate change might also be of advantage or even importance for them. Chapter 8 will carefully investigate these issues and will also provide various practical examples.

Not all of these issues can be addressed straightforwardly, however. The reason is that policies have to be directed to groups of consumers and producers for example, without knowing their preferences or their production possibilities. This may then lead to ineffective regulations, or to excessive reactions, or even to unexpected reactions. Current regulations provide ample experience for this. The question, which is considered in this book, is then to refine the policies to avoid at least major inconsistencies. Again, conclusions from behavioural economics will play a role in this context.

Another aspect worth mentioning refers to the management of global environmental commodities such as climate change. In this case, the cooperation of many countries is required, which, as experience shows, can be tedious without a global government. Benefits derived from attempts to mitigate climate change will typically be different from the costs associated with any activities in this area.

With these excursions into the practice of the circular economy, these findings regarding attempts to approach a circular economy, conclude the first part of the book. The second part discusses the economic background of the circular economy in more detail. This is necessary in order to introduce and analyse the tools available for implementing a circular economy.

References

China (2008). *Circular economy promotion law of the People's Republic of China*. Retrieved from http://www.fdi.gov.cn/1800000121_39_597_0_7.html.

Cossu, R., & Masi, S. (2013). Re-thinking incentives and penalties: Economic aspects of waste management in Italy. *Waste Management*, 33(11), 2541–2547. https://doi.org/10.1016/j.wasman.2013.04.011.

EU. (2008). *Directive on waste*. Retrieved from https://eur-lex.europa.eu/legal-content/EN/TXT/PDF/?uri=CELEX:32008L0098&from=DE.

EU. (2011). *Directive on end-of-life vehicles*. Retrieved from https://eur-lex.europa.eu/legal-content/EN/TXT/PDF/?uri=CELEX:32011L0037&from=EN.

EU. (2012). *Directive on waste electrical and electronic equipment*. Retrieved from https://eur-lex.europa.eu/legal-content/EN/TXT/PDF/?uri=CELEX:32012L0019&from=EN.

EU. (2014). *2030 climate & energy framework*. Retrieved from https://ec.europa.eu/clima/policies/strategies/2030_en.

EU. (2015). *Closing the loop—An EU action plan for the circular economy*. Retrieved from https://eur-lex.europa.eu/resource.html?uri=cellar:8a8ef5e8-99a0-11e5-b3b7-01aa75ed71a1.0012.02/DOC_1&format=PDF.

EU. (2018). *Directive on packaging and packaging waste*. Retrieved from https://eur-lex.europa.eu/legal-content/EN/TXT/PDF/?uri=CELEX:32018L0852&from=EN.

Gallardo, A., Bovea, M. D., Colomer, F. J., Prades, M., & Carlos, M. (2010). Comparison of different collection systems for sorted household waste in Spain. *Waste Management*, 30(12), 2430–2439. https://doi.org/10.1016/j.wasman.2010.05.026.

Germany. (2019). *Jahresbericht über die Altfahrzeug-Verwertungsquoten im Jahr 2017. (in German)*. Retrieved from https://www.bmu.de/fileadmin/Daten_BMU/Download_PDF/Abfallwirtschaft/jahresbericht_altfahrzeug_2017_bf.pdf.

Gerrard, J., & Kandlikar, M. (2007). Is European end-of-life vehicle legislation living up to expectations? Assessing the impact of the ELV Directive on 'green' innovation and vehicle recovery. *Journal of Cleaner Production*, 15(1), 17–27. https://doi.org/10.1016/j.jclepro.2005.06.004.

IEA. (2019). *International energy agency. CO2 emissions from fuel combustion—Highlights*. Retrieved from https://webstore.iea.org/download/direct/2521?fileName=CO2_Emissions_from_Fuel_Combustion_2019_Highlights.pdf.

Lin, Y., & Wang, G. (2018). *20th anniversary for soil environment protection in China*: (pp. 55–63). Singapore: Springer Singapore. https://doi.org/10.1007/978-981-10-6029-8_4.

Lösch, S., Okhrin, O., & Wiesmeth, H. (2019). Awareness of climate change: Differences among Russian regions. *Area Development and Policy*, 4(3), 284–307. https://doi.org/10.1080/23792949.2018.1514982.

McDowall, W., Geng, Y., Huang, B., Barteková, E., Bleischwitz, R., Türkeli, S., … Doménech, T. (2017). Circular economy policies in China and Europe. *Journal of Industrial Ecology*, 21(3), 651–661. https://doi.org/10.1111/jiec.12597.

Sander, K., & Schilling, S. (2010). *Transboundary shipment of waste electrical and electronic equipment/electronic scrap. (On behalf of the German environment agency)*. Retrieved from https://www.umweltbundesamt.de/sites/default/files/medien/461/publikationen/3933.pdf.

Silver, B., Reddington, C. L., Arnold, S. R., & Spracklen, D. V. (2018). Substantial changes in air pollution across China during 2015–2017. *Environmental Research Letters, 13*(11), 114012. https://doi.org/10.1088/1748-9326/aae718.

UBA. (2010). *Export von Elektroaltgeräten: Fakten und Maßnahmen.* (in German)German Environment Agency. Retrieved from https://www.umweltbundesamt.de/sites/default/files/medien/publikation/long/4000.pdf.

UBA. (2017a). *Packaging consumption in Germany.* German Environment Agency. Retrieved from https://www.umweltbundesamt.de/en/press/pressinformation/level-of-packaging-consumption-in-germany-remains.

UBA. (2017b). *Packaging* (in Germany). German Environment Agency. Retrieved from https://www.umweltbundesamt.de/en/topics/waste-resources/product-stewardship-waste-management/packaging.

UBA. (2018). *Indicator: Renewable energy.* German Environment Agency. Retrieved from https://www.umweltbundesamt.de/en/indicator-renewable-energy#textpart-1.

Yuan, Z., Bi, J., & Moriguichi, Y. (2006). The circular economy: A new development strategy in China. *Journal of Industrial Ecology, 10*(1–2), 4–8. https://doi.org/10.1162/108819806775545321.

Zhu, J., Fan, C., Shi, H., & Shi, L. (2019). Efforts for a circular economy in China: A comprehensive review of policies. *Journal of Industrial Ecology, 23*(1), 110–118. https://doi.org/10.1111/jiec.12754.

Integrating the economy and the environment

The second part of the book attends to the embedding of the circular economy (see also Definition 2.1) and its various perceptions into economics in general and environmental economics in particular. The goal of Part II is therefore to carefully link the circular economy to a regular economic system. Then basic economic concepts such as the allocation problems, feasible allocations as solution of these problems, and allocation mechanisms to achieve a solution can be considered in the more challenging context of a circular economy and its implementation. Economics of the circular economy is, however, anything else than a mere extension of the traditional economic procedures to a new, perhaps somewhat different field. To the contrary, the usual context of a market system, for example, does not work for a circular economy, at least not without any augmentations. Other allocation mechanisms have to replace or support the market system.

With this background, Chapter 6 discusses the concept of an environmental commodity, which is a natural extension of the concept of a regular economic commodity to the environmental context. Of importance is the fact that scarcity of environmental commodities need not be perceived sufficiently to raise environmental awareness for an environmentally friendly solution of the allocation problems. Similarly, a missing feedback or an insufficiently perceived feedback from environmentally friendly actions can retard or even prevent such actions.

The usual optimality concept for feasible allocations is Pareto-optimality characterising equilibria in pure market economies. The market system is the most prominent allocation mechanism for solving the fundamental allocation problems. Also central planning can lead to efficient or optimal outcomes, although the practical focus remains clearly on market systems after the failures of major centrally planned economies. Nevertheless, most economic systems need some planning, not least because of the increasing relevance of environmental issues, in particular the allocation of environmental commodities – equivalent to measures protecting the environment.

Chapter 7 focuses then in more detail on the allocation of environmental commodities. There are challenges solving the allocation problems in a circular economy, external effects lead to missing markets and methods to internalise these effects, to complete the market system need to be discussed. The Tragedy of the Commons and the Prisoners' Dilemma are two mechanisms of relevance for implementing a circular economy. They are simple to explain, but difficult to control, and they can also be triggered by an insufficient perceived feedback from environmentally friendly actions – also by inappropriate regulations in environmental policies, which are proposed as alternative allocation mechanisms for implementing a circular economy. The case study on the German Refillable Quota Issue demonstrates this "policy failure".

Chapter 8 turns towards a rather new field, nevertheless of importance for implementing a circular economy: behavioural environmental economics. This field deviates from classical economics in a variety of aspects. There is, for example, the insufficient capacity to gather and process relevant information, and there are social norms, which affect our behaviour in a multitude of ways. Especially social norms seem to play an important role for implementing a circular economy, as not all individuals can be monitored and controlled regarding their decisions with relevance for the environment. So the question arises, how to establish appropriate social norms to raise the perception of possible feedbacks from environmental actions, to raise environmental awareness, and to outwit the Tragedy of the Commons and the Prisoners' Dilemma. A case study on attempts to protect biodiversity shows some of the efforts to change behaviour.

Chapter 9 returns to the "economics of implementing a circular economy". It characterises feasible allocations for a circular economy, focusing on sustainability and, in particular on the waste hierarchy. The literature shows that characterising waste prevention in environmental regulations as priority goal is not sufficient: it seems to vanish behind collection and recycling activities, usually justified by reducing waste that goes to landfills and the various recovery procedures to save resources. This is, however, also in view of basic laws pf physics not in accordance with Definition 2.1 of the circular economy.

These obvious challenges to define and achieve "optimal" solutions of the allocation problems in the context of implementing a circular economy position technological advancements quite naturally as "allocation mechanism": new environmental technologies determine the path towards a circular economy. The advantage is that these technologies help to reduce current and future environmental pollution. However, the development and implementation of these technologies depend on societal path dependencies, for example on profitable business activities, and generate technological path dependencies. This leads to further interesting issues such as "vested interests" or "lock-in effects", which will be explored in Part III and in Part V.

Summarising, this second part of the book considers the bridge to the exciting world of the circular economy – across turbulent rapids with many challenges.

6

Economic foundation of a circular economy

This chapter presents fundamental economic concepts, which are necessary to grasp a circular economy and its implementation from an economic point of view: economics is about understanding and guiding the economic behaviour of individuals – for example, towards a circular economy in the context considered here. The starting point is the conceptualisation of the circular economy as presented in Definition 2.1 and proposed in Pearce and Turner (1989). In accordance with this picture of the circular economy, there is the requirement and the necessity to fully integrate the fundamental functions of the environment into all relevant economic activities in a sustainable way. Then a transition towards a circular economy, an implementation of a circular economy, implies a development, or at least an adjustment of the economic activities in this direction.

The first section reviews the basic concepts of commodities and scarcity with applications to the environmental context and references to the circular economy. "Perceived" scarcity and the necessity of a feedback from one's own action play some role in this context. The next section discusses the allocation problems including optimal solutions, followed by a brief review of possible allocation mechanisms. This section introduces and describes both the market system and methods of central planning – as contrasting examples. A summary on the economic foundation of a circular economy concludes the chapter.

The following presentations are kept short, just to the extent the material is needed for a sufficient understanding of the economics of implementing a circular economy. More details and examples regarding this basic economic context can be found in textbooks in economics and environmental economics (see, for example, Wiesmeth, 2011, Ch. 4).

6.1 Environmental commodities and scarcity

The first concept has to be, by necessity, the concept of an environmental commodity or an environmental good. This extension of the fundamental economic concept of a commodity is, on

65

the one hand, quite natural, but, on the other hand, also decisive: it indicates that environmental issues affect the wellbeing of humans, that various environmental commodities can be "produced" to protect the environment, and that economic tools and instruments can be used to stimulate production of these commodities, thus to stimulate an environmentally friendly behaviour.

6.1.1 Environmental commodities

The concept of a "good", or equivalently, a "commodity", comprising both a physical commodity, also a resource, or a service, is fundamental for any economic system and can be extended to include environmental goods or environmental commodities. Like any other commodity, environmental commodities touch the human sphere in one way or the other, affect the wellbeing of mankind in general, and of individuals in particular. Examples are the prevention or reduction of a polluting substance, the view of a spectacular landscape, protection from hazardous ultraviolet (UV) radiation or assimilation of waste – just to name a few.

Usually environmental "goods" include also pollutions of all kinds, measured by the amount and the environmental characteristics of the polluting substance, the location and the time period of the pollution. In this case it would certainly be better to talk about environmental "bads". However, there is always the possibility to switch from the discussion of environmental bads to environmental goods and vice versa by talking about pollution on the one hand, and the reduction of the pollution on the other.

One remark of utmost importance refers to the (economic) scarcity of goods in general and environmental goods in particular. A good or commodity becomes an economic entity only, when it is scarce, meaning that the total amount available at a particular location in a particular period of time is less than demanded in aggregate, without constraining demand in one way

or the other – through prices, for example, in the context of a market system (see Section 6.3.1). If a certain commodity is not scarce, then there is no need to care about it from an economic point of view – any amount required or desired is obviously available, by definition.

There is a difference between the economic scarcity of a commodity and its naturally given physical scarcity. Scarcity in the context considered here refers always to economic scarcity. Physical scarcity can, but need not, induce economic scarcity: for many regions, fresh air provides an example. Moreover, new technologies can reduce economic scarcity: this happened, for example, with the large-scale introduction of fracking to mine natural oil and gas. And this is also conceivable regarding the scarce assimilative services of the environment, of course.

Thus, economics deals with scarce commodities, in particular with scarce resources, which induce scarcity of the commodities produced with these scarce resources. Now, what about scarcity of environmental commodities? On a daily basis we learn more about the factual scarcity of various environmental goods – when we hear about the increasing pollution of the oceans with plastic waste, for example, about fine dust in the air, or about the continuing high emissions of greenhouse gases with the imminent risks of irreparable damages to the environment, if the assimilative capacity of the atmosphere is exhausted. In conclusion, environmental commodities affect the wellbeing of mankind and are typically scarce – they are inherently economic entities.

But then, of course, the question arises, why it took such a long time to realise this intrinsic economic nature of environmental commodities, to integrate the environmental commodities into the economic system? Most current environmental degradations could have been avoided, if the scarcity of these environmental commodities would have been respected earlier and more seriously, for example, at the beginning of the industrialisation some 250 years ago. Perhaps even more surprisingly, there are still quite a few countries without any major attempts to care about the environment or to restore the vital functions of the environment. What are then possible reasons for these observations?

First, decades or even centuries ago only little was known on environmental pollution, and relevant environmental goods such as clean water were not really scarce. There was perhaps soil, water and air pollution in the cities due to insufficient sanitary installations and limited knowledge on the health risks of these conditions, but there were often possibilities to escape this locally polluted environment, to find recreation in the pristine nature. The tradition of a "summer retreat" to recover from the pollution in the cities seems to be outdated today, but this opportunity, usually taken by the more affluent part of the population, was quite important not such a long time ago. Today's tourism is playing a somewhat different role.

Even man-made environmental disasters affected countries and their populations in earlier times. Some 300 years ago, for example, the area of what is today approximately the Free State of Saxony in Germany, was almost completely deforested due to all kinds of mining activities, demanding enormous amounts of wood before coal replaced wood as a source of energy. As this shortage, or increasing scarcity of wood, threatened to destroy the local industry and other more pleasurable activities such as hunting, every effort was made to reafforest the country in combination with a more thoughtful harvesting of wood. Not surprisingly, the concept of "sustainability" has one of its roots in Saxony, it was recorded in writing by Hans von Carlowitz, the then Director of Mines, in 1713 (Hamberger, 2013).

Despite these and other interesting examples, it is probably safe to say that known environmental commodities may have been locally scarce from time to time, and that there certainly were various man-made environmental

catastrophes, but for most of these occurrences there were possibilities for at least part of the population to "escape" local environmental degradations.

However, these historical developments, as interesting they may be in hindsight, do not yet answer the question raised earlier: why did it take such a long time to integrate economics and environment in a more careful way, and why is it apparently difficult to focus on this task – even today?

This gives rise to the important issue of "perceived scarcity" of environmental commodities. The following subsection introduces and discusses this concept – of high relevance for implementing a circular economy.

6.1.2 Perceived scarcity of environmental commodities

Slum dwellers in major cities in various developing countries face adverse circumstances regarding their daily life and often have to cope with all kinds of environmental pollution in their immediate neighbourhood. However, just because of a sheer lack of knowledge, or because of other issues, which are for whatever reason more important to them, these people seem not to care much about the environmental situation in their vicinity – in this sense, they do not really "perceive" scarcity of the environmental commodities. If they deplore these miserable circumstances, they do perceive scarcity, but then there are other items on their daily agenda, which are for the time being and for the circumstances given of more importance to them, the level of their "environmental awareness" and/or their financial means are not sufficient to direct more attention to environmental pollution.

Clearly, "environmental awareness", already addressed in Chapter 4, is related to "perceived scarcity": whereas the latter refers more to the general economic context, the former, environmental awareness, describes more an attitude, perhaps also a willingness towards improving the environmental situation or keep it on an acceptable level – resulting from perceived scarcity. In this sense, environmental awareness is dependent on perceived scarcity of environmental commodities.

Perceived scarcity is not a 0–1 issue: scarcity of some environmental goods might be perceived to a lesser degree in one country in comparison to another – not to speak of varying levels of environmental awareness between citizens in the same country. In fact, perceived scarcity and environmental awareness seem to vary a lot between different countries, and even wealthy industrialised countries need not always be leaders in this context.

The lack of perceived scarcity of most environmental commodities in earlier times likely contributed, together with the as well perceived unlimited supply of natural resources, towards the century-long conceptualisation of the economy as a linear take-make-dispose economy (Ellen MacArthur Foundation, 2017, Fig. 1.2). The functions of the environment were simply "perceived" as available – without any constraints.

There remain the general questions, how to measure perceived scarcity or environmental awareness and, in particular, how to raise the levels of these concepts? Section 4.2.3 pointed already to this aspect and referred in particular to the investigations of Grossman and Krueger (1995), Khakimova, Lösch, Wende, Wiesmeth, and Okhrin (2019), Lösch, Okhrin, and Wiesmeth (2019), Weber and Wiesmeth (2018). This is, as also indicated in Chapter 4, the issue of the Environmental Kuznets Curve (EKC), addressed by Dinda (2004), Stern (2004), and Huang, Lee, and Wu (2008) and Chen, Huang, and Lin (2019) among others, trying to establish, more or less successfully, a link between economic welfare and some concept of environmental awareness.

Many additional efforts to measure and apply environmental awareness directly can be found in the literature (see, for example, Altin, Tecer, Tecer, Altin, & Kahraman, 2014; Mei, Wai, & Ahamad, 2016; Teng & He, 2020). However, due to large conceptional differences, it is difficult to come up with a broadly applicable, precise concept of environmental awareness. For implementing a circular economy, raising environmental awareness as some kind of a pro-environmental attitude remains nevertheless important.

The social movement "Fridays for Future", active in various European countries, wants to raise environmental awareness regarding climate change. Time will tell, whether such unconventional activities can help not only to raise awareness of climate change sustainably, but also to initiate necessary behavioural changes to reduce greenhouse gas emissions.

The level of environmental awareness seems to be correlated to research activities, in particular in medicine, in natural sciences and in technology. New scientific results regarding hazardous substances and advanced technological possibilities to detect new environmental threats, can affect the (perceived) scarcity of certain environmental commodities with further challenges for adequately handling the ensuing issues. Vice versa, an increasing environmental awareness might also help to postulate and start more research activities regarding certain environmental issues.

For example, the ecological importance of the earth's ozone layer and its limited capacity to store chlorofluorocarbons (CFCs) only became known with the advancement of science and the development of sophisticated instruments to measure and document the relevant chemical processes. Similarly, the emission of nitrous oxides, which is currently under rigorous discussion in Germany and elsewhere, turned only recently into a serious environmental issue, due to various studies pointing to health risks associated with these emissions (Leopoldina, 2019).

There is yet another issue of relevance for environmentally friendly behaviour: a feedback from one's own environmental actions.

6.1.3 Relevance of a (perceived) feedback from actions

Behavioural aspects seem to be of significance for dealing with environmental issues, for mitigating climate change, for example. There is perceived scarcity, there is high environmental awareness, and yet individual efforts to reduce greenhouse gas emissions are not sufficient for successfully mitigating climate change. Even various industrialised countries, member states of the European Union (EU) among them, are not unanimously or only halfheartedly behind the initiatives of the United Nations Framework Convention on Climate Change (UNFCCC).

Of course, an individual or a small country alone cannot sustainably change this situation: there is, thus, no direct, perhaps not even a "perceived" feedback from one's own environmentally friendly actions. In such a situation, social mechanisms such as the Tragedy of the Commons and/or the Prisoners' Dilemma (see Section 7.1) with their adverse effects on environmentally friendly decisions gain relevance. Behavioural environmental economics provides some more insight into these relationships and is therefore of importance for implementing a circular economy (see Chapter 8).

In general, with a low level of environmental awareness there likely are no significant incentives for protecting the environment – the feedback from one's own action does not matter in this case. Similarly, even with a high environmental awareness there need not be much motivation to protect the environment, if the individual possibilities to change the situation are limited, if there is insufficient perceived feedback from one's own actions to protect the environment.

A missing or insufficient (perceived) feedback from environmentally friendly actions can therefore greatly affect attempts for the implementation of a circular economy: the many efforts to raise environmental awareness (see, for example, Robina-Ramírez & Medina-Merodio, 2019) are certainly good, but need

not be sufficient to foster environmentally friendly activities. And technological innovations, although they can have high relevance in this regard (see Section 11.3), cannot always compensate for a low environmental awareness or a missing feedback.

A complex network of perceived scarcity, environmental awareness and perceive feedback emerges, which affects the motivation for environmentally friendly behaviour. And this network is itself dependent on social norms and other features of behavioural environmental economics.

So far this extended discussion of environmental commodities in the context of (perceived) scarcity, environmental awareness and (perceived) feedback. The following section considers the allocation problems, of relevance for any economic system, and introduces the concept of an optimal allocation. Needless to say, that these basic concepts are also of importance for a circular economy.

6.2 Feasible allocations

Any society has to decide, in one way or the other, what to do with its scarce resources and the ensuing scarcity of commodities, including the environmental commodities. This points to the allocation problems, fundamental to any economic system, be it a decentralised market economy or a centrally planned economy, be it an industrialised or a developing country. Feasible allocations are the solutions of these allocation problems.

6.2.1 The allocation problems

In a simplified, way, answers to the following basic economic issues are required for any economic system for a given period of time:

- **Allocation Problem 1:** Which commodities shall be produced: e-vehicles, three-room apartments, universities, waste incineration plants, services to clean the environment, …? How many units with which characteristics are required?

- **Allocation Problem 2:** How shall these commodities be produced: labour-intensive, capital-intensive, environmentally friendly, …?
- **Allocation Problem 3:** Who shall have access to these commodities? Under which conditions will access be granted: prices, bonus system, personal characteristics, free access, …?

These allocation problems have to be solved and they are solved in any economic system, in any period of time – also for the environmental commodities. For a specific example in the environmental context think about the "services" provided by the earth's ozone layer. It is well understood today that the ozone layer protects life on earth from the hazardous effects of UV radiation from the sun, and should therefore not be depleted, should have no "holes", no significant reduction in its concentration. A way to restore the functionality of the layer in the long-run is the ban on CFCs, and once restored, nobody can be excluded from the services of the ozone layer. In the other case, the population in some parts of the earth will have limited access or no access at all to these protective services.

This simple consideration shows clearly that the allocation problems play also an important role in the environmental arena: there has to be a commitment to what extent to protect the ozone layer (allocation problem 1), how to do it (allocation problem 2). In this case, the answer to allocation problem 3 is provided through allocation problem 1 and the natural properties of the ozone layer: access is free. If we do protect the environment, or if we don't protect the environment – any decision implies a solution of the allocation problems in the corresponding context. In fact, the level of environmental protection is mirrored in the state of the environment, representing a feasible allocation, a solution of the allocation problems with respect to the environmental commodities.

An issue of high importance in economics is the existence of a mechanism, an "allocation mechanism", which provides support for the solution of the allocation problems – of course, in some reasonable, "optimal" way. This leads immediately to a variety of questions, which will be dealt with in the following sections and subsections. There is, first of all, the concept of an optimal allocation, characterising feasible allocations with certain properties.

6.2.2 Optimal solutions

The search for an appropriate mechanism to provide an optimal solution of the allocation problems asks first of all for an appropriate optimality concept. Of course, "optimality" can mean many different things to different societies. All kinds of social policies attempt to modify the solution of the allocation problems in a certain way, which is considered to better correspond to social justness and fairness – according to the ideas and beliefs of a particular society. For a liberal, democratic society, it is necessary that an optimal allocation as a normative concept finds the support of a large part of the population.

Different perceptions of fairness are accompanied by different notions regarding the optimality of a feasible allocation. Important for the circular economy: although an "optimal" allocation includes both regular and environmental commodities, there nevertheless often are separate perceptions regarding an optimal protection of the environment and they differ also widely across countries, even regions. A "circular economy", or "implementing a circular economy" may therefore mean different things for people in different countries or regions – and this can also be implicitly related to other notions of optimal solutions of the allocation problems, not only with respect to environmental commodities. This will be detailed later, also in view of the similarly imprecisely definable sustainability concept.

In addition to varying perceptions and notions of fairness etc., other framework conditions influence the social accord on optimality: these can be geographic and climatic conditions, the density of the population, but also tradition and religion, and, last but not least, the political system and economic prosperity.

These considerations point to challenges defining optimal allocations for a circular economy. The discussions in various countries on the implementation of certain aspects of a circular economy seem to mirror these issues: there are critical discussions on shutting down coal power plants in Germany, there are critical discussions on further raising certain emission standards in the EU, and California wants to enforce higher emission standards against the suggestions of the federal administration. Of course, the level of environmental awareness and the degree of individual/governmental influence, the eventual or perceived feedback from the actions, affect these discussions.

In a general economic context, the concept of optimality, or efficiency, can be linked to a normative criterion, thus representing kind of a social accord if the criterion is widely accepted. In decentralised market economies, optimality is usually based on the "Pareto criterion": a feasible allocation is (Pareto-) optimal or (Pareto-) efficient, if there is no other feasible allocation, which improves the wellbeing, the "utility", of at least one individual (consumer or household), without diminishing the utility of any other. This criterion seems to be applicable in an envy-free society, and – very importantly – it integrates or respects, at least to some extent, the wellbeing of each individual of the society. This is a feature, whose relevance for liberal, democratic societies should not be underestimated.

A simple consequence of these considerations is the observation that a feasible allocation, to which there exists an alternative providing the same amounts of the private commodities at a lower environmental pollution, cannot be (Pareto-) optimal or (Pareto-) efficient, if all individuals prefer less pollution, which can be taken for granted. Such a situation can sometimes be observed, when new technology replaces old equipment, enabling not only higher production rates, but also less pollution. Cases, where a lower environmental pollution is associated with smaller amounts of certain private goods are, of course, more difficult to evaluate regarding optimality: reducing the pollution of the air in a city with nitrous oxides might at least temporarily require restrictions on transport activities involving cars and trucks with diesel engines. Allocation decisions of such kind have to be based on reliable studies, evaluating and weighing health risks and economic losses – not an easy thing to do (Leopoldina, 2019).

A central question is therefore: how to achieve an optimal allocation? Allocation mechanisms are meant to provide an answer to this question.

6.3 Allocation mechanisms

Provided there is an appropriate optimality concept, the ensuing question is, how to achieve such a distinct solution of the allocation problems, especially in the environmental field. This becomes particularly urging in the case of an optimality criterion, which respects the wellbeing of individuals, such as the Pareto criterion.

There are two dominant classes of allocation mechanisms: decentralised mechanisms, usually of relevance for market economies, and mechanisms based on central planning, variations of which are used in centrally planned economies. Both variants will be considered in more detail in the next subsections. Beyond that there are the mechanisms, which combine elements of both.

Practically speaking, completely decentralised or completely centralised mechanisms probably never existed. All observable allocation mechanisms include both features of decentralised and centralised procedures – with varying degrees, of course. In this sense the mechanism applied in Germany, for example, is certainly to a large extent more decentralised than the mechanism employed in China with many allocation decisions, also regarding environmental commodities, issued by the government. These differences become sometimes visible in so-called top-down or bottom-up approaches, as indicated in the context of China's circular economy strategy (Section 4.1.2).

The allocation mechanism employed in a particular country is clearly of importance for implementing a circular economy: the tools, which will be introduced and investigated in the chapters of Part IV, are either tools, such as statutory requirements, which better suit central planning, or rely upon a decentralised approach, such as a tax on carbon dioxide emissions, for example. Also, not surprisingly, combinations thereof are possible and applied, with further complications regarding the functioning of the allocation mechanism. This issue will be reconsidered in later chapters.

The following subsection briefly discusses relevant features of the market mechanism, which plays a role in most countries.

6.3.1 The market mechanism

The market mechanism, or synonymously, the price mechanism, is one of the prominent allocation mechanisms. It is characterised by a "decentralisation of the economic decisions by means of a price system". This simply means that in a pure market system, consumers and producers make their economic decisions individually and also independently of each other. These individual actions are coordinated through the price system, with the same prices

for all economic agents serving as appropriate signals, thereby guiding these decisions. Therefore, a market system requires the following "ingredients":

- **Markets:** It is necessary to have a functional market – a real or a virtual place, where buyers and producers meet – for each commodity.
- **Price System:** There is a (market) price for each commodity, which is only determined by supply and demand of the commodity in question. Demand of the consumers depends on their preferences or utilities, on the prices, and their income from private property, also a necessary ingredient of a market economy, and comprising labour and capital income, among other sources. Similarly, supply of the producers is dependent on the prices, also on the prices of the natural resources, on wages for labour, and on the available technologies.

The motivation for demanding certain commodities results from the utility derived from the consumption of these commodities, and the opportunity cost of spending part of the income on these commodities. The motivation for producing and supplying commodities results from the goal of making profits. As production companies are owned after all by consumers as individuals, it is also in their interest that producers generate profits, which then constitute part of the consumers' income.

A crucial issue is, of course, the independence of the price system from the actions of individual consumers and producers. Otherwise certain individuals, in the extreme case a monopolistic producer, for example, could distort the prices to earn higher profits. In such a situation with powerful individual consumers or producers not only pursuing, but forcing through their own selfish interests, there is naturally little chance for attaining an optimal solution of the allocation problems. These considerations lead to the important assumption of perfect competition and the concept of a market equilibrium:

- **Perfect Competition:** In a pure market economy individuals, consumers and producers, do not have any influence on the price system.
- **Market Equilibrium:** When all the above requirements are satisfied, a market system is assumed to lead to an equilibrium constellation, a "market equilibrium". This is then a solution of the allocation problems, a feasible allocation. This solution is characterised by the fact that each economic agent, consumer or producer, can buy or sell exactly that amount of a commodity, which corresponds to demand or supply at the price prevailing in this equilibrium, the "equilibrium price".

The interesting and perhaps surprising feature is that in a formal model of a market economy with the characteristics and properties given above, a market equilibrium represents a (Pareto-) optimal solution of the allocation problems. This is Adam Smith's famous and well-known metaphor of the "invisible hand", which guides individuals to further the interests of the society, although this is not their direct intention (Smith, 1776, Book IV, Ch. II).

Of course, there are quite a few commodities, such as "health", for which it is difficult, if not impossible, to organise regular markets. Similarly, the requirement of a perfect competition is mainly a theoretical construct, which can at best be approximated in practice. Thus, the functioning of the "invisible hand" is restricted, and leaves therefore enough room and justification to interfere with the market solution with proposals from central planning, by means of all kinds of social policies and redistributions to attain an outcome of the allocation problems, which is preferred in a particular society.

Just to mention: there are other, additional functions of a market economy, which are considered important, also in an environmental context. According to Friedrich A. Hayek prosperity of a society is particularly dependent on innovation and creativity, requiring a system with free markets, where all individuals can participate with their knowledge, and their ideas (Machlup, 1974). This is an issue for implementing a circular economy, where Designs for Environment (DfE) are required to reduce waste and facilitate recycling. Similar to a market system, it will prove necessary to integrate individuals in an appropriate way – to generate knowledge and to provide incentives for DfEs (see Section 17.3). Exactly for this reason it seems to be advisable to make extensive use of the market system when implementing a circular economy. This corresponds to the role assigned to business models in this context (see Section 3.1.1). However, a circular economy is not a straightforward extension of a regular market economy.

Returning to the above requirements for a fully functional market economy, it becomes immediately clear that these requirements do not automatically extend to the environmental context. Thus, the pure market mechanism alone is not adequate to satisfactorily handle environmental issues in a circular economy. This will be further addressed in the next chapters of this Part II.

6.3.2 Central planning

An allocation mechanism completely based on central planning assumes a state, a "socialist" state so to say, in which the control of all production activities is in the hands of the state itself. The state is thus the sole responsible producer, authorised to employ the natural resources of the economy for producing the consumption commodities. The state then buys the productive services, labour, of the consumers and sells the consumption commodities to them. This applies at least to the situation where consumers are free to choose and buy the preferred commodities. Various authors investigated the viability of socialist economies, mainly in the first half of the last century.

Fred Taylor (Taylor, 1929) characterises a possible allocation mechanism in such a situation. Consumers are assured a certain money income, which they can use to demand and buy preferred commodities. The selling price of the commodities has to be determined to cover the costs of producing them, "the drain on the economic resources of the community" (p. 3). Thereby "the problem of ascertaining the effective importance in the productive process of each primary factor" has to be solved (p. 6). This requires in general a method of trial-and-error, leading to a "rational allocation" of resources, if implemented carefully.

Also Oskar Lange (Lange, 1938) points to the possibility of "economic accounting under socialism", prior to this disputed by Ludwig von Mises (Mises, 1951), whose work appeared in German language already in 1922. Economists such as Ludwig von Mises and, later, Friedrich A. Hayek (Hayek, 1944) and others mention problematic aspects such as "the abolition of private enterprise and private ownership of the means of production" in a socialist economy. The apparent success of "capitalist" economies and the apparent failure of "socialist" economies in the last decades brought this academic exchange largely to a halt. Nevertheless, most economic systems are based on more or less significant parts of central planning – no longer by an official Central Planning Bureau (CPB), but by institutions of the public administration.

One must, however, not forget that private initiative, a characteristic of a market economy, is restricted or even excluded in these cases. In a centrally planned economy, economic decision-making is the task of some agency, a CPB, for example. A comparatively small group of people is in charge of fundamental economic decisions. Even with support through modern technologies, they can make errors as all humans, and they might try to promote their own preferred alternatives. The chance, however, that these deviations or errors are detected and corrected through insight and counterbalances available elsewhere in the society, seems to be smaller in comparison to the situation in a decentralised market system with its error-correcting structure – the knowledge available in the society is usually better taken into account in decentralised market economies. This will become important for various tools to be used in environmental policies to implement a circular economy.

In view of historical experiences, the state of the environment in centrally planned economies was typically worse in comparison to the environmental degradations in market economies. This was sometimes also due to an ideological background, which simply declared environmental degradation impossible in socialist systems (see, for example, Welfens, 1993, p. 107). However, when a sociologist declares that "the principle of capitalist production does not take into account the last stage of the production process, namely the need for waste disposal" (Ermolaeva, 2019, p. 62), then this should be a matter for reflection.

In conclusion, central planning seems not to provide a generally applicable and appropriate tool for better allocating environmental commodities for which there typically are no regular markets. Examples of socialist or communist countries seem to confirm this more negative assessment. Nevertheless, without ideological barriers, a more central approach to various environmental issues, such as renewable energies, e-mobility, or forestation programs can be observed in various countries and can be of advantage regarding goals of a circular economy. On the other hand, one should not forget technological path dependencies established in this context. Whether these "technological" efforts always reach the "bottom", i.e., integrate all people appropriately, remains to be seen. This is currently an interesting question in Germany with respect to promoting e-vehicles and e-mobility.

With this basic information on central planning, this text considers the market mechanism as the driving tool for solving the allocation problems, at least for regular "private" commodities. The allocation of the environmental commodities should therefore respect this basic assumption and the framework conditions associated with it. These considerations, pointing also to possible societal path dependencies, should be kept in mind regarding all matters of a circular economy and its implementation.

6.4 Summary on the economic foundation of a circular economy

A few remarks shall help to keep some important aspects in mind. First, a circular economy is an economic system for which it is necessary to solve the allocation problems in one way or the other, also for environmental commodities. In classical, market-oriented systems this task is accomplished by the market mechanism, leading under certain conditions to a Pareto-optimal allocation. The concept of Pareto-optimality is based on a normative criterion, which involves all individuals – in accordance with a liberal, democratic society.

Central planning mechanisms are contrasting the decentralised allocation mechanism of a market system. Although it is in principle possible to achieve an optimal allocation with central planning, information requirements and possible strategic behaviour of economic agents render this task difficult.

Turning to environmental commodities, some planning activities are required for environmental commodities, for which regular markets do not exists. Moreover, issues such as perceived scarcity, environmental awareness,

and (perceived) feedback affect the task of solving the allocation problems. On top of that, an appropriate allocation mechanism comparable to the market system is not available.

There remains a certain focus on innovative technologies on the one hand, and on business models on the other. Thus, due to the special characteristics of the commodities of relevance in a circular economy, technological and societal path dependencies have a substantial influence on the implementation of a circular economy. This will be explored in more details in Part III and Part V, in particular.

The following chapter investigates possibilities and challenges of allocating environmental commodities in more detail. It involves aspects of market systems, but also insight from behavioural environmental economics.

References

Altin, A., Tecer, S., Tecer, L., Altin, S., & Kahraman, B. F. (2014). Environmental awareness level of secondary school students: A case study in Balıkesir (Türkiye). In: *4th world conference on learning teaching and educational leadership (WCLTA-2013)*Vol. 141, (pp. 1208–1214). pp. 1208–1214. https://doi.org/10.1016/j.sbspro.2014.05.207.

Chen, X., Huang, B., & Lin, C.-T. (2019). Environmental awareness and environmental Kuznets curve. *Economic Modelling*, 77, 2–11. https://doi.org/10.1016/j.econmod.2019.02.003.

Dinda, S. (2004). Environmental Kuznets curve hypothesis: A survey. *Ecological Economics*, 49(4), 431–455. https://doi.org/10.1016/j.ecolecon.2004.02.011.

Ellen MacArthur Foundation (2017). *Circular economy and curriculum development in higher education*. Retrieved from https://www.ellenmacarthurfoundation.org/assets/downloads/EMF_HE-Curriculum-Brochure-03.10.17.pdf.

Ermolaeva, Y. (2019). Problems of modernization of the waste management sector in Russia: Expert opinion. *Revista Tecnologia e Sociedade*, 15, 56–77. https://doi.org/10.3895/rts.v15n35.8502.

Grossman, G. M., & Krueger, A. B. (1995). Economic growth and the environment. *The Quarterly Journal of Economics*, 110(2), 353–377. https://doi.org/10.2307/2118443.

Hamberger, J. (2013). *Hans von Carlowitz: Sylvicultura oeconomica*. München: Oekom Verlag. Retrieved from https://www.oekom.de/_files_media/titel/inhaltsverzeichnisse/9783865814111.pdf.

Hayek, F. A. (1944). *The road to serfdom*. University of Chicago Press. Retrieved from https://cdn.mises.org/Road%20to%20serfdom.pdf.

Huang, W. M., Lee, G. W. M., & Wu, C. C. (2008). GHG emissions, GDP growth and the Kyoto Protocol: A revisit of Environmental Kuznets Curve hypothesis. *Energy Policy*, *36*(1), 239–247. https://doi.org/10.1016/j.enpol.2007.08.035.

Khakimova, D., Lösch, S., Wende, D., Wiesmeth, H., & Okhrin, O. (2019). Index of environmental awareness through the MIMIC approach. *Papers in Regional Science*, *98*(3), 1419–1441. https://doi.org/10.1111/pirs.12420.

Lange, O. (1938). *On the economic theory of socialism. O. Lange, F. M. Taylor, & B. Lippincott (Eds.), On the economic theory of socialism*. Vol. 2. University of Minnesota Press (NED-New edition). Retrieved from http://www.jstor.org/stable/10.5749/j.ctttsbzm.

Leopoldina (2019). *Clean air. Nitrogen oxides and particulate matter in ambient air: Basic principles and recommendations*. Halle (Saale): German National Academy of Sciences Leopoldina. Retrieved from (2019). https://www.leopoldina.org/uploads/tx_leopublication/2019_Leo_Stellungnahme_Saubere_Luft_en_web_05.pdf.

Lösch, S., Okhrin, O., & Wiesmeth, H. (2019). Awareness of climate change: Differences among Russian regions. *Area Development and Policy*, *4*(3), 284–307. https://doi.org/10.1080/23792949.2018.1514982.

Machlup, F. (1974). Friedrich Von Hayek's contribution to economics. *The Swedish Journal of Economics*, *76*(4), 498–531. https://doi.org/10.2307/3439255.

Mei, N. S., Wai, C. W., & Ahamad, R. (2016). Environmental awareness and behaviour index for Malaysia. *Procedia—Social and Behavioral Sciences*, *222*, 668–675. https://doi.org/10.1016/j.sbspro.2016.05.223 ASEAN-Turkey ASLI

QoL2015: AicQoL2015Jakarta, Indonesia, 25–27 April 2015.

Mises, L. v. (1951). *Socialism: An economic and sociological analysis*. New Haven, CT: Yale University Press. Retrieved from https://cdn.mises.org/Socialism%20An%20Economic%20and%20Sociological%20Analysis_3.pdf.

Pearce, D. W., & Turner, R. K. (1989). *Economics of natural resources and the environment*. Johns Hopkins University Press.

Robina-Ramírez, R., & Medina-Merodio, J.-A. (2019). Transforming students' environmental attitudes in schools through external communities. *Journal of Cleaner Production*, *232*, 629–638. https://doi.org/10.1016/j.jclepro.2019.05.391.

Smith, A. (1776). *An inquiry into the nature and causes of the wealth of nations*. London: W. Strahan and T. Cadell.

Stern, D. I. (2004). The rise and fall of the environmental Kuznets curve. *World Development*, *32*(8), 1419–1439. https://doi.org/10.1016/j.worlddev.2004.03.004.

Taylor, F. M. (1929). The guidance of production in a socialist state. *The American Economic Review*, *19*(1), 1–8. Retrieved from http://www.jstor.org/stable/1809581.

Teng, M., & He, X. (2020). Air quality levels, environmental awareness and investor trading behavior: Evidence from stock market in China. *Journal of Cleaner Production*, *244*, 118663. https://doi.org/10.1016/j.jclepro.2019.118663.

Weber, S., & Wiesmeth, H. (2018). Environmental awareness: The case of climate change. *RUJEC*, *4*(4), 328–345. Retrieved from https://doi.org/10.3897/j.ruje.4.33619.

Welfens, M. J. (1993). *Umweltprobleme und Umweltpolitik in Mittel- und Osteuropa*. (in German)Springer-Verlag.

Wiesmeth, H. (2011). *Environmental economics: Theory and policy in equilibrium. Springer texts in business and economics*. Berlin: Springer Nature. https://doi.org/10.1007/978-3-642-24514-5.

Allocating environmental commodities

Whereas the last chapter introduced and discussed basic economic concepts and principles, also in relation to a circular economy, this chapter turns to some of the challenges associated with allocating environmental commodities. The market mechanism alone will in general not be sufficient to solve this task, it needs to be augmented or replaced by some other mechanism. Moreover, there is the issue of optimality: how to guarantee optimality of the resulting allocation in a market system with environmental commodities? In a pure market economy, Pareto-optimality results from the market mechanism – under certain conditions (see Chapter 6). This is no longer the case with environmental commodities.

This chapter presents and investigates some of these issues, which clearly are of relevance for implementing a circular economy. In the next chapters, the discussion continues with aspects of behavioural environmental economics (Chapter 8) and a consideration of the allocation problems in a circular economy (Chapter 9).

The first section analyses external or environmental effects as an issue of missing markets – the external effects are "external" to the market system. Thus, the market system has to be completed in one way or the other to "internalise"

the environmental effects. However, mechanisms such as the Tragedy of the Commons and the Prisoners' Dilemma, possibly triggered by an insufficient (perceived) feedback from one's own action, provide further challenges (see also Section 6.1.3). Thereafter, appropriately designed environmental policies are proposed as tools to solve the allocation problems for environmental commodities, thus preparing the implementation of a circular economy. A case study on the "German Refillable Quota Issue" closes the chapter.

7.1 External effects and missing markets

Often defenders of a free and liberal market economy argue that bans or other prohibitive rules to "motivate" people to protect the environment should not be applied. This is not only the case in the European Union (EU) with plans to ban certain single-use plastic items. One of the arguments in favour of such a ban is that we cannot leave this decision to the market, because the market economy brought us these items. This argument is, of course, not quite valid, because environmental awareness at that earlier time was probably lower. Scarcity of the associated

77

environmental commodities was less perceived, thus opening the doors for the resulting pollution. On the other hand, can we really put so much trust in the intrinsic motivation of consumers and producers to protect the environment, even with a presumably higher environmental awareness? The discussion in Section 6.1.3 shows that other mechanisms, such as an insufficient feedback from actions, may interfere with environmental awareness. It is here that behavioural environmental economics comes into the picture (see Chapter 8).

But before addressing behavioural aspects, there is another important issue: for most environmental commodities regular markets do not exist – for more or less obvious reasons. This is true, for example, for the already mentioned services of the ozone layer: anyone buying these services is also buying them for many other individuals without, however, receiving, in general, their financial contributions. Thus, the consumption or the production of most environmental commodities involves "environmental effects" or "external effects": actions of an individual affect directly the utility or production possibilities of another individual without compensation – these effects are not reflected in the market system (Wiesmeth, 2011, Ch. 6).

The emission of nitrous oxides or particles through transport activities provides another example: without regulatory interferences with the market system, nobody pays or has to pay for these emissions while driving a car, and it is difficult, if not impossible, to imagine a regular market for these emissions for reasons similar to those mentioned in the context of the depletion of the ozone layer.

From a formal, theoretical point, some markets are missing. Consequently, it is the goal of environmental economics to provide appropriate instruments and tools, which complete the market system and which can be applied in a practical context (Wiesmeth, 2011, Ch. 6). Some further challenges associated with this procedure will be presented and discussed in the chapters of Part IV.

External effects associated with environmental commodities lead to a gap between social and private marginal costs of using or producing these environmental commodities. From a private point of view the earth's ozone layer can be used as a repository to store chlorofluorocarbons (CFCs) without any costs; however, it is well-known today that the social costs of such a behaviour, exhausting the assimilative capacity of the higher atmosphere, are significant. Completing the market system is therefore equivalent to reducing or closing such a gap between social and private marginal costs.

The market system can be completed in various ways. One possibility is establishing "artificial" markets, such as markets for tradable emission certificates or pollution rights. Pigou Taxes can be interpreted as market prices on some missing markets and can help to "internalise" environmental effects, to bring them back under the roof of the market system. In these cases, the property rights regarding the environmental commodities often remain with the government: public institutions allocate the emission certificates or the pollution rights, and they collect the revenue from the Pigou Taxes. Sometimes, however, negotiations and simple contracts between the relevant parties can be used to internalise externalities, especially if transaction costs for negotiating the contract are low. This approach is related to the Coase Theorem and requires the explicit assignment of property rights to individuals or groups of individuals, thereby explicitly pointing to the "two-sided" nature of external effects (Wiesmeth, 2011, Ch. 6.6).

The various possibilities to complete the market system are largely theoretically relevant. In a practical context it is necessary to gather the required information to allocate the correct number of certificates or to choose the correct value of the Pigou Tax, for example. As this is in general not possible, all practical attempts in this regard constitute only approximations. Nevertheless, these "market-based" instruments integrate and motivate the individuals to make use of their own information – an obvious advantage of these instruments. The chapters in Part IV provide more detailed information on relevant properties of command-and-control policies and market-based tools.

The services of the earth's ozone layer are an example of a "public commodity": exclusion of somebody from the consumption of these services is not feasible, and total supply is not affected by the number of consumers. Similarly, the available supply of "clean air" is (almost) not affected by the decision of an additional individual consumer to use a private car for commuting instead of the public transport: this decision has only negligible effects on traffic congestion and the state of the environment. This is but one example of the "Tragedy of the Commons", which affects decisions in a wide variety of environmental issues, including the case of single-use plastic items.

7.1.1 The Tragedy of the Commons

Consider the modal split in an industrialised country: the distribution of commuters to the various means of transportation, such as public transport or private cars. Despite a higher level of pollution (also in the form of greenhouse gases), and despite daily traffic congestion, especially during rush hours, many commuters continue to use their own car to get to and from work, although public transport alternatives are available and the general level of environmental awareness is presumably high.

The Tragedy of the Commons provides an explanation: the additional (or marginal) pollution of a commuter in a private car is negligible, as is the marginal effect on the overall traffic situation in the city. So why switch to public transport, which is for many people for a variety of reasons less comfortable, although it constitutes a feasible alternative? And if sufficiently many commuters take the public buses or trains, then the streets will be less crowded, there will be less

congestion and less need to take the bus. The consequence is clear: for environmental reasons alone, nobody has much of an incentive to change his or her commuting behaviour. Of course, there are other reasons to use public transport for commuting: cost saving, no need to search and pay for a parking space etc.

Scarcity of the associated environmental commodities is probably perceived individually, but it can hardly be changed through individual decisions (Section 6.1.2 and Wiesmeth, 2011, Section 5.3.3). There are those aspects of a too small feedback from one's own environmentally friendly decisions (see Section 6.1.3), bringing in considerations of behavioural environmental economics. Appropriate framework conditions, ideally supported by a rising level of environmental awareness might help to change the behaviour and the modal split, though.

The Tragedy of the Commons is indeed quite "common" and of high relevance for implementing a circular economy. The examples extend to all parts of the society and include often conspicuous consumption activities such as driving big cars with high greenhouse gas emissions – although there can be good reasons for driving a big car, of course. They include attempts to prevent transmission lines behind one's own backyard, although the transport of energy from renewable sources is generally supported, and they include the still observable habit – in spite of various awareness campaigns – of throwing small WEEE, such as batteries, into the household garbage, with the self-justification that one small battery does not really damage the environment.

The problematic issue with the Tragedy of the Commons is that it is simple, simple to explain, highly prevalent, but difficult to overcome. The typical approach in many countries seems to be to wait for a reaction from the government: the public administration should propose and provide solutions, which are acceptable for everyone. And this is, at least in many cases, difficult, if not practically impossible.

Bezin and Ponthièsre (2019) investigate issues of the Tragedy of the Commons in a dynamic, overlapping generations economy, thereby examining the dynamics of moral behaviours. The question, whether "positive framing" can help to solve the Tragedy of the Commons is investigated by Isaksen, Brekke, and Richter (2019), and Patt (2017) considers its "reframing" for mitigating climate change. The results and conclusions of these papers point to an interesting context in behavioural environmental economics (see Chapter 8).

These examples indicate that the Tragedy of the Commons affects not only environmental issues in general, but also many aspects regarding the implementation of a circular economy. By far not all individuals succumb to its temptations, though: Brown, Adger, and Cinner (2019) point to "evidence that collective action can be mobilised at various scales to avoid tragedies in population, in overfishing, in resource consumption, and in land degradation", obviously aspects of behavioural environmental economics. Nevertheless, because of its high prevalence it is indispensable to deal with the Tragedy of the Commons through appropriate policy tools, also methods of behavioural economics. The "Prisoners' Dilemma", presented in the next subsection, poses a similar challenge for implementing a circular economy.

7.1.2 The Prisoners' Dilemma

Consider the issue of mitigating climate change, referring to the global environmental commodity "limiting global warming". In the Kyoto Protocol of 1997 and the Paris Climate Change Conference of 2015 participating countries agreed to take appropriate measures to reduce greenhouse gas emissions, to fight climate change. Many countries have meanwhile delivered their Nationally Determined Contributions (NDCs), indicating "their best efforts" regarding the mitigation of climate change and "to strengthen these efforts in the years ahead".

Nevertheless, a couple of years after the Paris Agreement of 2015, overwhelming progress is not really visible. There are still various industrialised countries, which will not achieve their goals for 2020. Germany is among those countries, despite a reduction of greenhouse gas emissions of 28% between 1990 and 2016. Not surprisingly, also global greenhouse gas emissions keep rising, putting the delicate climate goals at risk.

In his opening remarks at the 2019 Climate Action Summit in New York, UN Secretary-General Gutteres said: "My generation has failed in its responsibility to protect our planet. That must change. The climate emergency is a race we are losing, but it is a race we can win. The climate crisis is caused by us – and the solutions must come from us. We have the tools: technology is on our side. Readily-available technological substitutions already exist for more than 70 per cent of today's emissions. And we have the roadmap: the 2030 Agenda for Sustainable Development and the Paris Agreement on climate change." Thus, why is it not happening? Why are countries obviously hesitating to reach their climate goals? Should it not be in their own interest?

One piece for explaining these "failures" could be the strategies associated with the Prisoners' Dilemma: as each country will profit from the corresponding efforts of all other countries, it might see less pressure on its own obligations and be less inclined to reduce its own efforts. This might in particular happen, when this country is small enough to have only a limited or negligible influence with its actions on the global climate, if there is no or only little perceived feedback. Government could get support of the population for such an attitude. In particular businesses might gain a temporary competitive edge by not having to immediately transform their production activities. This behavioural attitude could even be stronger in countries where there is some doubt regarding the extent of anthropogenic climate change, where

TABLE 7.1 The Prisoners' Dilemma: A simple example.

	C2: Payoff for 1	C2: Payoff for 2	N2: Payoff for 1	N2: Payoff for 2
C1: Payoff for 1	+1		−2	
C1: Payoff for 2		+1		+3
N1: Payoff for 1	+3		0	
N1: Payoff for 2		−2		0

Payoffs for two countries 1 and 2 for cooperative (C) and non-cooperative (N) behaviour relative to the status quo (0,0).

sufficient feedback from actions to mitigate climate change is not really perceived.

However, when more countries are thinking and acting like this, the global effect of the agreement on mitigation of climate change will be limited, will slow down, and in the worst case, it might be doomed to fail. There are likely other or additional explanations for such a development, but economic reasons are always a good guess (see, for example, Wiesmeth, 2011, Sections 5.3.1 and 5.3.2).

Table 7.1 depicts a simple Prisoners' Dilemma situation. The table shows the benefits of two countries (1 and 2) when they cooperate (C1 and C2) and when they do not cooperate (N1 and N2). Interestingly, country 1 is always better off with strategy N1, regardless of the action of country 2. And this holds vice versa for the other country, too. Thus, non-cooperation is a "dominant" strategy leading to the status quo as equilibrium with payoffs (0,0), although a cooperative behaviour would bring them payoffs (+1,+1) above status quo. Thus, the dominance of non-cooperation, regardless the action of others, drives this result. In case a specific country does not perceive any major feedback from its own actions, the motivation for non-cooperation will yet increase.

Another issue needs to be observed. As both the Kyoto Protocol and the Paris Agreement proved less successful than expected and needed for successfully mitigating climate change, more and more countries start to adapt to climate change. And this development is also supported by the establishment of the Green Climate Fund, which "pays particular attention to the needs of societies that are highly vulnerable to the effects of climate change". This retarding and deviating behaviour is in accordance with the Prisoners' Dilemma and reminds one of the situation in Saxony some 300 years ago when the country was almost deforested because many mining activities required lots of wood. Only then, when businesses went bankrupt, when jobs got lost, interest for the commons increased (see Section 3.1.3). The forests in Saxony recovered gradually – whether a similar prospect applies to climate change, remains doubtful.

The Prisoners' Dilemma is quite common and of relevance for implementing a circular economy – especially in a global context such as the climate change negotiations. Further examples refer to a Design for Environment (DfE) and to the issue of setting and raising environmental standards (see Chapters 15 and 17).

Of course, the issue of a gradually developing cooperation in such a context is of great relevance and has been addressed in the literature in a multitude of contexts. "Rational conformity behaviour", updating strategies with respect to certain factors, is investigated by Niu et al. (2018), and Szolnoki and Chen (2020) apply methods of evolutionary game theory to explore "the vitality of competing strategies in different social dilemma situations".

Interestingly, not all countries, not all producers succumb to the rationality of the

Prisoners' Dilemma. As indicated there is more than one element of behavioural economics influencing decisions – and behavioural economics might also provide tools to deal with mechanisms such as the Prisoners' Dilemma (see Chapter 8).

The next subsection proposes environmental policies as allocation mechanisms for environmental commodities. The cases presented in Chapter 5 point already to the requirement of carefully designed policies and to the challenges associated with this task.

7.2 Environmental policies as allocation mechanism

According to these considerations, the allocation of environmental commodities requires some special attention. In principle, it seems to be necessary to deal separately with the allocation of each environmental commodity: air, soil and water pollution need to be treated differently from mitigation of climate change, or handling all kinds of waste and end-of-life vehicles. Nevertheless, these issues are partially connected: transport activities, for example, are related to emission of greenhouse gases and air pollution through fine dust. Similarly, experience (see Chapter 5) shows that innovative recycling activities, for example, may affect efforts to prevent waste etc.

That markets for regular commodities are connected is no surprise and often clearly visible. The market mechanism usually solves these interdependencies silently and in the background. Without the market mechanism, however, these separate issues have to be linked in one way or the other. Moreover, due to the efficiency properties of the market mechanism (see Section 6.3.1) there is no critical necessity to think about the goal of the market process: there is at least the possibility that a Pareto-optimal allocation can be reached.

These are, therefore, once more the challenges associated with allocating environmental commodities: to set appropriate goals for solving the allocation problems and respect all these interdependencies – a holistic approach is, thus, needed. Mechanisms such as the Tragedy of the Commons and the Prisoners' Dilemma tend to complicate these tasks. Perceived scarcity, environmental awareness, a perceived feedback, and aspects of behavioural environmental economics play a role, too.

In view of a multitude of already existing environmental policies and the ample experience collected with them (see Chapter 5), their application with appropriate instruments and tools to implementing a circular economy seems natural. There is no single instrument for all needs, as attempts to prevent or reduce waste require different approaches in comparison to fostering DfEs or enhancing recycling activities or strategies to mitigate climate change. Moreover, as efforts to reduce waste on the one hand, and efforts to motivate DfEs on the other are to some extent related, holistic approaches aiming at more than just one issue are required to take these dependencies into account for the allocation of the environmental commodities in the context of a circular economy.

There is an additional remark in favour of environmental policies: the reference to a DfE directs the attention to the Porter hypothesis, the relationship between the stringency of an environmental policy and the rate of innovation activities. Martínez-Zarzoso, Bengochea-Morancho, and Morales-Lage (2019) find that there are indeed indications for such a positive relationship, a view, which is also supported by Wang, Sun, and Guo (2019) and Liao (2018). However, this effect likely depends on other framework conditions, too: on efforts to monitor and control compliance with these regulations, on possibilities to export these technologies etc. This implies that the design of environmental policies also has to take into account such possible "side-effects".

Observe that the market mechanism has the characteristics of a holistic allocation mechanism: any interdependencies between markets is respected and reflected in the equilibrium prices. Consequently, the attempt to construct or, rather, "design" holistic environmental policies is nothing but an attempt to imitate the market mechanism. The main challenge with these policies is to make appropriate use of the relevant knowledge, which is available in a decentralised way – an important feature characterising a market economy. That the level of environmental awareness plays a role regarding the design of environmental policies is analysed in Weng, Hsu, and Liu (2019) to "investigate how governments should adjust their environmental policies when worldwide environmental consciousness increases". As a consequence, these environmental policies should always take into account the specific situation in a country, also regarding environmental awareness.

Designing environmental policies will be one of the main tasks of Part V in the context of implementing a circular economy. The case studies of Chapter 5 indicate that current policies are not yet optimal with respect to meeting the goals.

The following section shows that environmental policies can themselves be the source of the Tragedy of the Commons and the Prisoners' Dilemma.

7.3 Case study: The German refillable quota issue

This case study refers to a peculiar environmental policy enacted in Germany at the beginning of the 1990s. This was the time when more and more one-way drinks containers including plastic bottles came into circulation in Germany. In 1991 the share of refillable drinks containers was close to 72% and the government wanted to keep this share at that level or even raise it.

For this purpose, the first version of the German Packaging Ordinance was enacted – and then, surprisingly, just the opposite happened, the refillable quota decreased substantially. Table 7.2 shows this development for the last years, which has not come to an end since 1991 (see also Wiesmeth & Häckl, 2016).

With respect to the portfolio of case study research designs this study could be considered an "anomaly" (see Ridder, 2017, p. 292), as the environmental policy did not produce the expected and desired result. Even worse, the policy led exactly to that outcome it wanted to avoid in the first place, and there is a lock-in effect due to sophisticated technologies, which have meanwhile been installed in Germany.

What happened? The purpose of the German Packaging Ordinance (first enacted 1991) and the later Packaging Act (enacted 2019) is to avoid or reduce the environmental impacts of waste arising from packaging. Regarding drinks packaging, the ordinance tries to regulate mainly the market behaviour of drinks producers.

In the first versions of the Packaging Ordinance, there is a general obligation to charge a deposit on non-reusable drinks packaging (Art. 8). However, there was an exemption from

TABLE 7.2 Quota of reusable beverage packaging: Germany 2015–2017.

Quota of reusable beverage packaging in Germany			
Year	2015	2016	2017
Water	39.7%	38.7%	38.4%
Beer	82.9%	82.1%	81.9%
Soft Drinks	30.6%	28.8%	27.1%
Mixed Alcoholic Drinks	6.4%	6.4%	6.3%
All Beverages	45.5%	44.2%	43.6%
Reusable	**44.3%**	**42.8%**	**42.2%**

Source: Data from Gesellschaft für Verpackungsmarktforschung mbH (https://gvmonline.de/wp/wp-content/uploads/2019/09/2019_09_EWMW2017_en.pdf).

this obligation as long as, roughly speaking, the combined proportion of drinks packaged in reusable packaging stayed at or above 72%, the actual share in 1991 (Art. 9 (2)).

These regulations focused clearly on the producers: with the threat of a mandatory deposit, they should be "motivated" to increase the share of refillable drinks containers. Unfortunately, these regulations did not directly address the consumers, although their drinking habits, as it turned out later, had a significant influence on the combined share of reusable drinks containers.

As already indicated, the policy failed, the refillable quota has dropped far below 72% since 1991 to less than 50%, despite the traditionally high share of refillable glass bottles for beer in Germany. In accordance with the regulations of the Packaging Ordinance the German government had to implement the deposit scheme in January 2003. Germany is now locked into a system it did not really want in the first place. This "lock-in effect" results from the development and installation of quite a sophisticated machinery to take back empty drinks packages and return the deposit fee. It would simply be too costly to abandon this system or replace it by a fundamentally different system – there are technological path dependencies.

One of the main reasons of this "anomaly" was the requirement of a "combined" quota for refillable drinks packages, which triggered the Prisoners' Dilemma. A producer of drinks, who alone cannot significantly affect this combined share, will raise the quantity of drinks offered in refillable containers, if customers are expected to follow suit and increase accordingly their demand for drinks in refillable containers. However, there is the Tragedy of the Commons waiting around the corner: even if consumers are environmentally conscious, they understand quickly that there is no sizeable feedback from their action to buy drinks in the perhaps not preferred refillable containers (see Section 6.1.3).

Thus, producers cannot expect to attract many more customers by increasing the share of drinks in reusable packaging.

Offering more drinks in refillable containers means investments in new equipment and higher variable costs for additional logistics services. This implies that waiting some time and learning from the experiences of other producers seems to be a feasible option for a drinks producer. The argument that the producer could help to prevent the mandatory deposit fee is not decisive: it is always better to let other producers go ahead with their risky investments and profit perhaps from their decisions. This explanation corresponds to the Prisoners' Dilemma, resulting from the environmental regulations.

This does not imply that the deposit fee does not affect drinks producers, only the development of the combined share does not play a major role in decisions regarding the individual share, producers focus more on the preferences of their customers (see also Ferrara & Plourde, 2003). As mentioned, this case is also related to an insufficient or missing feedback from the actions of both producers and consumers. And this "anomaly" is more an anomaly regarding the – in Germany – unexpected development, but it is not really an anomaly, if the Tragedy of the Commons and the Prisoners' Dilemma are taken into account as mechanisms driving this result.

Again, this development does not mean that the deposit system for one-way drinks packaging in Germany is not environmentally reasonable (see Chapter 20). The case only wants to point out that designing policies for allocating environmental commodities poses some challenges, that mechanisms such as the Tragedy of the Commons and the Prisoners' Dilemma, for example, can be triggered through inappropriate regulations of an environmental policy.

The following chapter focuses on relevant aspects of behavioural environmental economics.

References

Bezin, E., & Ponthièsre, G. (2019). The tragedy of the commons and socialization: Theory and policy. *Journal of Environmental Economics and Management, 98*, 102260. https://doi.org/10.1016/j.jeem.2019.102260.

Brown, K., Adger, W. N., & Cinner, J. E. (2019). Moving climate change beyond the tragedy of the commons. *Global Environmental Change, 54*, 61–63. https://doi.org/10.1016/j.gloenvcha.2018.11.009.

Ferrara, I., & Plourde, C. (2003). Refillable versus non-refillable containers: The impact of regulatory measures on packaging mix and quality choices. *Resources Policy, 29*(1), 1–13. https://doi.org/10.1016/j.resourpol.2004.04.001.

Isaksen, E. T., Brekke, K. A., & Richter, A. (2019). Positive framing does not solve the tragedy of the commons. *Journal of Environmental Economics and Management, 95*, 45–56. https://doi.org/10.1016/j.jeem.2018.11.005.

Liao, Z. (2018). Environmental policy instruments, environmental innovation and the reputation of enterprises. *Journal of Cleaner Production, 171*, 1111–1117. https://doi.org/10.1016/j.jclepro.2017.10.126.

Martínez-Zarzoso, I., Bengochea-Morancho, A., & Morales-Lage, R. (2019). Does environmental policy stringency foster innovation and productivity in OECD countries? *Energy Policy, 134*, 110982. https://doi.org/10.1016/j.enpol.2019.110982.

Niu, Z., Xu, J., Dai, D., Liang, T., Mao, D., & Zhao, D. (2018). Rational conformity behavior can promote cooperation in the prisoner's dilemma game. *Chaos, Solitons & Fractals, 112*, 92–96. https://doi.org/10.1016/j.chaos.2018.04.034.

Patt, A. (2017). Beyond the tragedy of the commons: Reframing effective climate change governance. *Energy Research & Social Science, 34*, 1–3. https://doi.org/10.1016/j.erss.2017.05.023.

Ridder, H.-G. (2017). The theory contribution of case study research designs. *Business Research, 10*(2), 281–305. https://doi.org/10.1007/s40685-017-0045-z.

Szolnoki, A., & Chen, X. (2020). Gradual learning supports cooperation in spatial prisoner's dilemma game. *Chaos, Solitons & Fractals, 130*, 109447. https://doi.org/10.1016/j.chaos.2019.109447.

Wang, Y., Sun, X., & Guo, X. (2019). Environmental regulation and green productivity growth: Empirical evidence on the Porter Hypothesis from OECD industrial sectors. *Energy Policy, 132*, 611–619. https://doi.org/10.1016/j.enpol.2019.06.016.

Weng, Y., Hsu, K.-C., & Liu, B. J. (2019). Increasing worldwide environmental consciousness and environmental policy adjustment. *The Quarterly Review of Economics and Finance, 71*, 205–210. https://doi.org/10.1016/j.qref.2018.08.003.

Wiesmeth, H. (2011). *Environmental economics: Theory and policy in equilibrium. Springer texts in business and economics.* Springer Nature. https://doi.org/10.1007/978-3-642-24514-5.

Wiesmeth, H., & Häckl, D. (2016). Integrated environmental policy: A review of economic analysis. *Waste Management & Research, 35*(4), 332–345. https://doi.org/10.1177/0734242X16672319.

Behavioural environmental economics

The implementation of a circular economy requires the motivation and the active support of all individuals, likewise of consumers and producers. Efforts regarding waste prevention or waste reduction, innovations regarding a Design for Environment (DfE) and technologies for recycling, participation in all kinds of activities mitigating climate change and others, all those activities depend on the integration of the relevant stakeholders and their cooperation.

This is, however, as practical experience shows (see the cases in Chapter 5), not always easy to accomplish: simply writing these requirements into laws and other legal regulations, is certainly an important first step. But, often the compliance with these regulations is low, at least not sufficient to be called a success story. Closely monitoring and controlling the behaviour of the individuals is politically not feasible beyond a certain magnitude, at least not in democratic, liberal societies. Whether China's "Social Credit System", which extends also to environmental issues, helps in this regard, remains to be seen.

Moreover, socially relevant mechanisms such as the Tragedy of the Commons or the Prisoners' Dilemma, make clear that individually rational behaviour may lead to results, which are inferior to what a society can achieve, if there is some cooperation among the countries or the individuals in the form of a higher rate of compliance regarding the environmental regulations in consideration. This refers explicitly, but not only, to the tedious, global efforts to mitigate climate change. In Section 6.1.3 an insufficient perceived feedback from one's own action was identified as one of the possible reasons for decisions corresponding to these mechanisms.

On the other hand, these last examples of the Tragedy of the Commons and the Prisoners' Dilemma provide some hope: in most cases there are stakeholders, countries or individuals, depending on the concrete case, who deviate from the usually expected behaviour in these situations. As these mechanisms are based on the rational behaviour of a utility-maximising consumer, a profit-maximising producer, or a government, which wants to protect its consumers and producers, these deviations also mean a deviation from the traditional model of the homo oeconomicus, which is formative for classical economics.

Thus, these observable deviations point to intrinsic motivations to act in an environmentally friendly way, even if this means, or, rather, seems to mean, sacrificing utility or profit.

Because of their obvious relevance for the implementation of a circular economy, this behaviour and the related contexts have to be studied more carefully. The still young economic branches of "Behavioural Economics" and "Behavioural Environmental Economics" investigate these issues.

The following section presents an introduction into behavioural economics, pointing in particular to various principles of possible relevance for environmental economics and a circular economy. The then following section considers more closely attempts in the even younger area of behavioural environmental economics. Thereafter, environmental awareness and the issue of a perceived feedback are reconsidered from a behavioural point of view, followed by some concluding remarks. Parts of the presentation in the following section are taken from Samson (2014).

8.1 An introduction to behavioural economics

Neoclassical economics is based on the completely rational homo oeconomicus, who is characterised by preferences on consumption bundles and utility maximising behaviour, given the budget. The preferences are usually considered to be stable, and the individual agent has full information on all aspects of relevance for decision-making and the capability to process this information in order to achieve optimal results. Becker (1976) applied rational choice theory to all kinds of human behaviour, ranging from marriage to crime, thereby proposing to other disciplines in the social sciences such as sociology to adopt the paradigm of rationality.

However, at more or less the same time, mostly psychologists challenged this prevailing economic thinking based on the rationality postulate. In particular Kahneman and Tversky (1979) demonstrated with their "prospect theory" that individual decisions are not always optimal, that

decisions, also the willingness to take risks, could depend on the way they are "framed".

Moreover, other research showed that there are restrictions to our information processing capabilities, resulting in the concept of "bounded rationality" (Simon, 1985). Thaler and Sunstein (2008) also mention adequate information and feedback as a result of one's own action, in particular, as key factors for decision-making. But fact is that all individual endeavours mitigating climate change do not lead to an, again individually, noticeable feedback or change. Scarcity of the underlying environmental commodities can be perceived, environmental awareness can be high, but this insufficient feedback can prevent or retard necessary individual actions – according to these findings.

There are additional complications: individuals might not be interested in the relevant up to date information regarding a particular issue. This "information avoidance" (Golman, Hagmann, & Loewenstein, 2017) refers to situations, where people prefer not to obtain the otherwise freely available knowledge on a certain issue of relevance for them, such as, perhaps, mitigating climate change.

Another aspect investigated in behavioural economics is the "psychology of price", also related to bounded rationality. The concept refers to the price of a commodity and the perception of value, which might be substantially different, depending on the chosen framing (Ariely, Loewenstein, & Prelec, 2003). This is of importance for environmental policies and the circular economy in various ways, as it might affect "demand" for certain environmental commodities.

Considering social dimensions of decision-making represents another issue, which is of relevance regarding the economic behaviour of individuals. This means that individuals do not make their decisions in complete isolation, but will usually take into account the social context. As a consequence, the social environment might exert some pressure regarding certain decisions, it might "shape" these decisions.

Trust is a cornerstone of social life, makes social life possible and valuable, and should therefore be expected to affect also human decision-making in profound ways. In this context, reciprocity plays an important role, when we extend trust to other people, whom we do not know so far, or whom we will not meet at all, but who may affect our situation with their decisions (Berg, Dickhaut, & McCabe, 1995).

Reciprocity, considered a "generalised moral norm" by Gouldner (1960), belongs therefore to the social norms, which are part of these social dimensions of decision-making and are of high relevance in environmental economics, in particular in many policies and other practical applications. To be more concrete, social norms represent explicit or implicit behavioural expectations for members of a society (Dolan, Hallsworth, Halpern, King, & Vlaev, 2010), and our preferences are influenced by these norms. This refers, in particular to norms associated with religion, traditions, and culture, of course, where they indicate adequate behaviour or actions, which are taken by many others, if not the majority of people in the corresponding group (see also Feirrera & van den Wijngaard, 2019). For obvious reasons, social norms play an essential role in a variety of issues in waste management, such as collection, segregation, and reduction of waste.

The following section discusses these fundamental aspects of behavioural economics in an environmental context.

8.2 Behavioural environmental economics

Although behavioural environmental economics, as a separate subject, is still relatively new, there is already a sizeable literature dealing, in an environmental context, with all the deviations from the model of the rational decision-maker, the homo oeconomicus. Shogren and Taylor (2008), and Croson and Treich (2014), for example, review and reflect the literature on bounded

rationality. Particular applications refer to "framing" of environmental developments and the consideration of low-risk, high-impact catastrophes in the context of the prospect theory. Beshears, Choi, Laibson, and Madrian (2008) consider further informational requirements, which crucially affect decision-making of individuals, again regarding environmental issues. Finally, social norms can provide guidance in all kind of environmental matters, in waste management, but also in mitigation of climate change. Buchholz, Falkinger, and Rübbelke (2014), Kesternich, Reif, and Rübbelke (2017), and Feirrera and van den Wijngaard (2019), among others, refer also to the literature in this area.

It becomes obvious that various issues of high relevance for implementing a circular economy require input and, even more importantly, further research results from behavioural economics, and will, therefore, hopefully inspire further research activities in this field. The following subsections point to a variety of challenges in the context of implementing a circular economy, which need to be addressed with more insight from behavioural economics. Finally, also the relationship between environmental awareness and a perceived feedback from environmentally friendly actions and behavioural economics will be reconsidered, providing some new insight.

8.2.1 Prospect theory

As already indicated, Kahneman and Tversky (1979) investigate situations, in which the risk-taking behaviour of individuals depends on the "framing". In this context, Gsottbauer and van den Bergh (2011), while explaining various situations of bounded rationality in an environmental context, explicitly address the possibility of gaining widespread support for a policy regarding a potential environmental catastrophe. The chances may depend on the framing, the way the policy is presented to the general public: if the focus is on saving lives, then support could be higher in comparison to a presentation with a focus in terms of deaths (p. 271). Thus, the traditional expected utility model might not be the perfect guide to evaluate environmental risks.

Shogren and Taylor (2008) also refer to this observation, when they point to the uncertain outcome of most environmental policies with possible consequences regarding the acceptance of the policies by the economic agents. Climate change, for example, is likely associated with quite a few disastrous developments, some of them, such as a climate induced shift in the Gulf Stream, perhaps with comparatively low probability (p. 31).

So far, it seems to be difficult to estimate, how people are going to deal with such low-probability, high-consequence risks. Interestingly, Rheinberger and Treich (2017) find "evidence in favour of catastrophe accepting attitudes" (p. 614). But these risks are of relevance, and the fact that global efforts to mitigate climate change have not been overly successful till now, emphasises the need for corresponding investigations with the tools and the approaches of behavioural economics.

But, of course, these considerations are of relevance also for other environmental issues and policies, such as levying a pollution tax, or restricting or preventing certain activities with some damaging potential, or raising environmental standards with unclear avoidance behaviour of the economic agents (see Part IV).

8.2.2 Informational requirements

Bounded rationality in combination with the limited information processing capabilities of the decision-makers affect possibilities of implementing a circular economy in various ways. Beshears et al. (2008) refer to factors such as passive choice, complexity and limited personal experience, which are also of relevance for environmental economics.

First of all, passive choice implies that economic agents might just follow the "default" actions of others. This can be of advantage for environmental issues, but can also prevent or at least delay the achievements of certain policy goals – depending on the default option. Thus, if waste segregation becomes the default, then this might convince further households to follow suit. The question remains, how to develop a certain alternative into the default, into a social norm?

Clearly, this aspect of passive choice is of relevance for quite a few other relationships in environmental economics. In principle, all contexts affected by the Tragedy of the Commons or the Prisoners' Dilemma are amenable to environmentally friendly default options. This refers in particular to efforts regarding the mitigation of climate change. Moreover, the strategic relationship between leaders and followers in some environmental context seems to be of relevance for making use of the factor of passive choice. Patt (2017) points to the possibility of "framing climate change in evolutionary terms" to allow for better policy options regarding the Tragedy of the Commons in this context.

Then, the complexity of certain environmental topics need not stimulate the interest of economic agents in dealing with such a topic, mitigating climate change, for example. On the other hand, however, this might induce people to accept a default option. Again, the questions remain, how to arrive at an environmentally friendly default option? How to turn an existing default option into an environmentally friendly one? Who is going to define an environmentally friendly default option?

Limited personal experience affects environmental decision-making in the following ways: if economic agents are not much familiar with certain environmental issues, in particular in the absence of social norms, then the emotional distance might not prove advantageous for supporting any initiatives. The case of mitigating climate change provides again an example regarding this context with an additional complication: there is no timely feedback from one's actions. With appropriate measures, the goal should therefore be to make economic agents at least "perceive" such a feedback.

The psychology of price seems to play a role in this context of bounded rationality, too. There is the example of deposit schemes for drinks packaging in a variety of countries, in Germany, for example. The interesting thing is that with a deposit fee of 0.25 Euro on one-way bottles Germany succeeds in collecting about 98% of all one-way plastic bottles. This is surprising, because the vast majority of German households is financially not dependent on returning these empty packaging to reclaim the deposit fee.

High environmental awareness does play a role regarding this observation, of course. But environmental awareness is for sure not lower in neighbouring Austria. But there is no such deposit scheme in Austria, and the collection rate of empty drinks packaging in Austria is with approximately 75% lower than in Germany. So, the question arises, whether this difference can nevertheless be attributed to the existence of the deposit system. Perhaps, this small fee does provide some extrinsic motivation, but at the same time generates an intrinsic motivation to return empty bottles.

There is, thus, the very interesting question: do these 0.25 Euro really contribute to changing habits? Or do they rather serve as a "cheap excuse": fulfilling one environmentally friendly action may "suffice for keeping a high moral self-image in agents", and render it easier to ignore other actions of environmental relevance (Engel & Szech, 2017).

This issue should be further investigated, because this or similar kinds of deposit fees could play a decisive role in other areas of waste management, for example in the context of waste electronic equipment – if this can be justified by means of appropriate studies in behavioural economics.

8.2.3 Social norms in environmental economics

According to the above reflections on bounded rationality, how the economic agents – consumers, producers, institutions – respond to environmental policies, seems to depend, at least to some degree, on what they consider a "default". Regarding the implementation of a circular economy, it is therefore important to make sure that these defaults are in favour of the environmental goals to be achieved with the policies.

It is in this context that social norms come into the picture. After all, they indicate the behaviour of a majority of people regarding particular issues. Among these issues count, of course, many with environmental relevance. There are, as already mentioned, aspects of preventing, collecting and segregating waste, which also depend on social norms.

However, the effectiveness or the strength of these norms seems to depend on the possibilities to observe the behaviour of other people. Consequently, putting small electronic equipment such as batteries into household waste, is still an issue in waste management, despite existing social norms (van den Bergh, 2008). On the other hand, in developing countries, littering can perhaps be reduced, if appropriate social norms affect the behaviour, in particular, if children are present (Moqbel, El-tah, & Hassad, 2019).

Social norms are of relevance in other environmental contexts, too. There is, for example, the provision of public goods, which is, with the assumption of rational decision-makers, usually governed by mechanisms such as the Tragedy of the Commons or the Prisoners' Dilemma leading to non-optimal outcome. The question thus arises, whether perceptions of fairness and equity influence or can influence these decisions, to achieve, for example, an egalitarian-equivalent equilibrium (Moulin, 1987). Perhaps an appropriate framing could be used to establish an appropriate social norm (see again Patt, 2017).

Regarding efforts to mitigate climate change, Pittel and Rübbelke (2013) investigate how, besides costs and benefits, fairness aspects can influence a negotiating party's willingness to join an international agreement on climate change. Their considerations point to the relevance of schemes such as the Green Climate Fund, which is a financial mechanism under the United Nations Framework Convention on

Climate Change (UNFCCC) and "was established to limit or reduce greenhouse gas emissions in developing countries, and to help vulnerable societies to adapt to the unavoidable impacts of climate change". Brown and Hagen (2010) discuss other issues related to fairness, referring in particular to the possibly differing environmental concerns of firms and their workers.

One of the central questions regarding social norms, refers, not surprisingly, how to enforce social norms? The literature (see, for example, Buchholz et al., 2014; Reif, Rübbelke, & Löschel, 2017, and other literature cited in Kesternich et al., 2017) presents various mechanisms. Whether they are applicable in practical contexts, in implementing a circular economy, for example, remains to be seen. At least Feirrera and van den Wijngaard (2019) are convinced "that social influence could be harnessed to develop effective strategies to encourage more sustainable decisions and climate resilient behaviour" (p. 122).

Fig. 8.1, taken from Feirrera and van den Wijngaard (2019), points out that Pro-Environmental Behaviour (PEB) is driven by personal values and beliefs, and by norms. The influence of social norms and the actions of other people are shown in dark-blue boxes. The authors find out that peers' PEB is by far the strongest motivator of individual environmentally friendly behaviour.

In this context, also White, Habib, and Hardisty (2019) come to the conclusion that "consumers are more inclined to engage in pro-environmental behaviours when the message or context leverages the following psychological factors: Social influence, Habit formation, Individual self, Feelings and cognition, and Tangibility".

In addition to that, Buchholz and Sandler (2017) refer to trust building and reciprocity regarding a successful leadership, motivating

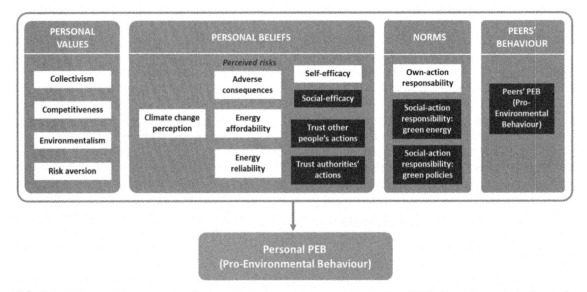

FIG. 8.1 Value-belief-norm theory. Factors driving pro-environmental behaviour (PEB). *From Feirrera, M., & van den Wijngaard, R. (2019). Pro-environmental behaviour: We care because others do. In Samson, A. (Ed.), The behavioural economics guide 2019 (p. 121). Retrieved from https://www.behavioraleconomics.com/the-be-guide/the-behavioral-economics-guide-2019/.*

followers to join the efforts to mitigate climate change, and Gneezy (2019) points to "four ways to use incentives to change behaviour": creating habits, breaking habits, providing upfront incentives, and removal of barriers. These issues, which will be reconsidered in other chapters of the book, seem to be important not only for global warming, but also for other areas of environmental economics, as indicated in Chapter 4 on leaders and followers in circular economy strategies.

8.3 Environmental awareness and perceived feedback

In order to briefly review Section 6.1, behavioural (environmental) economics can add further insight regarding the relationship between environmental awareness and environmentally friendly behaviour. As already indicated, a missing or simply non-available direct feedback, or also a non-perceived feedback from one's own environmentally friendly action or behaviour might as well play a role as various aspects of framing certain contents and goals of environmental policies: without a (perceived) feedback, there need not be an emotional attachment to the issue in consideration, and environmentally friendly actions need not follow. Similarly, the wording of relevant paragraphs and articles of policies could have an effect on this issue and should be taken into account. This refers in particular to fines or rewards associated with concrete regulations of a policy. Here comes the prospect theory of Kahneman and Tversky (1979) into the picture. Thus, there remains the questions, how and with which policy tools to adequately address the various groups of stakeholders in an environmental policy to raise "perceived feedback", perhaps with a suitable framing.

Bounded rationality with its multitude of features affects perceived feedback, too. Complexity of many environmental issues, aggravated by still incomplete scientific investigations, can simply reduce the interest of economic agents, thereby redirecting interest and associated actions to an available "default". It might then take a while and many efforts regarding creating or breaking habits in a society, till an environmentally friendly behaviour eventually becomes the new default.

This points finally to the important role of social norms. "We care because others care" (Feirrera & van den Wijngaard, 2019) shows the relevance of these norms to perceived feedback: an established social norm regarding the actual scarcity of an environmental commodity, for example, can raise perceived feedback from adequate and environmentally friendly behaviour for individuals.

As a further example, which indicates also the fragility of emerging social norms, the following case study investigates scientific efforts to understand and propagate biodiversity.

8.4 Case study: Sustainable use of the earth's biodiversity

The goal of a sustainable use of the earth's biodiversity gets more and more attention with research devoted to "the role of biodiversity in regulating ecosystem functioning and providing services to humanity", quoting from the Charter of the German Center for Integrative Biodiversity Research (iDiv) Halle-Jena-Leipzig. Consequently, one of this centre's role is to develop or strengthen social norms with the goal to raise perceived scarcity and awareness regarding biodiversity.

iDiv addresses, among others, the following questions: what is the role of biodiversity in regulating ecosystem functioning and provisioning of services to humanity? How should biodiversity be integrated in the management of our planet's resources and how can we safeguard biodiversity?

Undoubtedly, biodiversity plays a vital role in the ecosystem and its functioning needs to be protected: implementing a circular economy should thereby also integrate biodiversity. This brief case study considers iDiv and some aspects of the research activities to achieve its goals. The case study research design in this case is "no theory first" (Ridder, 2017, p. 291), pointing to the existence of a tentative research question, namely how to address and solve these tasks.

In order to spread the understanding of biodiversity as a valuable and scarce resource, iDiv researchers focus also on Biodiversity and Society. In this context they are "studying the impact of indirect socio-economic and direct drivers on biodiversity, ecosystem services and thereby human well-being". Among the key elements of their activities is also the engagement "of multiple stakeholders in monitoring, along with research on management and governance of biodiversity and ecosystem services". The "knowledge exchange between scientists, environmental resource managers, civil society

organisations, policy-makers and other stakeholders" is central to iDiv's work with an additional strong focus on citizen science.

Some of iDiv's publications point to methods of behavioural economics, applicable to the environmental context. In this sense, Wilson, Broughan, and Marselle (2019) "explore the potential for attempts to encourage student engagement to be conceptualised as behaviour change activity, and specifically whether a new framework to guide such activity has potential value for the Higher Education (HE) sector". And this new framework to be explored is the "Behaviour Change Wheel" (BCW), introduced in Michie, van Stralen, and West (2011).

Fig. 8.2 represents this BCW, which is not a linear model. According to the authors, "components within the behaviour system interact with each other as do the functions within the intervention layer and the categories within the

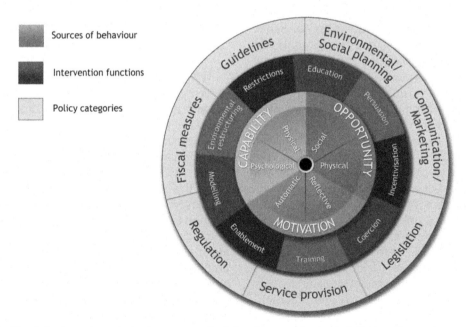

FIG. 8.2 The behaviour change wheel: a new method for characterising and designing change intervention. *From Michie, S., van Stralen, M.M., & West, R. (2011). The behaviour change wheel: A new method for characterising and designing behaviour change interventions.* Implementation Science, 6(1), 42. *https://doi.org/10.1186/1748-5908-6-42.*

policy layer". Behaviour change interventions are then "coordinated sets of activities designed to change specified behaviour patterns". Such interventions need to be designed carefully, including perhaps educational interventions, persuasion or even restrictions – supported by various policy measures. This complicated process is examined in more detail in Michie et al. (2011).

It remains to be seen, whether these attempts are successful with regard to changing the attitude towards biodiversity. The approach, attempting to change behaviour through appropriately coordinated policy tools and interventions seems to be applicable for the context of implementing a circular economy, too. But it seems that additional research activities are required to make better use of these tools.

8.5 Behavioural economics and implementing a circular economy

The considerations in this chapter show very clearly that behavioural economics has a lot to contribute to environmental economics in general, and the implementation of a circular economy in particular. What are the main reasons that traditional economic thinking, especially in a market context, does obviously not suffice to successfully internalise all external effects associated with environmental commodities?

An interesting exchange regarding this issue took place between economists who criticise behavioural economics and those who are open to what behavioural economics has to say about decisions of relevance for environmental policies. As Shogren and Taylor (2008) point out, a common criticism of behavioural economics is the argument that economic agents can learn to be rational "from a combination of market forces and evolution" (p. 28), with the market being more rational than the individual. Moreover, they refer to Gary Becker's critical opinion on the relevance

of "highly sophisticated" laboratory experiments in a market context (p. 29). According to Becker (2002), "economists have a theory of behaviour in markets, not in labs, and the relevant theories can be very different", with one reason being the division of labour.

One response from behavioural economists to this market-centred view is that this division of labour referred to by Becker (2002), does not always exist, in particular not for environmental commodities, for which regular markets usually do not exist. In fact, so the reasoning, we need environmental economics to deal with these missing markets (Shogren & Taylor, 2008).

The observations and the conclusions from existing environmental policies provide a mixed picture: there are obviously developments, which can better be explained by references to rational behaviour. Among those observations count, for example, the existence of vested interests, or decisions related to the mechanisms of the Tragedy of the Commons or the Prisoners' Dilemma. On the other hand, there are various aspects, which seem to require support from behavioural economics. It cannot be denied that social norms play a role in waste management. Moreover, not all people fall prey to the Tragedy of the Commons or the Prisoners' Dilemma – other or additional issues are of relevance for decision-making.

Therefore, implementing a circular economy should be open to respect economic decisions of a rational individual, but should also be ready for decisions deviating from the guidelines of the neoclassical theory. This constitutes an additional challenge because there is no closed and general theory of behavioural economics, comparable to the neoclassical equilibrium theory. Therefore, there is currently no possibility to make ample use of results from behavioural environmental economics for implementing a circular economy. Consequently, the functionality of various policies such as deposit systems, recommended for various circular economy policies (see Part V), needs to be observed and changed, if need be. Also, tools such as the behaviour change wheel need to be investigated regarding their applicability to the environmental context. The fact that these policies and these tools depend on local conditions does not make this situation easier.

The next chapter addresses the allocation problems in a circular economy. What are the challenges? What are the special requirements? What are the consequences for implementing a circular economy?

References

Ariely, D., Loewenstein, G., & Prelec, D. (2003). "Coherent arbitrariness": Stable demand curves without stable preferences. *The Quarterly Journal of Economics*, 118(1), 73–105. Retrieved from http://www.jstor.org/stable/25053899.

Becker, G. S. (1976). *The economic approach to human behavior.* Chicago, IL: The University of Chicago Press.

Becker, G. S. (2002). *Interview. The region.* Federal Reserve Bank of Minneapolis.

Berg, J., Dickhaut, J., & McCabe, K. (1995). Trust, reciprocity, and social history. *Games and Economic Behavior*, 10(1), 122–142. https://doi.org/10.1006/game.1995.1027.

Beshears, J., Choi, J. J., Laibson, D., & Madrian, B. C. (2008). How are preferences revealed? *Journal of Public Economics*, 92(8), 1787–1794. https://doi.org/10.1016/j.jpubeco.2008.04.010. Special Issue: Happiness and Public Economics.

Brown, G., & Hagen, D. A. (2010). Behavioral economics and the environment. *Environmental and Resource Economics*, 46(2), 139–146. https://doi.org/10.1007/s10640-010-9357-6.

Buchholz, W., Falkinger, J., & Rübbelke, D. (2014). Nongovernmental public norm enforcement in large societies as a two-stage game of voluntary public good provision. *Journal of Public Economic Theory*, 16(6), 899–916. https://doi.org/10.1111/jpet.12084.

Buchholz, W., & Sandler, T. (2017). Successful leadership in global public good provision: Incorporating behavioural approaches. *Environmental and Resource Economics*, 67(3), 591–607. https://doi.org/10.1007/s10640-016-9997-2.

Croson, R., & Treich, N. (2014). Behavioral environmental economics: Promises and challenges. *Environmental and Resource Economics*, 58(3), 335–351. https://doi.org/10.1007/s10640-014-9783-y.

Dolan, P., Hallsworth, M., Halpern, D., King, D., & Vlaev, I. (2010). *MINDSPACE: Influencing behaviour through public policy.* London: Cabinet Office. Retrieved from https://www.instituteforgovernment.org.uk/sites/default/files/publications/MINDSPACE-Practical-guide-final-Web_1.pdf.

Engel, J., & Szech, N. (2017). *Little good is good enough: Ethical consumption, cheap excuses, and moral self-licensing. Discussion paper No. 17-28 GEABA.* Retrieved from http://www.geaba.de/wp-content/uploads/2017/07/DP_17-28.pdf.

Feirrera, M., & van den Wijngaard, R. (2019). *Proenvironmental behaviour: We care because others do.* In A. Samson (Ed.), The behavioural economics guide 2019 (p. 121). Retrieved from https://www.behavioraleconomics.com/the-be-guide/the-behavioral-economics-guide-2019/.

Gneezy, U. (2019). *Introduction. Incentives and behaviour change.* In A. Samson (Ed.), The behavioural economics guide 2019 (p. VI). Retrieved from https://www.behavioraleconomics.com/the-be-guide/the-behavioral-economics-guide-2019/.

Golman, R., Hagmann, D., & Loewenstein, G. (2017). Information avoidance. *Journal of Economic Literature*, 55(1), 96–135. https://doi.org/10.1257/jel.20151245.

Gouldner, A. W. (1960). The norm of reciprocity: A preliminary statement. *American Sociological Review*, 25(2), 161–178. https://doi.org/10.2307/2092623.

Gsottbauer, E., & van den Bergh, J. C. J. M. (2011). Environmental policy theory given bounded rationality and other-regarding preferences. *Environmental and Resource Economics*, 49(2), 263–304. https://doi.org/10.1007/s10640-010-9433-y.

Kahneman, D., & Tversky, A. (1979). Prospect theory: An analysis of decision under risk. *Econometrica*, 47(2), 263–291. https://doi.org/10.2307/1914185.

Kesternich, M., Reif, C., & Rübbelke, D. (2017). Recent trends in behavioral environmental economics. *Environmental and Resource Economics*, 67(3), 403–411. https://doi.org/10.1007/s10640-017-0162-3.

Michie, S., van Stralen, M. M., & West, R. (2011). The behaviour change wheel: A new method for characterising and designing behaviour change interventions. *Implementation Science*, 6(1), 42. https://doi.org/10.1186/1748-5908-6-42.

Moqbel, S., El-tah, Z., & Hassad, A. (2019). *Litter control in developing countries.* Discussion Paper. The University of Jordan.

Moulin, H. (1987). Egalitarian-equivalent. cost sharing of a public good. *Econometrica*, 55(4), 963. https://doi.org/10.2307/1911038.

Patt, A. (2017). Beyond the tragedy of the commons: Reframing effective climate change governance. *Energy Research & Social Science*, 34, 1–3. https://doi.org/10.1016/j.erss.2017.05.023.

Pittel, K., & Rübbelke, D. (2013). International climate finance and its influence on fairness and policy. *The World Economy*, 36(4), 419–436. https://doi.org/10.1111/twec.12029.

Reif, C., Rübbelke, D., & Löschel, A. (2017). Improving voluntary public good provision through a non-governmental, endogenous matching mechanism: Experimental evidence. *Environmental and Resource Economics, 67*(3), 559–589. https://doi.org/10.1007/s10640-017-0126-7.

Rheinberger, C. M., & Treich, N. (2017). Attitudes toward catastrophe. *Environmental and Resource Economics, 67*(3), 609–636. https://doi.org/10.1007/s10640-016-0033-3.

Ridder, H.-G. (2017). The theory contribution of case study research designs. *Business Research, 10*(2), 281–305. https://doi.org/10.1007/s40685-017-0045-z.

Samson, A. (2014). In A. Samson (Ed.), *The behavioral economics guide 2014* Retrieved from https://www.behavioraleconomics.com/the-be-guide/the-behavioral-economics-guide-2014/.

Shogren, J. F., & Taylor, L. O. (2008). On behavioral-environmental economics. *Review of Environmental Economics and Policy, 2*(1), 26–44. https://doi.org/10.1093/reep/rem027.

Simon, H. A. (1985). Models of bounded rationality. Volume 1: Economic analysis and public policy. Volume 2: Behavioural economics and business organization: 1982 (reprinted 1983), Cambridge, MA: MIT Press. 478, 505 pages. *Organization Studies, 6*(3), 308. https://doi.org/10.1177/017084068500600320.

Thaler, R. H., & Sunstein, C. R. (2008). *Nudge: Improving decisions about health, wealth, and happiness.* Yale University Press.

van den Bergh, J. C. J. M. (2008). Environmental regulation of households: An empirical review of economic and psychological factors. *Ecological Economics, 66*(4), 559–574. https://doi.org/10.1016/j.ecolecon.2008.04.007.

White, K., Habib, R., & Hardisty, D. J. (2019). How to SHIFT consumer behaviors to be more sustainable: A literature review and guiding framework. *Journal of Marketing, 83*(3), 22–49. https://doi.org/10.1177/0022242919825649.

Wilson, C., Broughan, C., & Marselle, M. (2019). A new framework for the design and evaluation of a learning institution's student engagement activities. *Studies in Higher Education, 44*(11), 1931–1944. https://doi.org/10.1080/03075079.2018.1469123.

9

The economics of implementing a circular economy

Definition 2.1, following Pearce and Turner (1989), introduces "the" circular economy as the natural model of an economic system, integrating in particular the fundamental functions of the environment sustainably into the economy. The environment supplies natural resources, receives and assimilates waste and serves as a direct provider of utility. What does this mean from an economic point of view, from the point of view of solving the allocation problems for a circular economy?

These solutions obviously have to be in accordance with the requirements of the circular economy. Again, there certainly are many allocations satisfying these conditions. The goal is now to roughly describe and characterise these feasible allocations – with only brief hints to difficult optimality issues. The allocation mechanism in this context (see Section 6.3) seems to be largely technology-driven. This has certainly the advantage that a continuous stream of innovative technologies can facilitate and perhaps accelerate the implementation of a circular economy. There is, however, the risk of establishing significant technological path dependencies. This refers to all kinds of technologies handling waste, but also to new technologies such as e-mobility, e-commerce, or technologies for smart cities (see, for example, Leopoldina, 2017). Relevant issues associated with this strong orientation on technologies will be addressed in more detail in Part III.

Returning to the feasible allocations for a circular economy, they should first of all support a sustainable development: natural resources should be saved and the assimilative capacity of the environment regarding waste must not be exhausted. Together with sustainability the focus is therefore on all aspects of handling waste. In combination with a technology-driven development this implies that there should always be incentives for further technological innovations to reduce all negative issues associated with waste and to save resources through material recovery. Nevertheless, this remains a critical issue, ranging between technological path dependencies and degrees of freedom necessary to allow for as much knowledge as possible for implementing an adequate model of a circular economy.

Implementing the Circular Economy for Sustainable Development
https://doi.org/10.1016/B978-0-12-821798-6.00009-0

The following subsections present features of the allocation problems in a circular economy, and discuss the relevant concepts "sustainability" and the "waste hierarchy", also in the context of the literature. The final section summarises consequences for implementing a circular economy.

9.1 The allocation problems for a circular economy

Implementing a circular economy means solving the associated allocation problems adequately or even optimally in accordance with the framework conditions of a circular economy. But what exactly are the goals, what are the targets? What are "adequate" or even "optimal" solutions? How to characterise them practically? And how to combine the requirements of a circular economy with the optimality criterion? From a purely formal point of view it is, of course, possible to look for "bounded optimality", for feasible allocations which are Pareto-optimal under the constraint that they are also feasible for a circular economy.

This leads, not surprisingly, to various theoretical questions: how can circular economy allocations be formally characterised? Given an optimality concept, what is then the relationship between the set of optimal allocations and the set of allocations which are feasible for a circular economy? And last, but not least the question for an allocation mechanism: how to attain an optimal allocation for a circular economy, how to implement a circular economy?

If the vision of the Ellen MacArthur Foundation comes true that after having transformed all the elements of the linear take-make-waste system, we can "create a thriving economy that can benefit everyone within the limits of our planet", then there is a chance to arrive at an "optimal" allocation, benefitting everyone.

Nevertheless, these are assumptions, even visions, and the future will show their relevance.

And the question remains, how to transform "all the elements of the linear take-make-waste system"? How to motivate consumers and producers to manage resources, to make and use products accordingly?

In a pure market economy, the market mechanism guides economic decisions to a Pareto-optimal allocation, at least under the conditions discussed in Section 6.3.1. Such an allocation has features of a socially acceptable outcome, which can be further adjusted through suitable income redistributions etc. Thus, the market mechanism can guide the economy to an equilibrium, which is, in general, accepted and acceptable from a social point of view.

Regarding the allocation of environmental commodities, there is but one important difference: other tools have to augment or replace the market mechanism (see Chapters 7 and 8). There are environmental policies of all kinds, business models, and behavioural aspects, which have to be taken into account, combined and linked in order to implement a circular economy. Market solutions continue to play a role in this context – profitable business models can help (see Section 3.1.1). But Adam Smith's "invisible hand" (Section 6.3.1) is no longer fully functional, and it might also turn out that alternative optimality concepts have to replace or supplement Pareto-optimality. So far only little is known on the relationship between Pareto-optimality and allocations which are feasible for a circular economy.

The following subsections consider sustainability and waste management as core concepts of feasible allocations for a circular economy.

9.2 The concept of a sustainable development

The concept of sustainability has some of its roots in forestry: for a "sustainable development" no more wood should be harvested than the volume that regrows in the same period of time.

This was important at a time when, for example, parts of Germany, in particular Saxony, were deforested due to the immense consumption of wood in mining precious ores. At this time, more than 300 years ago, coal was not yet much in use, not discovered as valuable and almost unlimited source of energy – and long-term storage of carbon (see also Section 6.1.1).

Meanwhile there are many perceptions of sustainability, but practically all of them refer to a situation, in which human activity "conserves the function of the earth's ecosystems" (Geissdoerfer, Savaget, Bocken, & Hultink, 2017, p. 758). The Brundtland Commission proposed a widely accepted definition, likely emerging from ideas and actors from the "South" (Fukuda-Parr & Muchhala, 2020): "Sustainable development is development that meets the needs of the present without compromising the ability of future generations to meet their own needs." Importantly, the report refers to limits, also defined by the available technologies: "The concept of sustainable development does imply limits – not absolute limits but limitations imposed by the present state of technology and social organisation on environmental resources and by the ability of the biosphere to absorb the effects of human activities. But technology and social organisation can be both managed and improved to make way for a new era of economic growth." However, "sustainable development is not a fixed state of harmony, but rather a process of change in which the exploitation of resources, the direction of investments, the orientation of technological development, and institutional change are made consistent with future as well as present needs" (Brundtland, 1987).

Steps towards a sustainable development depend therefore on the concrete situation in a particular country or region – both regarding institutional frameworks and available technologies with significant differences between the countries (see also Section 2.1). As most

economies are linked through international trade and through global environmental commodities, such as mitigation of climate change, to achieve a global sustainable development is certainly one of the challenges of mankind in the years to come. Considerations in the context of the Tragedy of the Commons or the Prisoners' Dilemma might also play a role (see Section 7.1): which industrialised and highly developed countries are the first ones to adjust, perhaps even reduce their more or less affluent lifestyles in favour of the planet's ecological means? How can countries be motivated, extrinsically or intrinsically, to join the efforts to mitigate climate change?

Finally, according to Pearce and Turner (1989), "sustainable development thinking reflects both our greater understanding of what natural environments do for us as inhabitants and users of those environments, and stresses the uncertainty that results from our continuing ignorance of the many other ways in which we depend on life-support systems" (p. xii). This is a clear reference to the functions of the environment, which should be preserved, as "there are no substitutes for many natural assets: the oceans, the moors, the fells, the mountains, rivers and all their diverse wildlife. It is because a great many people want to experience these assets, and to feel reassured that they are being protected as far as seems reasonable, …, that the social science devoted to the efficient satisfaction of their wants is relevant" (p. xii).

In the context of the circular economy introduced in Section 2.1, the issue of sustainability is of relevance for each one of the fundamental functions of the environment. The supply of natural resources has to be taken care of in a sustainable way, the assimilative capacity of the environment should not be exhausted in the years to come, and also the direct provision of utility should be sustainably available.

In a regular market context, scarce natural resources tend to become more expensive thereby initiating the search for better mining technologies. Alternatively, new production technologies or a focus on new commodities can help to reduce or replace the input of these resources. As experience shows, market economies can provide substantial support to this regard. However, issues such as corruption (Erum & Hussain, 2019), conflicts (Vesco, Dasgupta, De Cian, & Carraro, 2020) or the resources curse (Henri, 2019) in countries endowed with resources tend to lessen the enthusiasm. Thus, recovery and recycling activities are gaining increasing importance in order to establish some independence. In addition to efforts to prevent waste they help to save valuable resources.

In September 2015, countries from all over the world adopted the UN 2030 Agenda for Sustainable Development and the 17 Sustainable Development Goals (SDGs), with a concrete list of actions. These SDGs address relevant global challenges, among them various of direct importance for a transition to a circular economy. There is Goal 7: "Affordable and Clean Energy" with its reference to air pollution and renewable energies, there is Goal 8: "Decent Work and Economic Growth", pointing to economic activities without harming the environment, there is Goal 11: "Sustainable Cities and Communities" addressing rising air pollution in cities and the need for adequate municipal waste collection, there is Goal 12: "Responsible Consumption and Production" with explicit reference to energy efficiency, thereby reducing resource use, degradation and pollution along the whole life cycle, and there is Goal 13: "Climate Action", postulating coordinated actions at the international level, to help in particular developing countries.

In conclusion, most of the SDGs are relevant for the path towards a circular economy. Moreover, the concept of sustainability is also important in related concepts, such as that of a "Smart City". According to various publications (see, among others, Ahvenniemi, Huovila, Pinto-Seppä, & Airaksinen, 2017; De Guimarães, Severo, Felix Júnior, Da Costa, & Salmoria, 2020; Hajduk, 2016; Martin, Evans, & Karvonen, 2018), the concept of a smart city refers, first of all, to sustainability. Nevertheless, Martin et al. (2018) also point to certain tensions between smart city visions and the goals of a sustainable development, and Ahvenniemi et al. (2017) study differences between smart and sustainable cities. Not surprisingly, that there are initiatives with the goal to bring the two concepts in close contact. This refers, in particular to Circular Economy in Smart Cities of the European Union (EU) and Circular Economy in Cities of the Ellen Mac Arthur Foundation.

In summary, a sustainable development is an integral part of a circular economy. However, the concept is dependent on many framework conditions, which differ substantially across countries. For this reason, each country or region should choose its own path towards sustainability, although this may, in view of the interdependencies mentioned, lead to controversies, accompanied by behaviours controlled by the Tragedy of the Commons or the Prisoners' Dilemma. We are, thus, to some extent back at the question Segerson (1991) asked after the publication of Pearce and Turner (1989) on "what sustainability would mean in terms of either theory or practical implementation" (p. 273). Relating sustainability to the feasible allocations remains therefore a challenge – not to speak of finding and employing an allocation mechanism to achieve an appropriate allocation.

The following section addresses another issue of utmost relevance for a circular economy: waste and its management – also an issue of a sustainable development.

9.3 Waste management

Pearce and Turner (1989) in their characterisation of the circular structure of an economy point to the fact "that natural environments are the ultimate repositories of waste products:

carbon dioxide and sulphur dioxide go into the atmosphere, industrial and municipal sewage goes into rivers and the sea, solid waste goes to landfill, chlorofluorocarbons go to the stratosphere, and so on" (p. 35). Natural systems such as forests, for example, produce their own waste, which is in general recycled in a natural way. This is, also in general, not true of economic systems, there is only a limited assimilative capacity for certain types of waste, even for bio-waste, for example.

Of course, waste can be collected and recycled, at least to some extent. The Second Law of Thermodynamics, however, "places a physical obstacle, another 'boundary', in the way of redesigning the economy as a closed and sustainable system" (p. 37f). Georgescu-Roegen (1971) strongly emphasises the relevance of this law to economics, in particular to dynamic economic analysis. Ayres (1998) refers, however, to the "the flux of available low-entropy energy (exergy) from the sun ...", which can sustain economic activity for a long period of time, "even though fossil fuel and metal ore stocks may eventually be exhausted" (p. 189). Nevertheless, fundamental functions of the environment, as a direct provider of utility, for example, can be harmed. Moreover, excessive waste may itself seriously disturb economic activities and cleaning up the environment may require significant resources. Thus, observing the waste hierarchy is inherently related to implementing a circular economy.

The physical reason is that the "materials that get used in the economy tend to be used entropically – they get dissipated within the economic system" (Pearce and Turner, 1989, p. 37). In general, it will be simply too costly or technically impossible to collect and recycle all materials used in the production process. This puts a temporary cap on economically reasonable collection and recycling quota, which are now ubiquitous in environmental regulations, especially in waste management. Further technological innovations may, of course, allow more stringent

caps. But whether it will be possible to "design out waste and pollution", as argued by the Ellen MacArthur Foundation, remains to be seen.

Therefore, returning to Pearce and Turner (1989), what are the consequences for preserving the functions of the environment in the circular economy? Waste management has to take into account various possible ways of handling waste.

9.3.1 Possible ways of handling waste

The following possibilities of handling waste are far beyond simple landfilling, which is, nevertheless, still a common way of "treating" waste in many countries. Turning towards a circular economy implies then a substantial rethinking, which starts with preventing waste (see Luttenberger, 2020 for the challenges waiting in Croatia to this regard). But also countries with more advanced methods of waste management need to adjust their strategies. Societal path dependencies with a strong focus on market economies need to be carefully redirected to make room for preventing waste, for reusing and recycling old and discarded commodities (see Section 12.2).

Prevention of Waste: Because it is, due to insight from the natural sciences, technically impossible and/or economically too costly to collect and recycle all waste products, waste should be prevented, wherever and whenever this is possible. The more waste we produce in terms of packaging waste, for example, of plastic waste, in particular, the more waste stays in the environment. Therefore, prevention of waste should be the ultimate goal, and this goal should be mirrored in the feasible allocations. Suitable designs of the products, a Design for Environment (DfE), could help to reduce material input in production. But there are other possibilities to reduce waste (see EU, 2008, Annex IV), which could be applied, at least in principle. The measures mentioned by the EU "affect the framework conditions related to the generation of waste", they "affect the design and production and distribution phase", and they "affect the consumption and use phase".

In general, however, efforts in this regard seem to be difficult and time-consuming to implement. This observation is true for many countries in all parts of the world (see also Chapter 5). Referring to the allocation problems: implementing feasible allocations regarding waste prevention is a challenge. In the context of a market economy efforts to prevent waste might seriously interfere with business interests. In the simplest case, this might mean higher costs due to costly design changes etc.

Reusing Products: Possibilities of reusing products exist in a manifold of ways. Old cars, textiles, and electronic equipment can be reused. Other commodities, such as houses or apartments, are typically reused. The gradual development with support from digital platforms towards a "sharing" society with less ownership regarding cars etc. and more sharing activities points to the increasing importance reusing is about to gain.

The possibilities of reusing older cars or other equipment is not always without problems. For example, when old cars or old electric and electronic equipment, which might still be reusable from a technical point of view, are exported to countries, which do not have the tools and knowledge to maintain or recycle these cars and other equipment, then environmental pollution associated with health risks due to improper handling might be one of the consequences. Thus, reusing is important in the context of a circular economy, but not without appropriate framework conditions. Referring to the allocation problems: how to integrate reusing activities adequately into the solutions? Again, in the context of a market economy, the challenge is to "compensate" business for a possibly reduced production of certain commodities, electronic equipment, for example, due to a prolonged life of the equipment.

Recycling of Waste: Due to the limited capacity of the environment to assimilate waste from the various kinds of economic activities, it is necessary to collect waste commodities and to recycle them. Reusing waste products could provide an intermediate alternative before recycling, and, again, an appropriate DfE could also help to facilitate recycling activities. There are, however, many recycling activities which have to be combined with reasonable collection procedures for the waste commodities. Technological path dependencies provide then solutions to the second allocation problem in this context: how to recycle, which methods, techniques can be applied in accordance with the technological framework conditions?

In contrast to efforts to prevent waste and also to reuse old products, recycling activities already gained widespread and increasing attention in recent years. One of the reasons is probably that recycling opens new business activities and is therefore more attractive than prevention of waste, for example. However, profitable recycling activities provide incentives for extending these activities – possibly to the disadvantage of efforts to prevent waste and reuse old equipment. Consequently, these societal path dependencies need some special attention when implementing a circular economy.

All these considerations lead to the waste hierarchy, a core concept in the context of implementing a circular economy.

9.3.2 The waste hierarchy

The waste hierarchy is of crucial relevance for a circular economy and its implementation. As is well-known, the waste hierarchy refers to the following priority order regarding waste management: prevention of waste; preparing for reuse; recycling of waste; other recovery, e.g. energy recovery of waste; and, finally, disposal of residues. In view of the functions of the environment and its limited assimilative capacity, the residues, which are to be landfilled in general, should be treated such that they are chemically and biologically inert.

The waste hierarchy is meanwhile part of most legislations on waste management, although interpretations, also in the literature, and practical handling vary quite a lot. Increasingly, collection and the recycling of waste are considered more or less equivalent to preventing waste, or the goal of preventing waste is simply neglected. This points to the societal path dependencies with a focus on profitable business activities. Kirchherr, Reike, and Hekkert (2017) mention that among their review of 114 definitions of a circular economy "some authors entirely equate circular economy with recycling ..." (p. 229). In various cases, "prevention of waste" is also replaced by "prevention of pollution" (see, for example, Ma, Wen, Chen, & Wen, 2014), which, in general, means an application of the polluter pays principle or similar measures (see the review of Reike, Vermeulen, & Witjes, 2018, p. 248). Also Prieto-Sandoval, Jaca, and Ormazabal (2018) point to two different groups of publications on the circular economy: the first group adheres to the principles of the waste hierarchy (see, for example, Wang, Che, Fan, & Gu, 2014 as one representative in this group), whereas the second group proposes so-called "sustainable design strategies" (SDSs) as the core principles of a circular economy. Among those SDSs counts in particular eco-design or, closely related, a DfE. The Ellen MacArthur Foundation is a proponent of these strategies (Prieto-Sandoval et al., 2018, p. 610).

There is, thus, a discrepancy between formal requirements regarding feasible allocations and practical implementations. What are possible reasons for this deviation from the waste hierarchy, with waste prevention being – as following from Pearce and Turner (1989) – an indispensable part of waste management for a successful transition to a circular economy? According to Kirchherr et al. (2017), "practitioners frequently neglect 'reduce' in their circular economy definitions, though, assumingly since this may imply curbing consumption and economic growth" (p. 229). Moreover, in Fig. 3 in their paper they document the definitions coded on the "4R framework" (see p. 227). Just for the sake of completeness, Fig. 9.1 shows a 9R framework with certain strategies associated with the linear respectively the circular model of an economy. Strategies which extend the lifespan of a product and its parts are in between.

There might be an additional reason for this observation: reducing waste or preventing the generation of waste requires the cooperation of many "stakeholders", in particular of consumers and producers, of all those, who are generating waste. This, as experience shows, is not a straightforward thing to do (see Chapter 5), and statistics provide in fact a mixed signal: total municipal waste generation in the countries of the partnership network of the European Environment Agency (EEA) declined by 1% in absolute terms and by 4% per capita from 2004 to 2012. According to a Briefing of the EEA, there was, however, an increase in per capita waste generation in 15, and a decrease in 20 out of 36 countries with an increase for Germany, for example, from 584 kg per capita in 2004 to 611 kg per capita in 2012. Regarding the EU countries, the latest Eurostat statistics states that waste generation per capita declined by 8 kg between 1997 and 2017 – a period of 20 years. Given that per capita waste generation in the EU is around 400 kg and that more and more plastics replaced other, heavier packaging material in this period of time, these numbers are not really impressive and convincing, and the reasons behind this development needs to be analysed. But, it has to be admitted that there is some decoupling of economic growth and waste generation.

On the other hand, collecting and recycling waste is much more technically oriented, and can be actively controlled by practitioners and the public administration. It leaves room for innovations, new jobs, and success messages

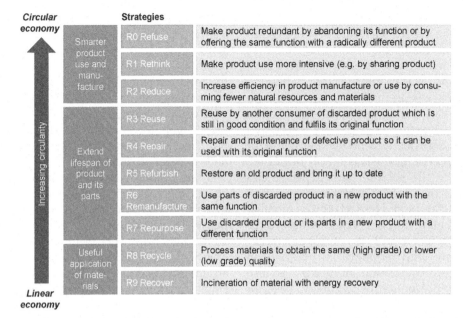

FIG. 9.1   The 9R framework regarding the waste hierarchy. Circularity strategies within the production chain. © *From Kirchherr, J., Reike, D., & Hekkert, M. (2017). Conceptualizing the circular economy: An analysis of 114 definitions.* Resources, Conservation and Recycling, 127, 221–232. https://doi.org/10.1016/j.resconrec.2017.09.005.

of increasing collection and recycling rates, thereby insinuating to the general public that the issue of waste management is completely under control, and there is no need to worry about the transition to a circular economy. Indeed, considering again the countries of the EEA partnership network shows a strong increase of average recycling rates from 22% in 2004 to 29% in 2012. According to the Briefing of the EEA, the numbers for Germany are 56% in 2004 and 64% in 2012.

This attitude, however, favours developments in waste management, which may at least retard the implementation of a circular economy. This refers in particular to the tendency to "misuse" the waste hierarchy for various other (business) purposes. In Germany, for example, chain stores, offering drinks in one-way packaging may enter the waste management and recycling business. Consequently, in addition to selling drinks, they have to provide for the collection and recycling of the empty

packaging, according to the environmental regulations. If the conditions for collecting and recycling plastics are reasonable, then they have another business case: by selling their products they also generate more packaging waste, which is specific for their brands, and then they gain additionally by recycling the waste drinks packaging.

At the first glance, this looks like a perfect implementation of the polluter pays principle, which is part of many environmental regulations: those, who generate waste are responsible for collecting and recycling it. However, it sounds too good to be true in view of the waste hierarchy: indeed, under certain conditions there is a clear incentive for more drinks in one-way plastic bottles, thus contradicting the priority goal of waste prevention.

Consequently, the currently observable increasing focus on collecting and recycling waste may deviate from the path towards a circular economy, it need not be incentive compatible regarding the prevention of waste – to the contrary, there might be incentives to produce more waste. In this sense, the implicit argument that recycled waste is equivalent to prevented or reduced waste, is treacherous. This becomes obvious, when this waste, in particular plastic waste, is exported to other countries, developing or emerging economies, with the official declaration, of course, that the waste will be recycled in these countries in an environmentally friendly way (see Chapter 23, for example).

9.3.3 Optimal allocations for the waste hierarchy

Returning to the economic background, due to the lack of a universally functioning mechanism for the allocation of the environmental commodities, there are only partial answers to structural properties of an optimal solution, to an optimal implementation of a circular economy. One of these properties is the priority order of the waste hierarchy: waste as an economic bad should first of all be prevented – at least to a level, at which it is possible to still have the same quantity of comparable private commodities: a solution to the allocation problems cannot be optimal, if it is possible to reduce waste and still enjoy the same quantity of the private commodities. Thereafter the other options come into the picture: reuse and recycling save resources and costs and help to reduce the pollution of the environment. To what extent they increase benefit is a priori difficult to judge – due to incomplete information.

Similarly, to what extent one should reduce, reuse or recycle waste, is difficult to say. To be more precise, it can happen that "reducing the environmental impact of a product at the production stage may lead to a greater environmental impact further down the line" (EU, 2012). This is also related to the Second Law of Thermodynamics signalling an increase in the entropy of the system, which can only be temporarily held back through increasing efforts regarding the collection and recycling of waste.

However, because there is only limited knowledge about the assimilative capacity of the environment regarding the various kinds of waste, and because the more waste is produced, the more tends to end up in the environment, it is always preferable to prevent waste in the first place. Moreover, why to generate waste, use valuable resources beyond the really unavoidable quantities? Thus, one should always check very carefully, if somebody, a business company, for example, plans to generate more waste, even if there are all promises to collect and recycle the waste products.

In EU (2012) various questions that can arise in this context in a local or regional setting are posed:

- Is it better to recycle waste or to recover energy from it? What are the trade-offs for particular waste streams?
- Is it better to replace appliances with new, more energy efficient models or keep using the old ones and avoid generating waste?
- Are the greenhouse gas emissions created when collecting waste justified by the expected benefits of a higher collection rate?

Answers to these and related questions also dependent on the local or regional context. Thus, universal answers can, in general, not be provided, making the design of environmental policies more challenging, but also more interesting. Clearly, in view of this lack of information, there needs to be a strong focus on the advancement of appropriate technologies. This leads to the conclusions of the following section.

9.4 Conclusions for implementing a circular economy

In order to respect and preserve the functions of the environment, sustainability is required regarding the use of, in particular but not only, non-renewable resources, the protection of the environment as direct provider of utility, and the waste management activities – observing the waste hierarchy, including the prevention or reduction of waste. This conclusion or, rather, recommendation, is in accordance with Definition 2.1 and with most perceptions of a circular economy in the scientific literature and in practical applications. Major deviations refer, however, to the requirement of preventing waste.

What does this imply for choosing a concrete path towards a circular economy? First of all, the possibilities of a sustainable development depend on the concrete situation, the detailed framework conditions in a particular country or region. These conditions include all kinds of exogenous facts, such as demographics, climatic and geographic conditions, level of economic development etc. The immediate consequence is that each country or region has to find its own way of a sustainable development with one country putting more effort into the social goals, the other more effort into the economic goals, and yet another one more effort into the environmental goals of a sustainable development.

As indicated, this can lead to difficulties, once the allocation of global environmental commodities requires attention. A coordinated procedure will then, under additional challenges from the Tragedy of the Commons and/or the Prisoners' Dilemma, ask for the unconstrained support of the various governments – with their probably quite different opinions regarding a sustainable development. The current global efforts to mitigate climate change are proof of such a difficult constellation.

Regarding the compliance with the waste hierarchy, the deviations from the priority goal of preventing waste, may lead to developments, which are not conform with the requirements of a circular economy. The challenge in this context seems to be to motivate all individuals, extrinsically or intrinsically, consumers and producers, to prevent waste. As experience shows, it is probably easier to extend and expand collection and recycling activities including success stories regarding rising recycling rates, thereby neglecting the obviously more difficult and less controllable aspect of waste prevention. This is also a

consequence of the focus and dependence on technologies which then largely determine the path towards a circular economy.

What remains to say regarding the allocation problems in the context of a circular economy? The issue of a sustainable supply of resources can be trusted to a market economy – with some proviso, though. So far, there have always been new commodities and new production technologies using less of the scarce resources, or new mining technologies better exploiting existing sources. Nevertheless, these activities should be supplemented by recovery and recycling activities, which also help to reduce landfilling of waste. A critical issue is thus a sustainable waste management, observing the waste hierarchy with a strong focus on preventing waste. Due to a lack of information not much more can be said on properties of feasible allocations for a circular economy.

This is, however, a clear signal for innovative technologies governing the development towards a circular economy. The thereby resulting technological path dependencies are supplemented by societal path dependencies with an orientation on a market system. For this reason, the chapters of Part III have to investigate the reaction or response of humans on all aspects related to technologies for a circular economy.

References

Ahvenniemi, H., Huovila, A., Pinto-Seppä, I., & Airaksinen, M. (2017). What are the differences between sustainable and smart cities? *Cities*, *60*, 234–245. https://doi.org/10.1016/j.cities.2016.09.009.

Ayres, R. U. (1998). Eco-thermodynamics: Economics and the second law. *Ecological Economics*, *26*(2), 189–209. https://doi.org/10.1016/S0921-8009(97)00101-8.

Brundtland, G. H. (1987). *Report of the World Commission on Environment and Development: Our common future. Retrieved from (1987).* https://sustainabledevelopment.un.org/content/documents/5987our-common-future.pdf.

De Guimarães, J. C. F., Severo, E. A., Felix Júnior, L. A., Da Costa, W. P. L. B., & Salmoria, F. T. (2020). Governance and quality of life in smart cities: Towards sustainable development goals. *Journal of Cleaner Production*, *253*, 119926. https://doi.org/10.1016/j.jclepro.2019.119926.

Erum, N., & Hussain, S. (2019). Corruption, natural resources and economic growth: Evidence from OIC countries. *Resources Policy*, *63*, 101429. https://doi.org/10.1016/j.resourpol.2019.101429.

EU (2008). *Directive on waste. Retrieved from(2008).* https://eur-lex.europa.eu/legal-content/EN/TXT/PDF/?uri=CELEX:32008L0098&from=DE.

EU (2012). *Life cycle thinking and assessment for waste management. Retrieved from(2012).* http://ec.europa.eu/environment/waste/publications/pdf/Making_Sust_Consumption.pdf.

Fukuda-Parr, S., & Muchhala, B. (2020). The Southern origins of sustainable development goals: Ideas, actors, aspirations. *World Development*, *126*, 104706. https://doi.org/10.1016/j.worlddev.2019.104706.

Geissdoerfer, M., Savaget, P., Bocken, N. M. P., & Hultink, E. J. (2017). The circular economy—A new sustainability paradigm? *Journal of Cleaner Production*, *143*, 757–768. https://doi.org/10.1016/j.jclepro.2016.12.048.

Georgescu-Roegen, N. (1971). *The entropy law and the economic process.* Harvard University Press.

Hajduk, S. (2016). The concept of a smart city in urban management. *Business, Management and Education*, *14*(1), 34–49. https://doi.org/10.3846/bme.2016.319.

Henri, P. A. O. (2019). Natural resources curse: A reality in Africa. *Resources Policy*, *63*, 101406. https://doi.org/10.1016/j.resourpol.2019.101406.

Kirchherr, J., Reike, D., & Hekkert, M. (2017). Conceptualizing the circular economy: An analysis of 114 definitions. *Resources, Conservation and Recycling*, *127*, 221–232. https://doi.org/10.1016/j.resconrec.2017.09.005.

Leopoldina (2017). *Pfadabhängigkeiten in der Energiewende. Das Beispiel Mobilität.* German National Academy of Sciences Leopoldina. Retrieved from(2017). https://www.leopoldina.org/uploads/tx_leopublication/2017_ESYS_Analyse_Pfadabhaengigkeiten.pdf.

Luttenberger, L. R. (2020). Waste management challenges in transition to circular economy—Case of Croatia. *Journal of Cleaner Production*, *256*, 120495. https://doi.org/10.1016/j.jclepro.2020.120495.

Ma, S. H., Wen, Z. G., Chen, J. N., & Wen, Z. C. (2014). Mode of circular economy in China's iron and steel industry: A case study in Wu'an city. *Journal of Cleaner Production*, *64*, 505–512. https://doi.org/10.1016/j.jclepro.2013.10.008.

Martin, C. J., Evans, J., & Karvonen, A. (2018). Smart and sustainable? Five tensions in the visions and practices of the smart-sustainable city in Europe and North America. *Technological Forecasting and Social Change*, *133*, 269–278. https://doi.org/10.1016/j.techfore.2018.01.005.

Pearce, D. W., & Turner, R. K. (1989). *Economics of natural resources and the environment.* Johns Hopkins University Press.

Prieto-Sandoval, V., Jaca, C., & Ormazabal, M. (2018). Towards a consensus on the circular economy. *Journal of Cleaner Production*, *179*, 605–615. https://doi.org/10.1016/j.jclepro.2017.12.224.

Reike, D., Vermeulen, W. J. V., & Witjes, S. (2018). The circular economy: New or refurbished as CE 3.0?—Exploring controversies in the conceptualization of the circular economy through a focus on history and resource value retention options. *Resources, Conservation and Recycling*, *135*, 246–264. https://doi.org/10.1016/j.resconrec.2017.08.027.

Segerson, K. (1991). Reviewed Work: Economics of natural resources and the environment by D. W: Pearce, R. K. Turner. *Land Economics*, *67*(2), 272–276. https://doi.org/10.2307/3146419.

Vesco, P., Dasgupta, S., De Cian, E., & Carraro, C. (2020). Natural resources and conflict: A meta-analysis of the empirical literature. *Ecological Economics*, *172*, 106633. https://doi.org/10.1016/j.ecolecon.2020.106633.

Wang, P. C., Che, F., Fan, S. S., & Gu, C. (2014). Ownership governance, institutional pressures and circular economy accounting information disclosure: An institutional theory and corporate governance theory perspective. *Chinese Management Studies*, *8*(3), 487–501. https://doi.org/10.1108/CMS-10-2013-0192.

PART III

The circular economy in a technological context

Innovative technologies are of significant importance for implementing a circular economy. These technologies are necessary for enabling a sustainable development, and for all aspects of waste management. But also design changes in favour of a design for environment (DfE) depend on innovations, as do many efforts of mitigating climate change. Moreover, according to the results of the chapters of Part II, the allocation of environmental commodities poses various challenges. The technological environment and its further development can therefore provide support for implementing a circular economy.

There are nevertheless some issues regarding this technological environment, which are of importance for implementing a circular economy – technology-guidance will prove challenging, too. The chapters of Part III address relevant topics and highlight and explain their role in circular economy strategies.

The first chapter characterises the technological environment. As important roots of the circular economy trace back to industrial ecology, it is obvious to explain some features of this 'school of thought' in its relation to economics. The discussion includes perceptions of industrial ecology and, in particular, its links to economics.

Beyond their relevance in industrial ecology, environmental technologies are also of interest for export-oriented countries. The global markets of environmental technologies are often critically dependent on environmental regulations which are in place in export and import countries. Governments then tend to promote certain technologies which fit best in their innovation policies. The two policy strains, implementing a circular economy and promoting environmental technologies, may thus interact with each other – not always fully supporting circular economy strategies: promoting a special technology with a focus on exports – technologies for renewable energy sources, for example, can establish critical technological path dependencies at home.

This influence of the governments on environmental technologies is affected by various informational issues addressed in the next chapter. There is, first of all, the complicated issue of economic reasonableness of certain technologies and environmental standards: when is the introduction of a certain technology or a certain environmental standard economically reasonable? Uncertainties and informational asymmetries affect in particular DfEs, as business companies have the necessary information regarding a DfE. And they will likely only use it to its full extent, if this is in accordance with their business interests. A case study on Germany's approach to promoting renewable energy sources demonstrates the relevance of informational requirements and path dependencies.

Rebound effects and path dependencies, investigated in the then following chapter, characterise certain responses of consumers and producers to environmental technologies and commodities. Rebound effects can be economically or psychologically grounded and adversely affect the behaviour of economic agents, thereby reducing the positive environmental impact of these technologies.

Technological path dependencies describe how – also due to an insufficient information – suboptimal choices of technologies can determine the technological environment possibly for a long period of time. They are of special relevance for implementing a circular economy, given the strong influence of the policy makers on various aspects of the further development of the environmental technologies. Moreover, societal path dependencies describing certain societal 'habits' may also interfere with the implementation of a circular economy. Path dependencies may trigger rebound effects.

The digital transformation with its manifold effects on the implementation of a circular economy is the topic of the last chapter. It addresses first commonalities between the digital and the circular transformation of an economy. After briefly referring to some aspects of establishing smart cities, which are related to sustainability and a circular economy context, a 'prospective' case study asks questions which deserve attention in view of the rapid development of e-commerce.

Thus, again, the implementation of a circular economy is substantially dependent on technologies and their further development, but, and this is the critical issue, providing these technologies may significantly affect the environmentally relevant behaviour of consumers and producers, and may, therefore, crucially interfere with the implementation of a circular economy. In the end, in Part V, it will turn out that a careful combination of economy-guidance and technology-guidance will be necessary for implementing a circular economy.

The technological environment of a circular economy

Besides economics, industrial ecology constitutes one of the roots of the circular economy. It helped to shape and develop the concept, and for various proponents it has always been an important source of inspiration. According to some supporters, "industrial" in industrial ecology should refer to all human activities, although the focus in the literature is clearly on the narrower concepts of product design and manufacturing processes. Nevertheless, the continuous progress of environmental technologies is considered essential for guiding the economy towards a circular economy (see also Chapter 9).

The chapters of Part II illustrated and discussed the economic challenges for implementing a circular economy. This is perhaps one reason for the outstanding role science and technology have in this context. "Technology-guidance" creates, however, other problems which need to be taken into account in this context. In the end, a careful combination of "economy-guidance" and "technology-guidance" will prove necessary for implementing a circular economy (see Part V). This chapter will first look into some details of industrial ecology before turning to the local and global markets for environmental technologies – to understand important aspects of the supply of these technologies.

10.1 Industrial ecology

With reference to the Ellen MacArthur Foundation, there are various "Schools of Thought", which have contributed to refining and developing the circular economy concept, among them the "Cradle to Cradle" concept with its special design philosophy, perceiving nature's 'biological metabolism´ as a model for a 'technical metabolism´. There is, further, the "Performance Economy" with its "closed loop" approach to production processes with the goals of product-life extension, long-life goods, reconditioning activities, and waste prevention, and its focus on selling services, rather than products, in view of a sharing economy. "Biomimicry" is based on nature's best ideas, and imitates these ideas to solve human problems, and, finally, "Industrial Ecology" is the

Implementing the Circular Economy for Sustainable Development
https://doi.org/10.1016/B978-0-12-821798-6.00010-7

"study of material and energy flows through industrial systems".

These thoughts materialised and led to applications already in the late 1970s, before Pearce and Turner (1989) introduced their economic concept of the circular economy. This time lag might also help to explain the role science, technology and engineering play in many aspects of implementing a circular economy. Economics with the explicit task and the goal to provide guidelines for motivating all kinds of stakeholders, in particular consumers, producers, government agencies, to contribute towards developing a circular economy, has therefore been on board only since the 1990s, at least with respect to both a fundamental and formative role.

10.1.1 Perceptions of industrial ecology

For Lifset and Graedel (2002) industrial ecology "focuses on product design and manufacturing processes" with "firms as agents for environmental improvement because they possess the technological expertise that is critical to the successful execution of environmentally informed design of products and processes" (p. 3). There is, thus, a systems perspective, and "eco-design", or a design for environment (DfE), is a "conspicuous element of industrial ecology" (p. 7), which helps to avoid or reduce environmental impacts by "incorporating environmental considerations into product and process design ex ante" (p. 7). Companies constitute a "locus of technological expertise", and are therefore "an important agent for accomplishing environmental goals" (p. 8), who should have an intrinsic motivation for accomplishing environmental goals. This implies a more active, more cooperative role for business in an environmental context, as a 'policy-maker' rather than a 'policy-taker' in the context of 'historic' command-and-control schemes.

There remains the important question for implementing a circular economy: how can firms be motivated to make optimal use of their technological expertise in this regard? How to develop their intrinsic motivation? The fact that they possess this expertise does not yet mean that they also have the intrinsic motivation to make unconditional use of it (Section 11.2.1). This throws then a new light on environmental policies, brings in more than a touch of behavioural economics, and poses interesting new questions regarding the design of policies, which assign companies a more active role and might help to turn them into policy-makers by further developing their intrinsic motivation.

According to Choudhary (2012), industrial ecology "explores the idea that industrial activities should not be considered in isolation from the natural world but rather as a part of the natural system" (p. 2). "Industrial", as already mentioned, thereby refers to all human activities with tourism, housing, medical services, transportation, agriculture and others among them, and "ecology" means the science of ecosystems (p. 2). Industrial ecology is therefore a branch of systems science and systems thinking (p. 3). Systems thinking is also inherent in the concept of a circular economy, and that all human activities should be considered as part of the natural system, is also a decisive feature of a circular economy. Again, the question arises, how to implement these features?

There is, thus, quite naturally, a strong focus on technologies in industrial ecology, and the concept of a circular economy promoted by the Ellen MacArthur Foundation, for example, adopts this position. The interesting view that waste is merely a "design flaw" points in this direction: with appropriate designs and new materials and technologies it should be possible to ensure that waste and pollution are not created in the first place. Moreover, according to this view, products should and could be designed such that they can be reused, repaired and remanufactured.

Regarding this last position, there is once more the immediate concern about the realisation in the context of a circular economy. If this technological approach to a circular economy in industrial ecology is considered the heart of a circular economy, then not only this question arises, but also the more general question regarding the role of economics at all. As mentioned, Lifset and Graedel (2002) refer to the doubts in industrial ecology circles regarding the role of environmental policies, in particular of traditional command-and-control policies. And, in fact, the review of environmental regulations in Chapter 5 has clearly indicated that environmental policies in the context of implementing a circular economy are still some way from functioning really satisfactorily. Thus is "technology-guidance" towards a circular economy the appropriate response?

10.1.2 The economics of industrial ecology

In industrial ecology the role of environmental policies, at least of command-and-control policies, is limited, the role of the market system in general is not particularly emphasised. This corresponds to the critical view some behavioural economists have regarding environmental issues to be considered in a market system (see Chapter 8). Companies, however, are assigned clear tasks in a circular economy: they are, among others, responsible for DfEs, and for a more effective approach to circular economy strategies. On the other hand, companies in general seem not yet to have grasped the active role of policy-makers expected of them, the role of "technology-guidance".

This observation points to the fact that stakeholders, consumers and producers, have their individual motivations to participate more or less actively in circular economy strategies. In the worst case, the goals of the strategies cannot be achieved, the regulations do not meet the interests of the stakeholders, other economic mechanisms, such as the Tragedy of the Commons or the Prisoners' Dilemma, for example, influence the behaviour (see Section 7.1).

In addition, public authorities can only partially control this behaviour due to information asymmetries and lack of information. Here, those researchers in industrial ecology, who view command-and-control regulation as inefficient, perhaps even counterproductive. Lifset and Graedel (2002) are right, at least to some extent: writing the requirement of an appropriate, environmentally friendly product design into environmental policies is, in general, not very helpful and not really goal-oriented (p. 8). Producers possess the required knowledge and expertise, and will only make adequate use of it, if it is in their business interests, which is, of course, legitimate in a free market economy. Nevertheless, such regulations seem to slow down the process of developing innovative and environmentally friendly designs, and the question arises, how to affect the behaviour of producers, how to motivate them for a DfE? Thus, how to interest companies for the role expected of them – not only by proponents of industrial ecology?

In fact, a variety of case studies, published by the Ellen MacArthur Foundation, supports the impression of viable, new business models, supporting and accelerating the transition to a circular economy. But these are still cases, which need not include all interesting DfEs. They might help to establish a different mindset in the near or not so near future, but are certainly not yet representative for larger parts of the economy. Establishing this different mindset required for a circular economy likely takes some time.

These profitable business cases refer probably to commodities and services, which are environmentally friendly, and/or which can be provided in a more sustainable, but also profitable way, for which there is, therefore, sufficient demand. Of course, this demand can also be "generated" through subsidies, driving competitive commodities, services or production processes out of the regular markets. For an example, consider the transition to an electric mobility system in the City of Shenzhen in China, which is presented by the Ellen MacArthur Foundation as one of the new business models enabling innovation with financial support. It remains, however, unclear, to what extent this business model would be financially and/or societally viable in the United States (U.S.) or in the European Union (EU).

There is also the possibility that some companies act out of pure altruism and turn to a DfE regarding their products or services. This would then correspond to findings from behavioural environmental economics and could play a role in creating new habits, presenting an intrinsic motivation for certain environmental issues. Beyond that, also a strategic behaviour might play a role with these "voluntary contributions": rushing ahead and adopting a DfE now could help a company to take advantage of a societal trend, create some leeway and gain some time against competitors, thus profiting from a "societal" first-mover advantage (see also Section 16.4).

The list of examples provided by the Ellen MacArthur Foundation includes various cases, which seem to fit into one or the other of these categories, but a more detailed investigation is needed to understand these business models. For sure, there are quite a few possibilities to deviate from mainstream economic behaviour, making it, however, still more difficult to predict the concrete development. Also the possibility of a "cheap excuse", of "greenwashing" should not be completely ruled out: consumers and producers might use one or a few environmentally friendly actions "to ease their moral conscience" (Engel & Szech, 2017).

No doubt, all these possibilities, all these behavioural aspects are important and have their place under certain framework conditions. Nevertheless, the transition to electric mobility in the City of Shenzhen and other examples are and remain "cases", initiated and enforced by public authorities, certainly with support from groups of engineers, architects and perhaps other knowledgeable people, or they result from strategic or some other behaviour. Therefore, the question remains, whether the characteristics of a market system, namely the integration of all stakeholders with their individual knowledge, could not be of an advantage with respect to the promotion and development of a circular economy. Choudhary (2012) points exactly to the necessity of such an integration, when he refers to the appropriate interpretation of "industrial" in industrial ecology. But, to integrate all these stakeholders, to employ them for these purposes, remains one of the challenges of a transition to a circular economy. Economics has to play an important role in this context, and economists have to contribute a lot – beyond cases and beyond "new" business models. Economic instruments are needed to design appropriate policies, which guide and motivate the stakeholders to make adequate use of their knowledge.

Therefore, industrial ecology is, for sure, an important and interesting school of thought, which has substantially influenced the concept of a circular economy and shaped its development. The idea that industrial activities should not be considered in isolation from the natural world, that these activities should be further developed to produce in a more environmentally friendly way, to provide more environmentally friendly commodities, corresponds perfectly to the academic concept of the circular economy introduced in Pearce and Turner (1989). In view of the various aspects raised above, a too strict focus on technology-guidance remains, however, questionable.

Given this strong technological background of the circular economy, it is necessary to investigate various features of local and global markets for environmental technologies. After all, the supply of these technologies, which are of relevance for a circular economy, is dependent on framework conditions, which are determined by actions and regulations of the governments.

10.2 Markets for environmental technologies

Environmental technologies are not just environmentally friendly technologies, they are rather meant to foster environmental protection and physical resource efficiency, to prevent or mitigate pollution, to manage or reduce waste streams, and to remediate contaminated sites, just to name a few characteristics of the definition from an industry perspective (ITA, 2017).

Innovations and new technologies are of relevance for many countries, in particular, export-oriented countries, which attempt to keep or even raise their competitive advantage by developing innovative goods and services, also in the environmental sector. Thus, not surprisingly, environmental technologies meanwhile turned into an important economic factor, with three countries, Japan, Germany, and the U.S., accounting for some 60% of total innovations regarding technologies to mitigate climate change in 2000–2005, for example (Dechezleprêtre, Glachant, Haščič, Johnstone, & Ménière, 2011).

What are, from a mostly economic point of view, special features of the markets for environmental technologies of importance for a circular economy, and what do they imply for circular economy strategies in countries around the world – exporters and importers of these technologies?

10.2.1 General aspects of markets for environmental technologies

The global market for environmental technologies is rapidly increasing and surpassed 1000 billion U.S.-Dollar (USD) in 2015, with the U.S. representing the single largest market accounting for approximately one third of the global market and exporting 47.8 billion USD worth of goods and services in the context of environmental technologies (see Fig. 10.1 for more details). Approximately 1.6 million people are employed in the U.S. industry producing environmental technologies (ITA, 2017).

Similarly, in Germany, "greentech" products accounted for a global trade share of 14% in 2016, and the markets for technologies for environmentally friendly energy generation, energy

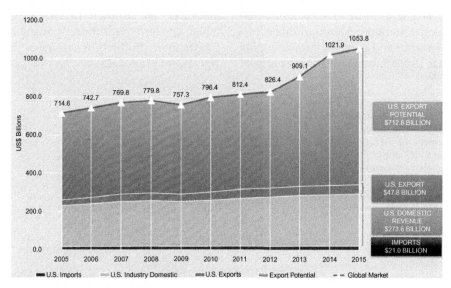

FIG. 10.1 **Global environmental technologies market overview.** The U.S. hosts the single largest market accounting for roughly a third of the global market. © *From ITA. (2017). 2017 Top Markets Report Environmental Technologies. Industry & Analysis. U.S. Department of Commerce—International Trade Administration. Retrieved from https://legacy.trade.gov/topmarkets/pdf/ Environmental_Technologies_Top_Markets_Report2017.pdf. International copyright, 2017, U.S. Department of Commerce, U.S. Government.*

efficiency, resource efficiency, sustainable mobility, sustainable water management, waste management and recycling represented some 15% of the German gross domestic product (GDP), expected to increase to 19% by 2025. Moreover, with an estimated average increase of almost 15% per year for the period from 2016 to 2025, the market for recycling technologies is expected to offer further fast-growing business opportunities in Germany – a great perspective for implementing a circular economy (GTAI, 2019). Fig. 10.2 shows the growth rates for the circular economy in general and for the sustainable water industry in particular. Germany is Europe's largest exporter of water treatment technologies.

No wonder then, that industrialised and export-oriented countries consider markets for environmental technologies as growth potentials for their economies. They tend to develop and test new technologies, in particular recycling technologies, for the local market, in order to offer them also on the global markets. Initiatives to stimulate innovations and the further development of these technologies and these markets help to establish these countries as leaders, perhaps in the global markets. Although this is certainly of importance and necessary for reducing environmental degradations in many parts of

the world, there are nevertheless some issues, which are also of relevance for implementing a circular economy, and which deserve therefore a closer look – both with respect to possible developments in the exporting and in the importing countries.

Markets for environmental technologies are, to a large extent, rather special markets. Of course, a certain share of these markets is characterised by regular demand and supply decisions from private households and private companies. This refers mainly to end-of-pipe environmental technologies used in private houses or factories: water purifiers, air cleaners, sound absorbing devices count among them. These markets certainly gain from a growing environmental awareness and a similarly growing health consciousness, which, supported by a continuous stream of new scientific insights, convinces, or at least motivates individual households to protect themselves and their families better against various kinds of environmental pollution.

These end-of-pipe technologies reduce the environmental effects of emissions by add-on measures, in contrast to cleaner production, which prevents pollution by using appropriate products or production processes. In their survey based on data of the Organisation for

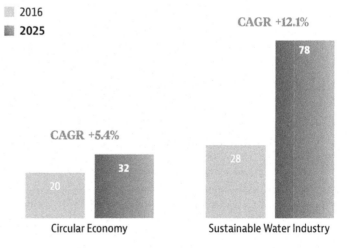

FIG. 10.2 **Germany's environmental technologies markets.** Market volume in EUR billion and Compound Annual Growth Rate (CAGR). © *From GTAI. (2019). German Trade and Invest. Environmental technologies in Germany. Fact sheet issue 2019/2020. Retrieved from https://www.gtai.de/resource/blob/64490/603917f069008c31cbf0 e732983b0427/fact-sheet-environmental-technologies-en-data.pdf.*

Economic Cooperation and Development (OECD), Frondel, Horbach, and Rennings (2007) point to the 76.8% of facilities investing predominantly in cleaner production technologies, with Japan leading with 86.5%, and with Germany lagging behind with only 57.5% (see Frondel et al., 2007, Fig. 3).

Although these are somewhat older data, the reasons for this observation seem to be related to the special structure of environmental policies and are, therefore, of relevance for the implementation of a circular economy. Frondel et al. (2007) refer to the command-and-control policies, which are largely used in environmental policies in Germany, and which "frequently impose technology standards that can only be met through end-of-pipe abatement measures" (p. 1). Nevertheless, as already indicated, only few industrialised countries seem to play a major role in developing innovative environmental technologies for the global markets.

10.2.2 Environmental technologies in global markets

The markets for environmental technologies, especially the global markets, are driven by various developments, which seem to affect, perhaps even partially enhance each other. The rising environmental awareness, likely supported by strong economic growth in the emerging economies but also in the developing countries in recent decades, puts some pressure on governments to reduce the air, water, or soil pollutions – often resulting from this economic growth – that severely reduce quality of life in many urban agglomerations. To some extent, this characterises the situation in many large cities in China or India, with their problems regarding the persistent air pollution, for example. Not surprisingly, these two countries, among other mostly emerging economies, are major buyers of environmental technologies in the global markets (ITA, 2017).

This rising environmental awareness is – and this is a crucial point – accompanied by more and more environmental regulations, again in particular in the emerging economies, and more restrictive environmental standards in the industrialised countries. Of course, these regulations, if they are sufficiently enforced by the governments, are of great importance for the industry manufacturing environmental technologies. Otherwise, due to aspects of the Tragedy of the Commons or the Prisoners' Dilemma, environmental technologies would likely only be used in situations, where individual health or individual wealth in a society is threatened directly (see Section 7.1).

However, for similar reasons, environmental regulations alone would not help to promote the development and the markets for environmental technologies. This can easily be seen by looking at the existing environmental regulations from countries all over the world. Many of these regulations contain rather precise provisions for pollution control, but as long as these provisions are not effectively enforced, pollution may and, as experience shows, does continue.

In this context, ITA (2017) refers to the case of air pollution control in China with the first pollution control law already passed in 1987, followed by revisions in 1995, 2000 and 2015 (see p. 11). Without an effective enforcement, however, the high legal standards could not prevent the widely reported and worsening air pollution problems in some of China's major cities. India presents another example: the Air (Prevention and Control of Pollution) Act was enacted in 1981, followed by various amendments. Nevertheless, various Indian cities continue to suffer from severe air pollution with emissions coming mainly from transport and industry, but also from biomass burning in rural areas.

In conclusion, governments have a major influence on the markets for environmental technologies: they can direct attention to certain environmental areas such as technologies for air pollution control, water technologies or waste management, just to name a few, and focus on particular issues such as renewable energies, for example. And, for the countries developing and exporting environmental technologies there seems to be a bright future ahead: with the increasing need to enforce environmental regulations in a stricter way, global demand for these technologies will increase.

Nevertheless, exporting environmental technologies raises a couple of questions, which are addressed in the following subsection.

10.2.3 Export promotions for environmental technologies

In view of the export-orientation of most countries producing environmental technologies, governments can and do support the development of these technologies through a variety of initiatives – beyond further environmental regulations. On the federal level in Germany, for example, the "Environment Innovation Programme for Projects Abroad" provides "financial support for projects which will have direct environmental effects on Germany or which will increase, via a 'philosophy transfer' and a maximum possible multiplier effect, the willingness to invest in climate action measures" in Central and Eastern European countries.

Similarly, the U.S. Department of Commerce is concerned about the successful export promotion of environmental technologies. Among the critical components are: a policy dialogue to identify and eliminate existing foreign trade barriers, technical assistance to foreign governments for the development of compatible regulatory approaches and for the creation of mechanisms for enforcement, maintenance and management of environmental systems, and the provision of financial vehicles for project development and export finance (ITA, 2017).

Not surprisingly, the governments have a clear view on the top markets – both countries and technologies – providing further export opportunities for their companies. Circular economy initiatives in various developing countries and emerging economies are of particular interest, as there is a potential for new business activities. Germany, for example, having been a first-mover in the development of the modern wind energy industry, is now looking at the potential regarding recycling components of wind power plants, which are soon going to reach their maximum funding period of 20 years. Moreover, recycling of carbon fibres, of waste electronic equipment, and electric vehicle batteries attract the interest of business companies, also of international companies (GTAI, 2019). This is interesting in so far as the design of these wind turbines and, probably of other technologies for renewable energy sources – meant for a circular economy – was not necessarily environmentally friendly. At least there are indications pointing in this direction.

Thus, just as an aside, there is the need for a DfE for environmental technologies, and an additional challenge for sustainably implementing a circular economy. The enormous quantities of photovoltaic modules, which need soon to be recycled, and, further into the future, also used batteries from all kinds of e-mobility, seem to require a new approach to developing environmental technologies.

Also, the U.S. Department of Commerce (ITA, 2017) is carefully analysing the needs of emerging economies such as China, India and others, and is targeting these countries with appropriate technology proposals containing all kind of technologies for air pollution control, waste water treatment and water management, and waste management, among others (see p. 17f).

Of course, there is nothing wrong with these promotional activities of the governments, which clearly focus on reducing environmental damage – at home and abroad. The fact that there is some support for the local industry is certainly a welcome add-on, and is used for justifying these policies, as the corresponding initiatives in Germany and the U.S. clearly show.

There are only a few remarks pointing to some possibly problematic issues: first, export promotion will, of course, focus on those fields and technologies, which are most promising for the world markets. This might put other environmental technologies, which are of more relevance for the domestic market, at a disadvantage, thereby establishing a certain technology path with ensuing path dependencies (see also Section 12.2), which might be difficult to control or handle. Secondly, circular economy strategies, which include compliance with the waste hierarchy and sustainability considerations, are dependent on the concrete situation of a country or a region (see Section 9.3). This refers to collecting, sorting and recycling technologies for all kinds of waste. The question arises, whether the specific needs of these countries are always sufficiently compatible with the strategies of the exporting countries and their companies, and whether these needs are adequately taken into account at all.

In this context, Broitman, Ayalon, and Kan (2012) draw the attention to schemes of waste management, which are often designed at a national level with local conditions neglected, and, interestingly, Arlosoroff and Rushbrook (1991) pointed to these problems already in the early years of integrated waste management activities. Thus, has nothing really changed since then? This issue becomes even more complicated in view of certain aspects of technological path dependencies, addressed in Section 12.2.

Perhaps one reason behind these observations can be found in those export strategies in combination with efforts to motivate foreign countries for compatible regulatory approaches. In this context, Agamuthu and Victor (2011) are referring to the "the pitfall of … the simplistic adoption of policies and legislations from other countries' 'purportedly' successful waste management system without taking into context the local, cultural and socio-economic waste management issues" (p. 952), and Bartl (2014) addresses the EU Waste Framework Directive (EU, 2008) with its occasionally unifying and standardising attempts. The same critique pertains to other EU Directives in the broader area of waste management.

This "power" of the governments to substantially influence business on the markets for environmental policies can assume monopolistic features. This aspect will be briefly addressed in the following subsection.

10.2.4 Technology policy – A journey into the unknown?

By making use of the well-known financial instruments, subsidies and export promotions, governments can significantly influence all kinds of activities on the local, but also on the global markets for environmental technologies. This quasi-monopolistic behaviour can, under certain circumstances, lead to problematic results, which are not always recognisable right away.

One of these policy decisions, which proved to be problematic in the end, refers to the promotion of renewable energies in Germany in the last 20 years. This is investigated in the case study on the "Promotion of Renewable Energy Sources in Germany" (see Section 11.3). Another decision, which might turn out problematic in near future, is the promotion of e-mobility, of e-vehicles to be more precise, in a variety of countries including Germany.

There is vast range of measures to increase the share of e-vehicles in Germany with the goal to make a significant contribution to meet the climate protection targets on the one hand, and to secure and extend the leading role of the German car manufacturing and parts supply industry (GTAI, 2015). This corresponds perfectly to the model of promoting globally relevant environmental technologies discussed above – according to the Association of the German Automotive Industry (VDA), "Germany wants to be the lead supplier and lead market for electric mobility".

First of all, the purchase of an e-vehicle has been subsidised in Germany since 2016. From July 2016, customers received a reduction on the listed price of up to 4000 €. This subsidy, initially limited till the end of 2020, is now extended to 2025 with a significantly higher subsidy of up to 6000 €. In addition to that there is a tax exemption for e-vehicles bought before end of 2020. The main reason for these adjustments is the still rather small number of e-vehicles sold: till the end of October 2019, only some 150,000 customers applied for the bonus on e-vehicles and hybrid cars. This is still considered too low regarding the climate protection goals of Germany.

There are also incentives for car manufacturers: the high environmental standards with respect to greenhouse gas emissions can only be reached, if sufficiently many e-vehicles are sold. Although right now, the electricity for these cars is mainly coming from fossil fuels, they are nevertheless factored into the emission standards to be observed by the car manufacturers with "super-credits": cars with an emission below $50g$ CO_2 per km, which includes e-vehicles with, by definition, $0g$ CO_2 per km, are counted twice in a car manufacturer's fleet fuel-economy in 2020, with the factor 1.66 in 2012, and with the factor 1.33 in 2023. Thereafter, there will be no "super-credits" for e-cars regarding the average greenhouse gas emissions of a car manufacturer's fleet. Therefore, the car manufacturers have a substantial interest in developing and selling as many e-cars as possible, and as soon as possible.

However, as of today it remains doubtful, whether this policy will be successful to promote e-mobility, also in the context of a circular economy. Potential customers seem still to be concerned about the cruising range of e-cars, which they deem – justifiably or not – as comparatively short. Moreover, they worry about not available charging stations and the long charging time. These are aspects, which need to be considered in the context of buying an e-vehicle. And, in addition to that, defenders of the diesel engine point to the economic and environmental advantages of the latest innovations, putting the diesel technology ahead of electric mobility with respect to CO_2 emissions, although the results of corresponding studies remain not undisputed (Buchal, Karl, & Sinn, 2019).

But, with the recent provisions, both of the government and the car manufacturers, it seems to be too late for a reversal in Germany. One has to admit that the export orientation of the German car manufacturers makes it necessary to pay attention to the situation in other countries. This applies in particular to China, which is an important market for German car manufacturers.

But, as of now, it remains doubtful, whether this policy will develop in the sense of a circular economy strategy. It could well happen that households turn towards an e-vehicle as a second or a third car, mainly for use in the cities for shopping and other small-range activities. This would certainly help to reduce air pollution in the cities, but need not replace cars with a gasoline or a diesel engine in a one-to-one way, thus leading to a perhaps significant rebound effect (see Section 12.1).

There is the bottom line that this export orientation for environmental technologies may have consequences, which need not necessarily correspond to all the goals of a circular economy strategy. To some extent, this field remains a journey into the unknown.

The fact that a large share of the markets for environmental technologies results from interventions of the governments in the form of export promotions on the one hand, and of new or stricter environmental regulations on the other, necessitates a more careful inquiry into certain new technologies, or also higher environmental standards, recycling quota, for example. This raises questions on the "economic reasonableness" of these technologies or standards for which, due to a lack of information, simple answers are not available. Chapter 11 investigates various informational issues in the context of environmental technologies, in particular DfEs.

References

Agamuthu, P., & Victor, D. (2011). Policy trends of extended producer responsibility in Malaysia. *Waste Management & Research*, *29*(9), 945–953. https://doi.org/10.1177/0734242X11413332.

Arlosoroff, S., & Rushbrook, P. (1991). Developing countries struggle with waste management policies. *Waste Management & Research*, *9*(1), 491–494. https://doi.org/10.1177/0734242X9100900169.

Bartl, A. (2014). Ways and entanglements of the waste hierarchy. *Waste Management*, *34*(1), 1–2. https://doi.org/10.1016/j.wasman.2013.10.016.

Broitman, D., Ayalon, O., & Kan, I. (2012). One size fits all? An assessment tool for solid waste management at local and national levels. *Waste Management*, *32*(10), 1979–1988. https://doi.org/10.1016/j.wasman.2012.05.023.

Buchal, C., Karl, H.-D., & Sinn, H.-W. (2019). *Kohlemotoren, Windmotoren und Dieselmotoren: Was zeigt die CO_2-Bilanz? Ifo-Schnelldienst*. Retrieved from https://www.ifo.de/DocDL/sd-2019-08-sinn-karl-buchal-motoren-2019-04-25.pdf.

Choudhary, C. (2012). Industrial ecology: Concepts, system view and approaches. *International Journal of Engineering Research & Technology*, *1*(9) Retrieved from https://www.ijert.org/research/industrial-ecology-concepts-system-view-and-approaches-IJERTV1IS9396.pdf.

EU (2008). *Directive 2008/98/EC of the European Parliament and of the Council on waste. Retrieved from (2008)*. https://eur-lex.europa.eu/legal-content/EN/TXT/PDF/?uri=CELEX:32008L0098&from=EN.

Frondel, M., Horbach, J., & Rennings, K. (2007). End-of-pipe or cleaner production? An empirical comparison of environmental innovation decisions across OECD countries. *Business Strategy and the Environment*, *16*(8), 571–584. https://doi.org/10.1002/bse.496.

GTAI (2015). *Electromobility in Germany: Vision 2020 and beyond. Issue 2015/2016*. German Trade and Invest. Retrieved from https://www.gtai.de/resource/blob/64490/603917f069008c31cbf0e732983b0427/fact-sheet-environmental-technologies-en-data.pdf.

ITA (2017). *ITA environmental technologies top markets report*. U.S. Department of Commerce—International Trade Administration. Retrieved from https://legacy.trade.gov/topmarkets/pdf/Environmental_Technologies_Top_Markets_Report2017.pdf.

Lifset, R., & Graedel, T. E. (2002). Industrial ecology: Goals and definitions. In R. U. Ayres, & L. W. Ayres (Eds.), *A handbook of industrial ecology*. Cheltenham: Edward Elgar Publishing. https://doi.org/10.4337/9781843765479.00009.

Pearce, D. W., & Turner, R. K. (1989). *Economics of natural resources and the environment*. Johns Hopkins University Press.

11

Technology and information

Whereas the last chapter focused on the technological environment of a circular economy, on industrial ecology and the (global) markets for environmental technologies, this chapter is about various informational issues: the role information, or rather the lack thereof, plays regarding the supply of certain technologies, designs for environment (DfE) in particular. As implementing a circular economy is dependent on the availability of appropriate technologies, informational requirements need to be investigated and taken seriously.

Markets for environmental technologies in general are to a large extent dependent on environmental regulations, monitored and controlled by the governments. Nevertheless, informational issues appear additionally in a variety of contexts and interfere with the necessities of implementing a circular economy:

- With their regulations, governments want to set and raise certain environmental standards, collection rates for waste, or they want to "force" the implementation of innovative technologies – for recycling or e-mobility, for example. To what extent are these measures "economically reasonable"? How to provide an answer to this question?

- For obvious reasons, business companies supply those technologies and commodities, which are profitable or are expected to become profitable under the given framework conditions. What about certain DfEs, which are not deemed profitable, but which are desired or even necessary for a circular economy? Observe that business companies have the required information in these cases, not the public authorities.

- A related context refers to new lead technologies such as technologies for renewable energies or e-mobility. Again, governments can enforce such technologies through subsidies and legal regulations. What about the acceptance through the population? How to integrate the population into these decisions?

This chapter tends to these issues which need to be adequately respected in all attempts to implement a circular economy, in particular in environmental policies. After all, these policies shall serve as allocation mechanisms for environmental commodities – together with the advancement of science and technologies. The first section explores the economic reasonableness of new technologies or setting and raising

Implementing the Circular Economy for Sustainable Development
https://doi.org/10.1016/B978-0-12-821798-6.00011-9

environmental standards – mainly from a theoretical point of view. Thereafter, the informational asymmetries, which might retard or even prevent the development of DfEs in certain contexts are investigated. The success of certain circular economy strategies depends substantially on whether and how consumers and producers adapt to such a modified environment: are households willing to accept innovative lead technologies? Finally, a case study on Germany's way to introduce and raise the share of electricity from renewable sources will be presented. This case study reveals consequences of incomplete and asymmetric information in the context of efforts to mitigate climate change.

11.1 Economic reasonableness of environmental technologies and standards

Of course, circular economy strategies depend and have to react on the advancement of science in general. Possibly harmful substances or hazardous processes, the damaging potential of certain activities, recycling technologies, for example, can only be detected and investigated through scientific research involving, among others, biology, chemistry, physics, and medical sciences. For this reason, the implementation of a circular economy will be an ongoing, never-ending project: advancements in the sciences will lead to new insights, also of environmental relevance, which will induce or even stipulate the development of new technologies, and then in turn help to advance academic research.

Chapter 10 discusses the dependence of markets for environmental technologies on government actions, on environmental regulations, on export promotions, which lead to the development of specific technologies to prevent or reduce environmental degradations. Thus, the decision to adopt specific technologies, to demand or supply certain technologies, does not always result from a "normal" market

environment with a multitude of individual sellers and buyers contributing to the market equilibrium. A similar consideration applies to setting and raising environmental standards, subsequently inducing the necessity to further develop environmental technologies.

For this reason, there is not only sometimes the question, whether it is necessary, both from an environmental and economic point of view, to introduce a certain advanced technology or to raise environmental standards in a particular country or region right now. This is the question of the "economic reasonableness" of some activity of environmental relevance, which meanwhile appears in a variety of environmental regulations, often in the context of implementing a circular economy.

11.1.1 The concept of economic reasonableness

Section 7 "Basic Obligations of Circular Economy" of the German Circular Economy Act (Germany, 2012) states in paragraph 4: "The obligation to recover waste shall be met, to the extent that this is technically possible and economically reasonable. … Waste recovery shall be deemed to be economically reasonable if the costs which recovery entails are not disproportionate to the costs that waste disposal would entail". And, similarly, the Circular Economy Promotion Law of the People's Republic of China (China, 2008) postulates in Article 4: "The promotion of circular economy shall be implemented on the basis of being feasible in technology, reasonable in economy …"

Comparable regulations on economic reasonableness or economic feasibility can be found in a variety of environmental policies from countries all over the world. These formulations refer usually to recovery or recycling quota for special kinds of waste, but refer, as is true for China, also in more general contexts to the implementation of a circular economy. Thus, the question of "economic reasonableness" affects circular economy strategies and should be taken as a serious issue. Requirements, which are considered non-reasonable from an economic or societal point of view, may lead to all kinds of avoidance behaviours (see the case study in Section 15.4, for example).

Postulating economic reasonableness sounds, of course, "reasonable". But, what does it practically mean? We know that many activities in waste management such as collection, segregation and recycling of waste are financed through fees from households and companies, perhaps supported through additional funds from various public sources. Therefore, raising these standards or quota can, in principle, be financed through higher fees etc., at least to some extent, and this may still be considered "reasonable" in view of the environmental issues associated with these activities.

But, where does this raising of standards end, where should it end, what is, therefore, the limit for "economic reasonableness"? A straightforward answer is not available and not easily attainable, and the formulation in the German Circular Economy Act that "waste recovery shall be deemed to be economically reasonable if the costs which recovery entails are not disproportionate to the costs that waste disposal would entail", is not really helpful, either. First of all, one would have to make precise, what "disproportionate" means in this context. Moreover, the costs of waste disposal are also difficult to evaluate with many possibilities to arrive at different estimates.

An interesting example in this context is provided by the German efforts for a permanent disposal of nuclear waste. For various political and societal reasons these costs are already so high that any future attempts to generating electricity from nuclear power stations in Germany are a priori ruled out. Thus, the question of economic reasonableness depends in this case on the costs of the disposal of waste, which can become, as this example shows, prohibitively high. For some further comments regarding this issue see Wiesmeth (2011, p. 95).

Of course, in such a context it is tempting to look and refer to profitable business activities: only those quotas, standards, technologies etc. are considered economically reasonable, which allow "reasonable", i.e., profitable business activities. This corresponds to the many attempts of governments to "outsource" waste management activities to private companies in the form of more or less complex public private partnership (PPP) projects, and this corresponds also to the list of profitable business models (case studies) provided by the Ellen MacArthur Foundation in support of circular economy strategies.

But, looking only at profitability, there is another problem: in this case, a regulation would not really be necessary, because those profitable business opportunities, if there are any, would sooner or later be discovered and taken by private business companies. The existence of such opportunities depends, of course, on other framework conditions, the level of wages, or the technical state of the facilities employed for these activities, or the concrete PPP project, rendered profitable through an appropriate share of the revenue resulting from fees. In the latter case, we are again back at the question of the extent of the "reasonable" share of the revenue.

Import and recycling of contaminated plastic waste through private companies in quite a few low-income countries with possibly low environmental awareness, is related to this context: whereas most of these activities are no longer profitable in industrialised countries, but rather dependent on additional financial support, they might still be interesting for these low-income countries, with the risk to result in additional environmental pollution (see Section 23.2.1). In their editorial, Ragossnig and Schneider (2017) ask the question about the right level of recycling of plastic waste, thereby pointing exactly to matters of product design, but in particular also to the dependence on financial issues (see p. 129).

Thus, we are more or less left with the first situation resulting in the question: to what extent are activities, in particular those related to implementing a circular economy, economically reasonable, to what extent is it economically reasonable to support or subsidise these activities by additional financial means? To put it in more concrete terms: is the transition of the City of Shenzhen to electric mobility (see Section 10.1.2) economically reasonable, or is it just the consequence of government decisions? Is the ambitious increase of the recycling targets in the German Packaging Act (Germany, 2019a) economically reasonable, or will it rather lead to avoidance practices? Are the German attempts to mitigate climate change at high costs economically reasonable, or are they, in view of the small share of German emissions on global greenhouse gas emissions, a more or less symbolic act? Or are they perhaps important and, thus, economically reasonable in order to motivate other countries to follow suit? This last aspect, which is, obviously, of relevance for the German position regarding efforts to mitigate climate change, reveals once more the difficulties to make a profound judgement on matters of economic reasonableness.

These and many other important, practical examples demonstrate the relevance of an answer to this basic question of economic reasonableness. However, a formal analysis of this issue shows that providing a practical solution is not simple. This might help to explain the comparatively low attention that is given to this question in most practical contexts.

11.1.2 Some formal considerations

Wiesmeth (2011) investigates this issue formally in the context of an economy with an activity, which pollutes the environment (see Wiesmeth, 2011, Section 9.3). The associated environmental or external effect is not taken into account by the market system, leading to an equilibrium allocation, which is not optimal, not Pareto-efficient (see Section 7.1). Engineers now develop a recycling technology, which abolishes these externalities and allows a remediation of the environment. This new technology is more expensive and more costly in operation and cannot be applied profitably in the given equilibrium situation. Nevertheless, it turns out that the implementation of this technology would allow a Pareto-superior allocation, both in comparison to the former market equilibrium with externalities, and to the former efficient allocation attainable by internalising the external effects in the economy with the old technology.

In view of the ultimate goal of all economic activities, namely to care about the economic welfare of the individuals, this result seems to justify the declaration of such a recycling activity as economically reasonable. As environmental commodities are also of relevance for economic wellbeing, this definition is of interest in the context of implementing a circular economy, too.

However, what does this mean in a practical context? To have this result in a comparatively simple formal model is one thing, but to make use of it in a complicated practical environment, such as e-mobility in the City of Shenzhen, another. There are the tools of the cost-benefit analysis, which can be used to provide an answer to this kind of questions. The decision criterion of a cost-benefit analysis, the "Criterion of Potential Pareto Improvement", compares those individuals, who gain from a certain activity, with those, who lose. If the winners can monetarily compensate the losers, such that all individuals are better off if the project would be implemented, then the project should be realised. As this compensation is usually only a potential compensation, one usually refers to this situation as a "potential" Pareto improvement (see Atkinson & Mourato, 2015, or OECD, 2018 for some recent applications of the cost-benefit analysis to environmental contexts).

However, even a cost-benefit analysis is not easy to undertake to clarify such a situation and depends itself on a substantial amount of information, which is usually not straightforwardly available, and which economic agents do not or cannot always provide without any bias. Thus, again, is the circular economy strategy of the City of Shenzhen economically reasonable? Are the other approaches mentioned above economically reasonable? A convincing and economically grounded answer will be difficult to obtain, although there is a societal need for a profound answer.

This shows also the analysis of the Italian packaging waste management system by Beccarello and Di Foggia (2018). They come to the conclusion that the benefits of higher recycling targets "are associated with positive effects on job creation, production and value added", and "there are positive outcomes due to both direct and indirect effects on the economy" (see p. 538). Although this is in fact a positive result, it might be tempting to use exactly this argument as an argument for more recycling activities, and less efforts to prevent waste in the first place. This could, thus, be another example of a "societal path dependency" to be introduced and discussed in Chapter 12.

There is one additional lesson to be learned from the approach to clarify economic reasonableness adopted in Wiesmeth (2011): the formal analysis shows that economic reasonableness depends not only on the technological issues, i.e., on the technical feasibility of a certain circular economy strategy, the technical and technological aspects of the transition to electric mobility in the City of Shenzhen, for example. It rather depends also on the perception of this issue through the consumers, the individuals. Thus, introducing some new technologies or higher standards affecting the wellbeing of individuals should be considered in a careful way. Are households, individuals really ready for these changes? Will significantly more households in Germany be willing to buy e-vehicles? Are households in low-income countries willing to support higher

collection rates of municipal solid waste? Should they? Are they willing to support the efforts of the governments to mitigate climate change similar to those in high-income countries?

Obviously, this dependence of the concept of economic reasonableness on the effect on the wellbeing of individuals implies that certain technologies, certain environmental standards, which are economically reasonable in one country need not be economically reasonable in another country. This applies in particular to industrialised countries on the one hand, and developing countries on the other. Although this seems quite plausible, this principle is probably sometimes violated when certain environmental technologies are presented or even sold to poorer countries. Moreover, this tricky issue is also somehow associated with technology policies of export-oriented countries, which are investigated in Section 10.2.3.

These considerations link this subsection to the remarks made in Chapter 8, indicating that the concept of economic reasonableness could also be of relevance from the point of view of behavioural environmental economics. Thereby, attempts to create new habits, to break old habits and to raise intrinsic motivation could also be helpful to "establish" economic reasonableness, to gain the support of the individuals in a society for new technologies by raising environmental awareness, perceived scarcity or perceived feedback.

There are other phrases, which often appear in environmental policies. One of them is the requirement of the application of the "best available techniques" in certain production processes. This formulation plays a role, for example, in the Clean Air Act of the U.S., or the Federal Immission Control Act of Germany, referring to production processes, which make use of environmentally friendly technologies (see also Section 15.3.1).

Therefore, these phrases can be regarded as some kind of a requirement for a DfE for production processes with similar considerations and challenges in view of informational requirements. For this reason, this issue will be considered more carefully in the context of designing environmental policies. Of course, the wording "best available techniques" already points to some information asymmetry, which will be addressed in the next section.

11.2 Information asymmetries

In a regular market economy, business companies have a strong incentive to stay ahead or, at least, to not fall behind their competitors. One means to accomplish this is to continuously develop new commodities and to make use of innovative technologies, which raise cost efficiency of the production process. In an international context, governments should then provide framework conditions, which allow companies to successfully compete on the international level. Beyond that governments do not really have to care much about innovations, expertise is with the companies and it is in their own interest to make optimal use of it.

Unfortunately, this might be quite different in an environmental context. No doubt, innovations play an important role there, too, as the discussion and the examples in Chapter 10 show. However, there are innovations of environmental relevance, which are not necessarily of interest for private business companies. Informational asymmetries can retard or even prevent the implementation of such innovations.

11.2.1 Informational requirements for a design for environment

DfEs can provide perfect examples for such a situation, although a continuous stream of DfEs is at the heart of a circular economy and in the focus of industrial ecology as one of the schools of thought of a circular economy (see Duchin, Levine, & Fath, 2014; Lifset & Graedel, 2002). Companies might think twice about a DfE, as they might find themselves in a framework quite different from a regular business environment with its clear focus on profitable innovations.

Environmental regulations, especially in the context of waste prevention and recycling, typically include a paragraph on "product design". Consider, for example, the Directive on WEEE (waste electric and electronic equipment) of the European Union (EU) of 2012, which states in Article 4: "Member states shall … encourage cooperation between producers and recyclers and measures to promote the design and production of electrical and electronic equipment (EEE), notably in view of facilitating reuse, dismantling and recovery of WEEE, its components and materials."

It is interesting to note for later considerations that consumers, or customers, are not explicitly integrated into the list of relevant stakeholders. This is surprising, beyond the fact that it does not correspond to the integrative approach of some proponents of industrial ecology (Choudhary, 2012). A DfE as a product, service or process innovation need not necessarily be in the interests of a business company, just because of possible adverse consequences for profits. It can be more costly, and the new design need not meet the taste of the customers, or is expected to attract less additional customers. If there is too much uncertainty about the reactions of the customers, it might be better to wait, to let other companies go ahead and learn later from their experiences. Thus, in some situations there seems to be a late-mover advantage in contrast to the first-mover advantage, which is usually associated with "regular" innovations, and a careful observation of consumer behaviour is thus essential.

Clearly, such a framework tends to slow down the innovative activities in certain sectors of the environmental field, and the governments cannot do much about it: the intrinsic reason for such a constellation has to be found in the information asymmetry: as with innovations in general, the technological knowledge is mainly with the companies, they know best, what could be done and what should be done, given their evaluation of the markets. But if there are risks, regarding the reaction of the customers on a

DfE, for example, then it might be better to wait and to not make use of this knowledge right away.

This leads to a straightforward, but nevertheless important remark: DfEs will be realised, if it is in the interest of the companies. This interest can have various reasons: certain DfEs might be profitable right away, companies might have an intrinsic motivation, or they might expect to profit from signalling environmentally friendly behaviour. Moreover, environmental policies might be designed in a way to make DfEs profitable in comparison to the alternatives available to the producers, or they might expect stricter regulations in near future, which force them to act now in the sense of a first-mover. Thus, considering these reasons, for the vast majority of the companies, an appropriate policy seems to be the right approach to motivate them for a DfE. This corresponds to Porter's Hypothesis (Wang, Sun, & Guo, 2019). To design such a policy, is not always an easy thing to do, but it seems to be a crucial issue for implementing circular economy strategies (see Part V).

11.2.2 Examples from the literature

There is, again for informational reasons, no immediate and straightforward possibility to examine the behaviour of companies regarding a DfE for their products. They will always claim that they did their best to improve reusability and recyclability of the commodities. No doubt that this is true, at least to a certain extent – but the issue is that likely more could have been done, had they fully employed their knowledge, and the challenge is that more should be achieved in the future.

Monteiro, Silva, Ramos, Campilho, and Fonseca (2019) provide a good example for this case. They analyse, how the Portuguese packaging industry deals with the issues eco-design and sustainability. In their survey they find that "quality and cost" are "very important or extremely important" for all those packaging manufacturers interviewed (p. 1745), and 94% of the companies admit "that there is customer guidance to opt for more ecological packaging solutions …" (p. 1747). Moreover, energy consumption and packaging weight are the two most important factors for purchasing products and raw materials. These are factors, which have an immediate effect on production costs (p. 1747, Fig. 6). In their conclusions they refer to 6% of the interviewed companies considering government incentives to eco-design as unimportant, and 15% giving not much attention to university and research partnerships in product development (p. 1749).

This study thus shows the focus of the companies on the market situation, on costs, on quality and the demand of the customers. Eco-design or a DfE come only, if the market develops in this way – largely independent of too simple government regulations. Policy-makers have only limited information and can hardly enforce a DfE, at least not by means of a traditional command-and-control policy. And whether many companies are ready to act as policy-makers, rather than as policy-takers, and have an intrinsic motivation to introduce "voluntarily" a DfE, remains to be seen (Lifset & Graedel, 2002). Behavioural economics (see Chapter 8) leaves ample room for interpreting this special intrinsic motivation, and provides suggestions to cultivate and develop it.

Of course, to complete the picture, DfEs can lead to cost reductions for business companies, especially when they help to save resources, and they can generate additional profits. Examples of such activities exist, the Ellen MacArthur Foundation provides a long list thereof.

The view indicated above that DfE will typically happen, if the market situation is in favour of a particular environmentally friendly design, is supported in Gupt and Sahay (2015). In their study they ascertain the most important aspects of cases of extended producer responsibility (EPR) from developed and developing countries and arrive at the conclusion, which is surprising, at least at the first glance, that the "success of EPR depends on the upstream management …", and that "changes in the design of the products to reduce environmental impacts (DfE) … do not seem to have a major role in the success of EPR …" (p. 609). Also, Massarutto (2014) concedes that "the impact of EPR on green design and product innovation has been much lower than expected" (p. 14).

This relationship between DfEs and EPR policies is of importance for implementing a circular economy and will be explored later (see Section 17.2). But these observations already point to some incomplete specifications of existing EPR policies, which do not automatically and in each situation motivate producers for a DfE (see also Wiesmeth & Häckl, 2016 and Wiesmeth, Shavgulidze, & Tevzadze, 2018 for further details on this issue).

For a more specific context, consider the manufacturing of cars in the EU. The end-of-life vehicle legislation (Section 5.1.3) seems to have some effect regarding a DfE for cars. Gerrard and Kandlikar (2007) show in their "evaluation framework" that "legislative factors and market forces have led to innovation in recycling …" and that "carmakers are also taking steps to design for recycling and for disassembly. However, movement towards design for re-use and remanufacturing seems limited". This last sentence points, not unexpectedly, again to possible improvements regarding a DfE.

Another area of concern is the textile industry, more precisely the handling of chemicals and additives (see also Chapter 24). Producers and consumers of textiles are often separated in the sense that many production activities, in particular environmentally sensitive ones, are located in developing countries. Darbra et al. (2012) argue that the "fate of their additives can be widely spread, and the potential impact of these chemicals on the environment and human health can, therefore, be present on a global scale", and a DfE for textiles, meaning less potentially hazardous chemicals, less additives,

seems to be in dire need, perhaps in relation to REACH, the European Regulation on Registration, Evaluation, Authorisation and Restriction of Chemicals (Wiesmeth & Häckl, 2015).

A DfE for textiles in this sense is of relevance for yet another aspect: Friege (2013) asks the following important questions regarding contaminants in secondary resources: "What is happening on the global market with respect to secondary resources? How do national governments regulate the recycling of potential valuables from waste? How are people protected against hazardous compounds in products? What is the significance of contaminants in products made from secondary resources? What do we know about the separation of hazardous compounds from industrial or household waste?" These questions play a role in the project RISKCYCLE (see also Bilitewski, 2012, the contributions in Bilitewski, Darbra, & Barceló, 2011). Of course, these questions do not only refer to the recycling of textiles, but touch also various other recycling processes, in particular recycling of plastics, and the more than occasional remarks in favour of "upgrading" recycled plastics in order to allow a better substitutability between virgin plastics and recycled plastics (see Section 23.1.2).

These remarks regarding a DfE can be directly extended to include the requirement of using the "best available technique" in environmental regulations in order to achieve a high level of protection of the environment. Again, public authorities can control and compare the techniques employed, but they usually do not have the information on improvements, which companies might not want to apply, because of cost considerations, for example.

Thus, it is and will remain one of the challenges of environmental policies to motivate producers to make use of their knowledge regarding a DfE or an improvement of the best-available technique for certain production processes.

11.3 Case study: Promotion of renewable energy sources in Germany

The data regarding the energy transition in Germany are impressive. In 2018 the application of renewable energy sources developed as follows (data taken from Germany, 2019b):

- Electricity: In 2018, electricity generation from renewable energy increased further by over 4% to 225 billion kilowatt-hours. Its share of total electricity consumption rose from 36.0% to 37.8%.
- Heat: The consumption of renewables-based heat in 2018 remained at the previous year's level. However, since the overall consumption of heat fell due to the weather conditions, the proportion derived from renewable energy rose from 13.8% to 14.2%.
- Fuels: The sale of biofuels increased by around 5% between 2017 and 2018. As a consequence, the share of renewables in the transport sector rose significantly, from 5.2% to 5.7%.

Fig. 11.1 shows the development of electricity generation from renewable energy sources in Germany from 1990 to 2019.

The use of renewable energy sources helped to save some 203 million tonnes of CO_2 equivalent in Germany in 2019. Fig. 11.2 shows that in particular wind energy contributed significantly towards the reductions of greenhouse gas emissions in Germany, followed by biomass and photovoltaics. With this development Germany has leading rankings within the EU: with respect to total installed photovoltaics capacity per capita Germany is heading the field, and with respect to total installed wind energy capacity per capita Germany is occupying the 4th place (see Germany, 2019b).

This energy transition is Germany's long-term strategy for transforming the "energy supply to make it economic and environmentally compatible", the share of renewables in electricity consumption has grown from 6% in 2000 to almost 38% in 2018 (Germany, 2019b).

Nevertheless, this remarkable development, part of a circular economy strategy, was not without challenges, resulting also from incomplete information and information asymmetries. How did it all happen?

Key to this development is the German Renewable Energy Sources Act (EEG), whose aim it is to increase the proportion of electricity generated from renewable energy sources as a percentage of gross electricity consumption to 40–45% by 2025, to 55–60% by 2035 and to at least 80% by 2050. The EEG entered into force in 2000 and has since been revised several times. The goal of the EEG was and is to help mitigating climate change, to become less dependent on the imports of fossil fuels, and to develop innovative technologies, also for the export markets: "The purpose of this Act is to enable the energy supply to develop in a sustainable manner in particular in the interest of mitigating climate change and protecting the environment, to reduce the costs of the energy supply to the economy not least by including long-term external effects, to conserve fossil energy resources and to promote the further development of technologies to generate electricity from renewable energy sources" (Germany, 2017, Section 1).

11.3.1 The original regulations for renewable energy sources in Germany

The first versions of the EEG from the early 2000s regulated priority connections to the public grid system for electricity supply of plants generating electricity from renewable energy sources and the priority purchase and transmission of, and payment for, such electricity by the grid system operators. With this regulation the government created "demand" for electrical energy from renewable sources: the grid system operators simply have to buy electrical energy from these providers.

According to these early versions of the EEG, the fees paid for electricity produced from landfill

Development of gross electricity production from renewable energy sources in Germany

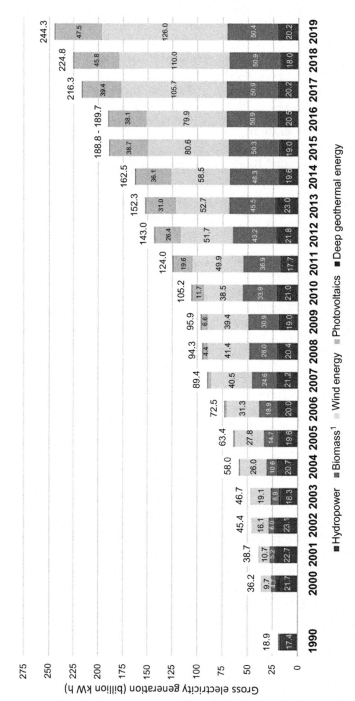

FIG. 11.1 Electricity from renewables in Germany. Development of gross electricity production from renewable energy sources in Germany. © *From BMWi. (2020). Development of renewable energy sources in Germany in the year 2019. Charts and figures based on statistical data from the Working Group on Renewable Energy-Statistics (AGEE-Stat). Berlin: Federal Ministry for Economic Affairs and Energy (BMWi). Retrieved from https://www.erneuerbare-energien. de/EE/Redaktion/DE/Downloads/development-of-renewable-energy-sources-in-germany-2019.pdf?__blob=publicationFile&v=25Chart 6*

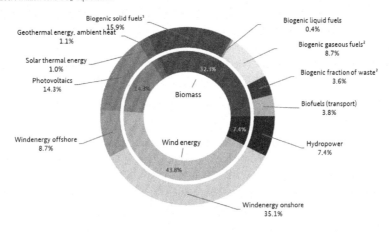

Greenhouse gas emissions avoided through the use of renewable energy sources in the year 2019

Total: 203.4 million tons CO_2-equivalents

¹ incl. sewage sludge, without charcoal; ² biogas, biomethane, sewage gas, and landfill gas; ³ biogenic fraction of waste in waste incineration plants estimated at 50%

BMWi based on AGEE-Stat using data of the German Environment Agency (UBA); as of February 2020.

FIG. 11.2 Avoided greenhouse gas emissions in Germany in 2019. The use of renewable energy sources helped to avoid emissions of greenhouse gases. © *From BMWi. (2020). Development of renewable energy sources in Germany in the year 2019. Charts and figures based on statistical data from the Working Group on Renewable Energy-Statistics (AGEE-Stat). Berlin: Federal Ministry for Economic Affairs and Energy (BMWi). Retrieved from https://www.erneuerbare-energien.de/EE/Redaktion/DE/Downloads/development-of-renewable-energy-sources-in-germany-2019.pdf?__blob=publicationFile&v=25Chart 34.*

gas, sewage treatment plant gas and from biomass were fixed: for installations commissioned in 2009, the fees for electricity from landfill gas and sewage treatment plant gas were up to 9.0 Euro cents per kWh, and the fees paid for electricity produced in plants exclusively using biomass were 11.67 Euro cents per kWh. For electricity from solar radiation (small installations attached to or on top of buildings) the feed-in tariff was reduced to 28.74 Euro cents per kWh in 2011 from 43.01 Euro cents per kWh in 2009.

The critical components of the EEG were therefore the obligation for priority purchases at guaranteed prices. These regulations induced private households and business companies to install solar technology or set up biogas plants, and the thereby perceived "demand" for this technical equipment should initiate research for advancing these technologies. The annual decrease in the guaranteed prices for electricity from new installations was meant to stimulate the development of more effective equipment.

11.3.2 Conclusions from the promotion of renewable energy sources in Germany

The promotion of renewable energy sources in Germany certainly built upon traditional economic theories, in particular the comparatively risk-free business opportunities, which were attractive for many households and business companies, perhaps more attractive than originally expected. However, it turned out that not all of the goals associated with the EEG could be achieved, issues of incomplete or asymmetric information, which were not fully taken into account, created some problems. For this reason, the research design of this case study could be classified "gaps and holes" according to Ridder (2017).

The "business case" established by the EEG was quite appealing. On the one hand, for a household planning to install photovoltaic modules on the roof of the house costs were easy to calculate. But on the other hand, there was "unlimited" demand at a fixed price for electricity generated from these modules. And it was similar for electricity from biomass, from wind turbines and other renewable sources. Thus, the EEG succeeded in motivating many small and large investors to accept this business risk to produce electricity from renewable sources. By means of these framework conditions it was possible to address economic agents in a positive way, and to interest a large number of them in the programme. It would have been difficult to achieve this result with a traditional command-and-control policy.

This unexpectedly successful development pushing Germany to the top ranks was perhaps one of the gaps of the policy. At the same time, when the feed-in tariffs had to be distributed to those generating electricity from renewable sources, fees had to be collected from basically all households for the EEG surcharge: in 2019, households had to pay 6.405 Euro cent per kWh. Fig. 11.3 shows the development of this surcharge contributing to the comparatively high price of electrical energy in Germany.

This high price for electricity has an additional effect, which could not be foreseen at the time of the implementation of the policy: households, in particular those, who do not directly profit from the energy transition, start to complain not only about high prices, but also about side-effects such as wind turbines or transmission lines in the vicinity. This observation corresponds to the missing perceived feedback from one's own willingness to accept these side-effects (see also Section 6.1.3). Of course, Germany's decision to shut down all nuclear power stations and all coal power plants in near future puts additional pressure on further raising the share of renewable energy sources.

Yet another observation is related to incomplete or asymmetric information: the companies producing the technologies for using renewable energy sources, were expected to further develop these technologies, also for making them available for the global markets. However, these companies profited from high demand for their products, such that a lack of necessary innovations opened the door for producers from China, Japan and South Korea: in 2010, half of the solar panels installed in Germany were produced in Asia and obviously proved to be of good quality. Thus, in the end the high feed-in tariffs supported producers in East-Asia and not so much the German companies. Due to asymmetric information it was difficult for the German policy-makers to interfere with this development in due time.

Another issue refers to biomass. Again due to the high feed-in tariffs many farmers set up biogas plants and cultivated energy crops such as maize. As this happened for various reasons on a large scale globally, the "Tortilla Crisis" emerged in 2006 together with the discussion "table or tank". The response of farmers on the EEG and similar regulations in other countries was not expected to this extent – the competition between energy plants and the food chain became clearly visible.

Additional aspects regarding Germany's efforts to mitigate climate change will be presented in Chapter 22. One of the lessons learned in 20 years of EEG is that fast reactions, quick responses are sometimes needed to cope with unwanted developments in this environment of incomplete or asymmetric information. This is, of course, problematic, when the regulations are legal acts of the legislature, which take time to get adjusted.

The following chapter addresses the interesting question, how individuals react on the availability of certain technologies. Moreover, technological and societal path dependencies are interfering with the path towards a circular economy.

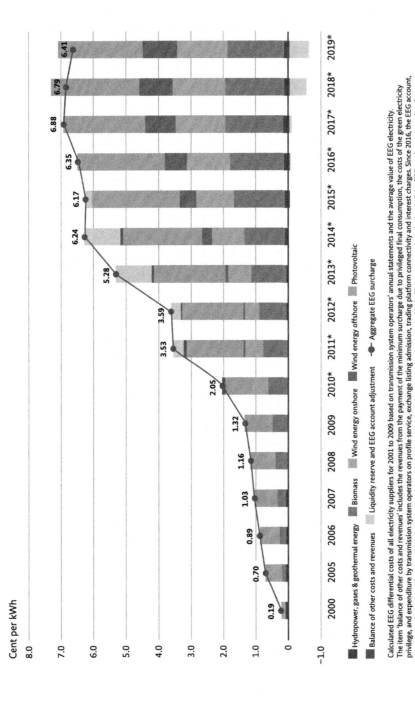

Cent per kWh

Hydropower, gases & geothermal energy ■ Biomass ■ Wind energy onshore ■ Wind energy offshore Photovoltaic
■ Balance of other costs and revenues ■ Liquidity reserve and EEG account adjustment —●— Aggregate EEG surcharge

Calculated EEG differential costs of all electricity suppliers for 2001 to 2009 based on transmission system operators' annual statements and the average value of EEG electricity.
The item 'balance of other costs and revenues' includes the revenues from the payment of the minimum surcharge due to privileged final consumption, the costs of the green electricity
privilege, and expenditure by transmission system operators on profile service, exchange listing admission, trading platform connectivity and interest charges. Since 2016, the EEG account,
which records revenues from the EEG surcharge and payments made to installation operators, has been in the black. These assets reduce the level of the EEG surcharge, meaning that
it can be lower than the total costs of technology-specific funding.
* From 2010 onwards, transmission system operators' forecast of EEG surcharge in accordance with the Renewable Energy Sources Ordinance, published on www.netztransparenz.de

FIG. 11.3 EEG surcharge in Germany. Development of the renewable energy surcharge 2000–2019. © *From BMWi. (2019). Renewable energy sources in figures: National and international development, 2018. Berlin: Federal Ministry for Economic Affairs and Energy (BMWi). Retrieved from https://www.bmwi.de/ Redaktion/EN/Publikationen/renewable-energy-sources-in-figures-2018.pdf?_blob=publicationFile&v=2 p. 31.*

References

Atkinson, G., & Mourato, S. (2015). *Cost-benefit analysis and the environment. OECD environment working papers.* OECD Publishing. https://doi.org/10.1787/5jrp6w76tstg-en.

Beccarello, M., & Di Foggia, G. (2018). Moving towards a circular economy: Economic impacts of higher material recycling targets. In: *International conference on processing of materials, minerals and energy (July 29th–30th) 2016, Ongole, Andhra Pradesh, India, 5(1, Part 1)*, pp. 531–543. https://doi.org/10.1016/j.matpr.2017.11.115.

Bilitewski, B. (2012). The circular economy and its risks. *Waste Management, 32,* 1), 1–2. https://doi.org/10.1016/j.wasman.2011.10.004.

Bilitewski, B., Darbra, R. M., & Barceló, D. (2011). Global risk-based management of chemical additives: Production, usage and environmental occurrence. In *The handbook of environmental chemistry.* Berlin: Springer-Verlag. https://doi.org/10.1007/978-3-642-24876-4.

China (2008). *Circular economy promotion law of the People's Republic of China. Retrieved from (2008).* http://www.fdi.gov.cn/1800000121_39_597_0_7.html.

Choudhary, C. (2012). Industrial ecology: Concepts, system view and approaches. *International Journal of Engineering Research & Technology, 1*(9) Retrieved from (2012). https://www.ijert.org/research/industrial-ecology-concepts-system-view-and-approaches-IJERTV1IS9396.pdf.

Darbra, R. M., González Dan, J. R., Àgueda, A., Capri, E., FAIT, G., Schumacher, M., … Guillén, D. (2012). Additives in the textile industry. In B. Bilitewski, R. M. Darbra, & D. Barceló (Eds.), *The handbook of environmental chemistry.* Berlin: Springer-Verlag. https://doi.org/10.1007/698_2011_101.

Duchin, F., Levine, S. H., & Fath, B. (2014). *Industrial ecology:* (pp. 352–358). Oxford: Elsevier. https://doi.org/10.1016/B978-0-12-409548-9.09407-0.

Friege, H. (2013). Waste—valuables—secondary resources—contaminants—waste again? *Environmental Sciences Europe, 25*(1), 9. https://doi.org/10.1186/2190-4715-25-9.

Germany (2012). *Act to promote Circular Economy and safeguard the environmentally compatible management of waste. Retrieved from (2012).* https://www.bmu.de/fileadmin/Daten_BMU/Download_PDF/Abfallwirtschaft/kreislaufwirtschaftsgesetz_en_bf.pdf.

Germany (2017). *Renewable Energy Sources Act (EEG 2017). Retrieved from (2017).* https://www.bmwi.de/Redaktion/EN/Downloads/renewable-energy-sources-act-2017.pdf%3F__blob%3DpublicationFile%26v%3D3.

Germany (2019a). *Packaging act (factsheet). Retrieved from (2019a).* https://verpackungsgesetz-info.de/wp-content/uploads/2018/06/20171019_landbell_verpackg-factsheet_en_final.pdf.

Germany (2019b). *Renewable energy sources in figures: National and international developments 2018.* Berlin: Federal Ministry for Economic Affairs and Energy (BMWi). Retrieved from (2019b). https://www.bmwi.de/Redaktion/EN/Publikationen/renewable-energy-sources-in-figures-2018.pdf?__blob=publication.File&v=2.

Gerrard, J., & Kandlikar, M. (2007). Is European end-of-life vehicle legislation living up to expectations? Assessing the impact of the ELV Directive on 'green' innovation and vehicle recovery. *Journal of Cleaner Production, 15*(1), 17–27. https://doi.org/10.1016/j.jclepro.2005.06.004.

Gupt, Y., & Sahay, S. (2015). Review of extended producer responsibility: A case study approach. *Waste Management & Research, 33*(7), 595–611. https://doi.org/10.1177/0734242X15592275.

Lifset, R., & Graedel, T. E. (2002). Industrial ecology: Goals and definitions. In R. U. Ayres, & L. W. Ayres (Eds.), *A handbook of industrial ecology.* Cheltenham: Edward Elgar Publishing. https://doi.org/10.4337/9781843765479.00009.

Massarutto, A. (2014). The long and winding road to resource efficiency—An interdisciplinary perspective on extended producer responsibility. *Resources, Conservation and Recycling, 85,* 11–21. https://doi.org/10.1016/j.resconrec.2013.11.005 SI:Packaging Waste Recycling.

Monteiro, J., Silva, F. J. G., Ramos, S. F., Campilho, R. D. S. G., & Fonseca, A. M. (2019). Eco-design and sustainability in packaging: A survey. In: *29th international conference on flexible automation and intelligent manufacturing (FAIM 2019), June 24-28, 2019, Limerick, Ireland, beyond industry 4.0: Industrial advances, engineering education and intelligent manufacturing, 38,* (pp. 1741–1749). https://doi.org/10.1016/j.promfg.2020.01.097.

OECD (2018). *Cost-benefit analysis and the environment: Further developments and policy use.* Paris: OECD Publishing. https://doi.org/10.1787/9789264085169-en.

Ragossnig, A. M., & Schneider, D. R. (2017). What is the right level of recycling of plastic waste? *Waste Management & Research, 35*(2), 129–131. https://doi.org/10.1177/0734242X16687928.

Ridder, H.-G. (2017). The theory contribution of case study research designs. *Business Research, 10*(2), 281–305. https://doi.org/10.1007/s40685-017-0045-z.

Wang, Y., Sun, X., & Guo, X. (2019). Environmental regulation and green productivity growth: Empirical evidence on the Porter Hypothesis from OECD industrial sectors. *Energy Policy, 132,* 611–619. https://doi.org/10.1016/j.enpol.2019.06.016.

Wiesmeth, H. (2011). *Environmental economics: Theory and policy in equilibrium. Springer texts in business and economics.* Springer Nature. https://doi.org/10.1007/978-3-642-24514-5.

Wiesmeth, H., & Häckl, D. (2015). Integrated environmental policy: Chemicals and additives in textiles. *Waste Management*, *46*, 1–2. https://doi.org/10.1016/j.wasman.2015.10.028.

Wiesmeth, H., Shavgulidze, N., & Tevzadze, N. (2018). Environmental policies for drinks packaging in Georgia: A mini-review of EPR policies with a focus on incentive compatibility. *Waste Management & Research*, *36*(11), 1004–1015. https://doi.org/10.1177/0734242X18792606.

Wiesmeth, H., & Häckl, D. (2016). Integrated environmental policy: A review of economic analysis. *Waste Management & Research*, *35*(4), 332–345. https://doi.org/10.1177/0734242X16672319.

The rebound effect and path dependencies

The introduction of new environmental technologies and innovative commodities, which are meant to replace or displace other, less efficient and less environmentally friendly technologies and commodities, is in various contexts accompanied by a "rebound effect", which partially offsets the impact of efficiency gains or other positive effects of new technologies and commodities on the environment.

Rebound effects emerge from economic and psychologic mechanisms: changes in the consumption and production decisions of households can be induced by new, more energy efficient commodities and technologies, but also by moral licensing and spill-over-effects. They therefore characterise a complex interdependency between the supply of technologies and the response of consumers and producers.

This is also true regarding technological path dependencies: the decision of a company or a government for a special (environmental) technology may constrain choice for future technologies in this area. Technologies for waste management, for mitigating climate change and for other circular economy strategies are concerned. These dependencies can, thus, lead to lock-in effects, and may provide further challenges for reusing and recycling activities. Societal path dependencies characterise, perhaps even determine the way we conduct our daily life and our businesses. Some of these dependencies or habits might not be in accordance with the systems change necessary for implementing a circular economy.

This chapter looks into these technological issues, which are related: both rebound effects and technological path dependencies refer to complex interactions between the supply of environmental technologies and the economic behaviour of producers and consumers. Societal path dependencies may generate rebound effects and additionally interfere with the implementation of a circular economy.

12.1 The economic rebound effect

Economic rebound effects arise from behavioural changes, typically in response to improvements of certain environmental technologies, improvements of energy efficiency, for example.

Implementing the Circular Economy for Sustainable Development
https://doi.org/10.1016/B978-0-12-821798-6.00012-0

12.1.1 Classification of economic rebound effects

In traditional microeconomic models, demand for consumption commodities depends on the preferences of the consumers, on the prices of the commodities and on the income of the consumers. Preferences of the consumers are usually taken as given, although it is clear that innovations or other new commodities have to make their way into consumers' preferences, an aspect of particular relevance in behavioural economics (Chapter 8). Prices of various commodities may change due to these innovations and efficiency gains or due to other developments in the economy. Moreover, the incomes of the consumers increase, in general, with economic growth. Thus, demand for consumption commodities is affected by a variety of factors, also through the introduction of new, more (energy) efficient technologies, and the continuous supply of innovative commodities.

Of importance is that quite a few of the "game-changing" factors result from or are connected with activities, which are of relevance for implementing a circular economy. With a view on sustainability, there are, first of all, innovations related to a design for environment (DfE) – increasing the efficiency of various technologies regarding the use of energy, water and raw materials, for example. These innovations allow products to be manufactured and services to be delivered with less resources. Associated cost decreases can also reduce prices, thereby stimulating demand for these commodities. This price effect, which is typically called the direct rebound effect (Zink & Geyer, 2017, p. 595), may be further strengthened through regular economic growth, raising income, which additionally tends to raise demand for commodities, at least in general.

In microeconomic theory, the effect of a price change of a commodity can be separated into a substitution effect and an income effect. A more precise analysis assigns the term "direct rebound" to the substitution effect (Borenstein, 2013), which is negative: a lower price of a commodity leads to higher demand for this commodity, after the income change resulting from the increasing purchasing power is compensated. If one takes further into account the income effect, resulting from the increase in purchasing power of the consumer due to the lower price for a commodity, then one arrives at "indirect rebound" (p. 6). This income effect leads normally to an increasing demand for other commodities, which can be further intensified through economic growth with additional income boosts.

Whereas this classification of direct and indirect rebound is precisely mirrored in basic microeconomics (see, for example, Pindyck & Rubinfeld, 2018), other concepts of direct, indirect or secondary rebound can be found in the vast literature. With a focus on the United Stated (U.S.), Greening, Greene, and Difiglio (2000) review the literature regarding definitions and introduce a classification of the rebound or "take-back" effect: there are, as already mentioned, the direct rebounds and secondary effects, there are additional economy-wide effects referring to "larger and largely unpredictable effects that increased efficiency has on prices and demand of other goods", and there are "transformational effects, referring to the potential of energy efficiency increases to change consumer preferences, societal institutions, technological advances, regulation, or other large-scale effects" (see Zink & Geyer, 2017, p. 595).

12.1.2 Empirical results for economic rebound effects

Of course, the numerical results obtained for the rebound effects, presented and discussed in the literature, depend on the precise definition used. Often, however, estimations of price elasticities, the percentage change of demand for commodities resulting from an increase of the price of 1%, play a role in the calculations of the rebound effects (see, for example, Frondel, Ritter, & Vance, 2012). As it is understood that rebound effects partially or even fully offset beneficial effects of new technologies and commodities, they are measured as percentage reductions from the initially expected environmental benefits.

Regarding empirical results, the German Environmental Agency (UBA) estimates the direct rebound effect for space-heating use, i.e., the rebound effect resulting from more energy efficient heating systems in Germany, anywhere from 10% to 30% (UBA, 2014). Similarly, Aydin, Kok, and Brounen (2017), investigating the rebound effect in residential heating in the Netherlands, find an effect of 26.7% among homeowners, and of 41.3% among tenants. Thus, the outcomes of energy improvement programs are largely determined by household behaviour.

Vehicle fuel efficiency seems to be characterised in Germany by a direct rebound effect of around 20% (UBA, 2014). Interestingly, Frondel et al. (2012), studying the heterogeneity of the rebound effect in private transport, arrive at estimates in the range of 57–62%. However, they also find that the magnitude of estimated fuel price elasticities, determining the rebound effect, depends inversely on the household's driving intensity. Thus, rebound effects tend to be significantly larger for households with low vehicle mileage than those for households with high vehicle mileage.

In a more general context, Walnum, Aall, and Løkke (2014) come to the conclusion that regarding sustainable mobility there are rebound effects associated with more efficient technologies, less polluting means of transport, and changing consumption patterns – all of them "offsetting expected savings in energy use and GHG [greenhouse gas] emissions in the transport sector".

Comparing these results, it seems to be difficult to arrive at comparable and reliable estimates. However, individuals react differently to any economic changes with the consequence,

that the estimated rebound effects might depend critically on the group of consumers participating in a survey (see also Aydin et al., 2017).

A similar conclusion holds for the rebound effects regarding lighting. The actual energy savings in Germany in this context may be up to 25% lower than the projected technically feasible savings. In this sense, rebound effects "undermine resource use reductions", and they need to be taken into account in targets for resource use (UBA, 2014).

Schleich, Mills, and Dütschke (2014), on the other hand, measuring the direct rebound effect in lighting as the elasticity of energy demand for lighting with respect to changes in energy efficient lamps, come up with further interesting details. They, in particular, point to the possibility of negative direct rebound effects, resulting from more energy efficient lamps with energy savings larger than expected.

The International Energy Agency (IEA) refers to the increase in global primary energy demand by 2.3% in 2018, the largest increase since 2010 – despite improvements in energy intensity of 1.7% in 2017 and 1.2% in 2018. China, India and the U.S. were responsible for 70% of this demand growth in 2018, up from 43% in 2017. Moreover, fossil fuels accounted for 70% of the growth in primary energy demand, by far outweighing the 24% increase from renewable energy sources in primary demand (see IEA, 2019, p. 13f).

The IEA contributes this rebound in global energy demand to several factors, but mainly to global economic growth, which was 3.7% in 2017, and to a higher demand for oil, mainly from the transport sector. The road transport sector accounts for 50% of global final oil consumption and 22% final energy consumption. In 2018, there was an additional increase in energy-intensive industries, such as steel production in China and petrochemical industry in the U.S. According to the IEA, this rebound in energy demand led to an increase in energy sector greenhouse gas emissions, "which reached an historic high of over

33 billion tons of CO_2" (see IEA, 2019, p. 15ff). Estimating economy-wide energy rebound effects in China, Yan, Ouyang, and Du (2019) find average rebound effects of 88.55% for all provinces, and an average long-run rebound effect of 77.5%. Interestingly, the rebound effect in the developed eastern part of China had decreased continuously during the research period 1997–2015.

12.1.3 Rebound effects in a circular economy

Beyond the direct rebound effects, there are, as already indicated, additional rebound mechanisms of even more relevance for a circular economy: secondary, economy-wide and transformational rebound effects. These mechanisms are in particular associated with "secondary production", which Zink and Geyer (2017) associate, in the context of a circular economy, with reuse at the product level ("repair", "refurbishment"), reuse at the component level ("remanufacturing"), and reuse at the material level ("recycling"), and which are therefore of high relevance for implementing a circular economy (p. 594).

Zink and Geyer (2017) continue to argue that secondary goods, often recycled commodities, may be insufficient substitutes for primary goods, because they are of inferior quality – think about recycled plastics, which must, in general, not be used in food packaging – or because they are otherwise less desirable for consumers. Consequently, these secondary goods, in particular recycled goods, "are likely to be produced in addition to, rather than instead of, primary materials, and the potential benefits of recycling will be reduced". Therefore, circular economy strategies may prevent some primary consumption, but not on a one-to-one basis (p. 598). For example, according to BUND (2019), in 2016 only 2.8% of the plastics products manufactured in Germany were from recycled plastics. This number is certainly still

far from the intentions associated with a circular economy (p. 37).

In this context, Allwood, Worrell, and Reuter (2014) draw attention to the fact that many materials cannot be recycled, that most materials that are recycled today are downgraded in the process, and recycling may be more energy intensive than new production. For this reason, recycling or material reuse should be seen as an option in a hierarchy of material management strategies. Observe that these considerations again touch the issue of the economic reasonableness (see Section 11.1) of certain activities proposed in the context of implementing a circular economy.

If the secondary goods compete directly with primary goods, then there might be a general pressure on the prices due to the resulting increase in the quantity. Consequently, demand for both secondary and primary goods might increase, and in the end, a larger quantity will be sold at a lower price (Zink & Geyer, 2017, p. 598). In addition, Allwood et al. (2014) point to the fact that lower prices in these markets leave income to be spent elsewhere with unpredictable results, also regarding circular economy strategies.

The possibility of a "backfire effect", i.e., of a rebound effect of more than 100% is addressed in Makov and Font Vivanco (2018). With a focus on imperfect substitution between "recirculated" and new products, and "respending" due to economic savings, they investigate rebound effects for life-cycle greenhouse gas emissions from reusing smartphones in the U.S. There is a range of 27–46% for specific smartphone models. Moreover, they argue that in other regions and under different consumer behaviour patterns it might be possible that all emission savings from smartphone reuse are lost due to the rebound effect.

Economy-wide and transformational rebound effects are, according to Zink and Geyer (2017), more speculative. They point to the possibility that an increased use of refillable drinks containers, for example, could lead to an increased production and operation of refilling stations. Even more interesting: an "increased emphasis on recycling could lead consumers to purchase more disposable products, believing they can erase their impact at the recycling bin", or "availability of cheaper materials attributed to increased recycling may change consumer tastes", and "repair occupations that have systematically disappeared over the past century may start to re-emerge with unpredictable effects on employment, affluence, immigration, and overall consumption levels and patterns" (p. 599). Some of these aspects play a role in the context of path dependencies, in particular societal path dependencies, to be discussed in Section 12.2, and some of these issues will be addressed in the context of designing policies for implementing a circular economy in Part V.

One of the consequence of all these considerations is that the economic rebound effects are difficult to estimate and, regarding economy-wide and transformational effects, difficult to predict. More studies, more investigations are required to throw some more light on these issues, which are of importance for implementing a circular economy.

12.1.4 Psychological rebound effects

In addition to rebound effects resulting from economic mechanisms, behavioural changes due to an improved energy efficiency, for example, rebound effects can also be induced by non-economic, psychological mechanisms. According to Merritt, Effron, and Monin (2010), "past good deeds can liberate individuals to engage in behaviours that are immoral, unethical, or otherwise problematic, behaviours that they would otherwise avoid for fear of feeling or appearing immoral". This "moral self-licensing" is one of the relevant cognitive processes in this context, "acting virtuously can make people feel

licensed to act less-than-virtuously" (Effron & Conway, 2015).

Of course, there are many possibilities with relevance for implementing a circular economy. Unfortunately, not many concrete examples have been investigated thoroughly so far. The following list provides therefore only possible psychological rebound effects in the environmental context (see also Dütschke, Frondel, Schleich, & Vance, 2018):

- There might be households carefully segregating waste, thereby justifying the purchase of drinks in one-way packaging.
- There might be consumers returning all empty one-way bottles to justify their preference for drinks in one-way containers.
- There might be producers advertising some measures to protect the environment just to continue or increase other environmentally doubtful activities.
- And there might be technological improvements motivating individuals to reduce efforts to mitigate environmental damage.

Especially this last bullet point draws attention to "behavioural rebound effects" investigated by Dorner (2019): firstly, technological improvements reduce marginal environmental damage and thereafter pro-environmental efforts, and secondly, endogenous technological change further reduces pro-environmental behaviour through moral self-licensing. Some of these examples seem to point to "cheap excuses": a focus on one environmentally friendly action might serve as an excuse for other more problematic activities, or "little good is good enough" (Engel & Szech, 2017).

It is problematic that only comparatively few empirical analyses regarding this issue exist so far. The implementation of a circular economy is dependent on environmentally friendly actions, on pro-environmental behaviour. The following subsection provides a few hints how to deal with rebound effects with a focus on economic rebound effects, however.

12.1.5 How to deal with the rebound effect?

Because of their detrimental impact on the environmental effectiveness of new technologies, rebound effects should be avoided or at least reduced, wherever and whenever this is possible. In particular, the goods produced and provided on the secondary markets should be genuine substitutes for primary production alternatives (Zink & Geyer, 2017, p. 599). Simply marketing and selling used products such as smartphones or cars, among others, to developing countries might be good for business in general, and quite effective regarding a simplistic reusability concept, but quite ineffective regarding the expected environmental benefits.

If, for example, consumers in one of these countries – developing countries or emerging economies, could not afford a new car, then the resulting rebound effect is clear: there will be a new car in an industrialised country and a used car in a developing country. This is reuse, of course, and as that it might be meaningful for a variety of reasons, in particular for the consumers in the developing countries. But it is not the reuse at the product level, intended and meaningful for a circular economy. A similar argumentation applies to the export of used smartphones and laptops, to the export of used electronic equipment to developing countries in general (see also Makov & Font Vivanco, 2018; Sovacool, 2019). This "issue of reuse" needs to be reconsidered in more depths in a separate chapter (see Section 19.3).

How to deal with the rebound effect? Font Vivanco, Kemp, and van der Voet (2016) address this question "for achieving absolute energy and environmental decoupling". They examine the extent the rebound effect is considered in environmental policies. Their conclusion is that an appropriate policy design and policy mix "are key to avoiding undesired outcomes". This issue will be further considered in the context of designing policies for implementing a circular economy in Part V.

Thus, avoiding rebound effects seems not an easy thing to do. The following section turns to technological and societal path dependencies, which are related to rebound effects.

12.2 Path dependencies

Path dependencies, in particular technological path dependencies, refer to sub-optimal or inefficient technologies which can become "locked-in" as industry standards and may therefore persist for extended periods of time (Stack & Gartland, 2003). The concept emerged from contributions of Arthur (1989), who relates the concept to two or more competing increasing-returns technologies. Insignificant events may by chance then give one of them an initial advantage of adoption. David (1985) on the other hand, investigates the "economics of QWERTY", referring to the dominant keyboard layout, which has been considered sub-optimal already for a long period of time.

In sociology and political sciences, formulations of path dependence typically capture "the idea that history matters". Castaldi, Dosi, and Paraskevopoulou (2011) refer also to the relevance of path dependence "in the evolution and patterns of decision making of organisations": persistent organisational choices emphasise the importance of past events for the future orientation of organisations (p. 3). Moreover, social scientists "frequently trace outcomes back to critical junctures when unexpected, random, or small events trigger deeply patterned sequences of subsequent events", thereby also pointing to QWERTY and the role of increasing returns (see Mahoney, 2006, p.130).

The investigation of both technological and societal path dependencies reveals interesting relationships   between   circular   economy strategies, technologies and the economic behaviour of consumers and producers. Similar to the case of rebound effects, the reaction of the economic agents, who themselves might be "guided" by societal path dependencies, is not always completely predictable. For sure, implementing a circular economy requires a careful consideration of possible behavioural reactions of the economic agents, pointing to aspects of behavioural environmental economics.

12.2.1 Technological path dependencies

In a market context, technological path dependencies are usually related to competing increasing-returns technologies, one of which has to be adopted – with the possibility that various accidental factors interact with the decision. Environmental technologies, such as water treatment technologies, waste incineration plants, or renewable energy technologies, exhibit similar characteristics, of course. Decisions regarding the adoption of certain technologies are typically made by bureaucracies or other organisations – the market mechanism is not really applicable to situations with increasing returns.

There is, however, an additional aspect: implementing a circular economy is dependent on the availability of appropriate technologies and their further development. In this sense, the technological environment of a circular economy has – with support from environmental policies – some relevance as an allocation tool, guiding the economic agents towards a circular economy (see Chapter 10). The design and development of environmental technologies results mainly from the regulatory framework (see Chapter 14). Thus, public authorities determine to some extent the direction of the technological development, and competition with its advantage to integrate many individuals into the decision-making process is relegated to the background. Therefore, technological path dependencies play a decisive role in the context of implementing a circular economy.

To begin with a simple case: the choice of a particular technology in waste management, for example in the context of collecting and segregating household waste, leads to typical path dependencies, which are usually caused by sunk costs. Due to a typical lack of financial means in municipalities, it will be, in general, difficult to adopt a completely new technology within a short period of time. Once a technology has been chosen, it will remain in place and can have a long-lasting relevance for various aspects of waste management, and it will frame or even restrict the way, consumers and producers can make their decisions regarding their part in waste management. The case study on the "German Refillable Quota Issue" in Section 7.3 shows how technological path dependencies can arise from environmental regulations.

In order to avoid inefficient or sub-optimal technologies it is therefore necessary to start with "optimal" technologies. As any optimality concept depends on local conditions it is not always easy to collect enough information for adequate decision making. Moreover, often the financial volume of these technologies does simply not allow testing different environments. Collection systems for various kinds of waste have, however, been investigated in the literature. There obviously is some variance regarding the systems in application:

Pires, Martinho, and Chang (2011) conduct a literature review of the pros and cons of waste management practices, and, similarly, Seyring, Dollhofer, Weißenbacher, Bakas, and McKinnon (2016) study the legal framework and the practical implementation of collection systems for various waste streams in the member states of the European Union (EU). Gallardo, Carlos, Colomer, and Edo-Alcón (2017) and Gallardo, Bovea, Colomer, Prades, and Carlos (2010) analyse the efficiency of collection systems in Spanish cities, and compare different waste collection systems in Spain with a focus on areas with different income levels.

Feil, Pretz, Jansen, and Thoden van Velzen (2016) compare separate collection of plastic waste with technical sorting from municipal solid waste, and finally Hahladakis, Purnell, Iacovidou, Velis, and Atseyinku (2018) focus on post-consumer plastic packaging waste with the conclusion that consumers tend to correspond differently to different collection systems, and Broitman, Ayalon, and Kan (2012) question a unified technical approach to waste management in view of relevant differences among regions.

These issues are particularly important for improving waste management in developing and transition countries, and emerging economies. These countries are, for example, experimenting with different approaches to reduce littering (see Moqbel, El-tah, & Haddad, 2019 for Jordan), or discussing the implementation of policies to reduce plastic waste (see Wiesmeth, Shavgulidze, & Tevzadze, 2018 for Georgia). Due to the fact that each country, each region, has to find those technologies, which are adequate given its specific needs for a sustainable development, it is certainly recommendable to reflect this situation and to choose carefully.

In this context, Collivignarelli, Sorlini, and Vitali (2013) refer to a variety of factors, which cause the failure of international cooperation projects promoting environmental technologies in developing countries. Marshall and Farahbakhsh (2013) point to the significant differences between the main drivers of solid waste management in industrialised and developing countries, limiting the applicability of approaches that proved to be useful in industrialised countries to developing countries. Finally, Guerrero, Maas, and Hogland (2013) collect data and investigate waste management in major cities in developing countries. The objective of their research is to analyse the behaviour of stakeholders and additional influential factors on waste management systems. Among other things, they arrive at a list of factors that reveal the most important causes for the failures of such systems.

12.2.2 Societal path dependencies

Path dependencies can, however, arise on a much deeper level of a society's dealing with waste or other aspects of a circular economy. One example is the prevention of waste, which is often considered the most important part of the waste hierarchy. The EU is working on guidelines for waste prevention and presents many examples of successful practices, but a real breakthrough is still missing. Collection, recovery and recycling quota still play the dominant role in all statistics regarding waste management. What is the reason for this general observation? After all, there is common consent that waste prevention is leading the waste hierarchy and recycling should come at the very end.

Wilts (2012) argues that "the prevention of waste as the top of the waste hierarchy … is much more than a simple amendment of ways of dealing with waste, but means nothing less than a fundamental change of the socio-technical system of waste infrastructures … " (p. 29). With a reference to a case study of the socio-technical waste regime in urban agglomerations in Germany, he focuses on "the relationship between physical waste infrastructures, actor constellations in waste governance and incentives for waste prevention" as "an aspect for the success or failure of waste prevention measures … " (p. 32). The important conclusion of this analysis is that this limited progress might be related to "path dependencies in the way a society deals with waste". It is certainly true for many countries that "treatment and prevention of waste are most often regarded as separate policy fields", which is, of course, correct from a static point of view: "waste that ends up in treatment facilities can no longer be prevented". From the perspective of a "socio-technical regime", however, "the prevention of waste has a fundamentally different function than the environmentally friendly, reliable and cheap treatment of waste". As the components of a socio-technical regime are connected, treatment and recycling of waste reduce the urgency to prevent waste (p. 32): waste that is treated and recycled is no longer considered as waste, is, rather, interpreted in large parts of the societies as prevented waste.

Thus, the prevention of waste is probably not yet sufficiently anchored in societal paths, at least not to the extent of the environmentally friendly collection, treatment and recycling of waste, which is currently dominating the discussion. Therefore, these path dependencies resulting from the ways a society is "used" to deal with waste, has significant consequences for the success of efforts regarding waste prevention and, therefore, for a successful implementation of a circular economy.

Unfortunately, appropriate technologies in waste management and their continuous further development may contribute to the stability of this societal path dependencies: hi-tech systems to collect and recycle waste seem to reduce the need for preventing waste. In fact, that is at least what is happening in Germany and likely also in other countries: large drinks producers raise their share of drinks in one-way containers exactly with the argument and the promise that the empty bottles will be collected and perfectly recycled with no harm to the environment. This refers, among others, to Coca-Cola with its "World without Waste" initiative, and to Lidl Deutschland, a chain-store company, with its clear position in favour of one-way PET bottles, given the well-functioning German deposit scheme for one-way drinks packaging and the high-quality recycling system.

These arguments, which seem to be accepted and supported in large parts of the societies, are probably not false regarding the promises of these companies. However, the tricky issue with these path dependencies is that they "can be understood as 'self-reinforcing feedback loops´ meaning that once a decision is made, this is favoured over all other, as well as future alternatives" (Wilts, 2012, p. 32). There is, thus, a disadvantage regarding the diffusion of possibly better solutions, with a "lock-in effect" limiting the system's capability for innovation.

Path dependencies, both in a technological and a societal context, are also important with respect to reusing certain commodities – another core issue for implementing a circular economy. Take vehicles, for example, or electronic equipment, which should be, according to many regulations, designed to allow for reuse and for a simpler and less costly dismantling or recycling. Reusing old vehicles or old electronic equipment is not much of an issue within industrialised countries, which have the technologies and the means to recycle end-of-life vehicles or waste electric and electronic equipment (WEEE), after a period of reuse. Once, however, used vehicles and electronic equipment are exported to low-income countries, then certain issues gain relevance from an environmental point of view and need to be respected when implementing a circular economy (see Chapter 21).

The issue behind this observation is again a kind of a societal path dependence. The carefully designed cars of the manufacturers in industrialised countries might lead to environmental issues in countries, for which these cars are not necessarily optimal, at least not without further precautionary measures such as mandatory and qualified checks etc. Interestingly, although exporting used vehicles and electronic equipment for reuse in low-income countries seems to correspond to circular economy strategies, these cases create environmental issues, which extend to recycling the waste equipment in these countries. Recycling of electronic waste in countries such as Ghana, among others, have often been documented in the literature (see Ongondo, Williams, & Cherrett, 2011; Sovacool, 2019). Clearly, these observations point to rebound effects.

Thus, regarding the implementation of a circular economy, is there an issue that these commodities and others are mainly and predominantly focused on consumers in industrialised countries? Would it not, from the point of view of implementing a circular economy, be better to design special versions of these products for customers in the low-income countries, or help them to establish their own production facilities focusing on appropriate models of cars and electronic equipment? So far, the "philosophy" of international trade, again a societal path dependence in this context, does not sufficiently take into account these environmental issues, at least not at large. Unfortunately, experiences with "home-made" means of transportation in some countries in South and South-East Asia indicate that such proposals might contribute more to the problem than to the solution, at least regarding air pollution.

These observations point clearly to a rebound effect: reusing old vehicles and old electronic equipment in this way likely does not really contribute to saving resources, and may lead to further environmental degradations, thus establishing a link between path dependencies and rebound effects.

How to deal with these technological and societal path dependencies? As the above examples demonstrate, new technologies can play an ambiguous role in the context of environmental economics in general, and in implementing a circular economy in particular: on the one hand, they provide necessary technical solutions for environmental issues, but on the other hand, they can create new environmental issues due to societal path dependencies. This, for sure, points to the necessity to extend the horizon of a circular economy beyond that of the industrial ecology with its focus on science and technologies. Of importance is also the reaction of consumers and producers on all aspects in the context of implementing a circular economy in general, and on the introduction of new technologies in particular. As societal path dependencies point to behavioural economics, special attention should be given to change these path dependencies, to break adverse habits.

Finally, one of the main societal path dependencies potentially interfering with various aspects of implementing a circular economy is the general focus on characteristics of a market

system. It is, of course, good to have business models for certain circular economy strategies as listed by the Ellen MacArthur Foundation, and it is also good to have profitable recycling activities with "positive effects on job creation" (Beccarello & Di Foggia, 2018). However, the discussions in the chapters of Part II addressed the limits of the market mechanism regarding the allocation of environmental commodities. Thus, the fact that a particular activity is not profitable does not a priori mean that this activity is not economically reasonable in the context of implementing a circular economy and vice versa, of course (see also Section 11.1.1).

This does, of course, not imply that markets should not play any role in implementing a circular economy – to the contrary. However, one should always make sure that by means of appropriate framework conditions incentives of the individuals are in line with the goals of the circular economy strategies. These important aspects will be reconsidered in Part V in the context of designing environmental policies – which try to make use of features of the market system.

The digital transformation affects most parts of the society in general, and our lives in particular, although what we see is probably, only the beginning. What role for digital technologies should we expect in the context of implementing circular economies?

References

Allwood, J. M., Worrell, E., & Reuter, M. A. (2014). *Squaring the circular economy: The role of recycling within a hierarchy of material management strategies*: (pp. 445–477). Boston: Elsevier. https://doi.org/10.1016/B978-0-12-396459-5.00030-1 (chapter 30).

Arthur, W. B. (1989). Competing technologies, increasing returns, and lock-in by historical events. *The Economic Journal*, *99*(394), 116–131. https://doi.org/10.2307/2234208.

Aydin, E., Kok, N., & Brounen, D. (2017). Energy efficiency and household behavior: The rebound effect in the residential sector. *The Rand Journal of Economics*, *48*(3), 749–782. https://doi.org/10.1111/1756-2171.12190.

Beccarello, M., & Di Foggia, G. (2018). Moving towards a circular economy: Economic impacts of higher material recycling targets. In: *International conference on processing of materials, minerals and energy (July 29th–30th) 2016, Ongole, Andhra Pradesh, India, 5(1, Part 1)*, pp. 531–543. https://doi.org/10.1016/j.matpr.2017.11.115.

Borenstein, S. (2013). *A microeconomic framework for evaluating energy efficiency rebound and some implications. NBER working paper series, (19044). Retrieved from (2013).* https://www.nber.org/papers/w19044.pdf.

Broitman, D., Ayalon, O., & Kan, I. (2012). One size fits all? An assessment tool for solid waste management at local and national levels. *Waste Management, 32*(10), 1979–1988. https://doi.org/10.1016/j.wasman.2012.05.023.

BUND (2019). *Plastics Atlas 2019: Facts and figures about a world full of plastics.* Berlin: BUND, Heinrich-Böll-Stiftung. Retrieved from (2019). https://www.boell.de/de/2019/05/14/plastikatlas?dimension1=ds_plastikatlas.

Castaldi, C., Dosi, G., & Paraskevopoulou, E. (2011). Path dependence in technologies and organizations: A concise guide. In *Laboratory of Economics and Management, Pisa.* Retrieved from https://ideas.repec.org/p/ssa/lemwps/2011-12.html.

Collivignarelli, C., Sorlini, S., & Vitali, F. (2013). *Sustainability of appropriate environmental technologies in developing countries: General framework. In: Imagining cultures of cooperation: Universities networking to face the new development challenges. Proceedings of the III CUCS congress* Retrieved from (2013). https://pdfs.semanticscholar.org/a76b/299988fc64c746f0cd5c39ea3de32d97fee7.pdf.

David, P. A. (1985). Clio and the economics of QWERTY. *The American Economic Review, 75*(2), 332–337. Retrieved from (1985). http://www.jstor.org/stable/1805621.

Dorner, Z. (2019). A behavioral rebound effect. *Journal of Environmental Economics and Management, 98*, 102257. https://doi.org/10.1016/j.jeem.2019.102257.

Dütschke, E., Frondel, M., Schleich, J., & Vance, C. (2018). Moral licensing—Another source of rebound? *Frontiers in Energy Research, 6*, 38. Retrieved from (2018). https://www.frontiersin.org/article/10.3389/fenrg.2018.00038.

Effron, D. A., & Conway, P. (2015). When virtue leads to villainy: Advances in research on moral self-licensing. *Morality and Ethics, 6*, 32–35. https://doi.org/10.1016/j.copsyc.2015.03.017.

Engel, J., & Szech, N. (2017). *Little good is good enough: Ethical consumption, cheap excuses, and moral self-licensing. Discussion Paper No. 17-28 GEABA. Retrieved from (2017).* http://www.geaba.de/wp-content/uploads/2017/07/DP_17-28.pdf.

Feil, A., Pretz, T., Jansen, M., & Thoden van Velzen, E. U. (2016). Separate collection of plastic waste, better than technical sorting from municipal solid waste? *Waste Management & Research, 35*(2), 172–180. https://doi.org/10.1177/0734242X16654978.

Font Vivanco, D., Kemp, R., & van der Voet, E. (2016). How to deal with the rebound effect? A policy-oriented approach. *Energy Policy, 94*, 114–125. https://doi.org/10.1016/j.enpol.2016.03.054.

Frondel, M., Ritter, N., & Vance, C. (2012). Heterogeneity in the rebound effect: Further evidence for Germany. *Energy Economics, 34*(2), 461–467. https://doi.org/10.1016/j.eneco.2011.10.016.

Gallardo, A., Carlos, M., Colomer, F., & Edo-Alcón, N. (2017). Analysis of the waste selective collection at drop-off systems: Case study including the income level and the seasonal variation. *Waste Management & Research, 36*(1), 30–38. https://doi.org/10.1177/0734242X17733539.

Gallardo, A., Bovea, M. D., Colomer, F. J., Prades, M., & Carlos, M. (2010). Comparison of different collection systems for sorted household waste in Spain. *Waste Management, 30*(12), 2430–2439. https://doi.org/10.1016/j.wasman.2010.05.026.

Greening, L. A., Greene, D. L., & Difiglio, C. (2000). Energy efficiency and consumption—The rebound effect—A survey. *Energy Policy, 28*(6), 389–401. https://doi.org/10.1016/S0301-4215(00)00021-5.

Guerrero, L. A., Maas, G., & Hogland, W. (2013). Solid waste management challenges for cities in developing countries. *Waste Management, 33*(1), 220–232. https://doi.org/10.1016/j.wasman.2012.09.008.

Hahladakis, J. N., Purnell, P., Iacovidou, E., Velis, C. A., & Atseyinku, M. (2018). Post-consumer plastic packaging waste in England: Assessing the yield of multiple collection-recycling schemes. *Waste Management, 75*, 149–159. https://doi.org/10.1016/j.wasman.2018.02.009.

IEA (2019). *Energy efficiency 2019.* International Energy Agency. Retrieved from (2019). https://www.iea.org/efficiency2019/.

Mahoney, J. (2006). *Analyzing path dependence: Lessons from the social sciences:* (pp. 129–139). London: Palgrave Macmillan UK. https://doi.org/10.1057/9780230524644_9.

Makov, T., & Font Vivanco, D. (2018). Does the circular economy grow the pie? The case of rebound effects from smartphone reuse. *Frontiers in Energy Research, 6*, 39. Retrieved from (2018). https://www.frontiersin.org/article/10.3389/fenrg.2018.00039.

Marshall, R. E., & Farahbakhsh, K. (2013). Systems approaches to integrated solid waste management in developing countries. *Waste Management, 33*(4), 988–1003. https://doi.org/10.1016/j.wasman.2012.12.023.

Merritt, A. C., Effron, D. A., & Monin, B. (2010). Moral self-licensing: When being good frees us to be bad. *Social and Personality Psychology Compass, 4*(5), 344–357. https://doi.org/10.1111/j.1751-9004.2010.00263.x.

Moqbel, S., El-tah, Z., & Haddad, A. (2019). Littering in developing countries: The case of Jordan. *Polish Journal of Environmental Studies*, *28*(5), 3819–3827. https://doi.org/10.15244/pjoes/94811.

Ongondo, F. O., Williams, I. D., & Cherrett, T. J. (2011). How are WEEE doing? A global review of the management of electrical and electronic wastes. *Waste Management*, *31*(4), 714–730. https://doi.org/10.1016/j.wasman.2010.10.023.

Pindyck, R., & Rubinfeld, D. (2018). *Microeconomics* (9th ed.). Pearson.

Pires, A., Martinho, G., & Chang, N.-B. (2011). Solid waste management in European countries: A review of systems analysis techniques. *Journal of Environmental Management*, *92*(4), 1033–1050. https://doi.org/10.1016/j.jenvman.2010.11.024.

Schleich, J., Mills, B., & Dütschke, E. (2014). A brighter future? Quantifying the rebound effect in energy efficient lighting. *Energy Policy*, *72*, 35–42. https://doi.org/10.1016/j.enpol.2014.04.028.

Seyring, N., Dollhofer, M., Weißenbacher, J., Bakas, I., & McKinnon, D. (2016). Assessment of collection schemes for packaging and other recyclable waste in European Union-28 member states and capital cities. *Waste Management & Research*, *34*(9), 947–956. https://doi.org/10.1177/0734242X16650516.

Sovacool, B. K. (2019). Toxic transitions in the lifecycle externalities of a digital society: The complex afterlives of electronic waste in Ghana. *Resources Policy*, *64*, 101459. https://doi.org/10.1016/j.resourpol.2019.101459.

Stack, M., & Gartland, M. P. (2003). Path creation, path dependency, and alternative theories of the firm. *Journal of Economic Issues*, *37*(2), 487–494. Retrieved from (2003). http://www.jstor.org/stable/4227913.

UBA (2014). *Rebound effects. Retrieved November 30, 2019, from (2014).* https://www.umweltbundesamt.de/en/topics/waste-resources/economic-legal-dimensions-of-resource-conservation/rebound-effects.

Walnum, J. H., Aall, C., & Løkke, S. (2014). Can rebound effects explain why sustainable mobility has not been achieved? *Sustainability*. https://doi.org/10.3390/su6129510.

Wiesmeth, H., Shavgulidze, N., & Tevzadze, N. (2018). Environmental policies for drinks packaging in Georgia: A mini-review of EPR policies with a focus on incentive compatibility. *Waste Management & Research*, *36*(11), 1004–1015. https://doi.org/10.1177/0734242X18792606.

Wilts, H. (2012). National waste prevention programs: Indicators on progress and barriers. *Waste Management & Research*, *30*(9_Suppl), 29–35. https://doi.org/10.1177/0734242X12453612.

Yan, Z., Ouyang, X., & Du, K. (2019). Economy-wide estimates of energy rebound effect: Evidence from China's provinces. *Energy Economics*, *83*, 389–401. https://doi.org/10.1016/j.eneco.2019.07.027.

Zink, T., & Geyer, R. (2017). Circular economy rebound. *Journal of Industrial Ecology*, *21*(3), 593–602. https://doi.org/10.1111/jiec.12545.

The digital transformation – An ongoing process

Economies today rely on digital and internet technologies to an extent and in ways people could not have anticipated even a few years ago, and we are probably only at the beginning of this "digital transformation". Vial (2019) considers it "a process where digital technologies create disruptions triggering strategic responses from organisations that seek to alter their value creation paths while managing the structural changes and organisational barriers that affect the positive and negative outcomes of this process". Mergel, Edelmann, and Haug (2019) provide an understanding of digital transformation in public administrations, and Verhoef et al. (2019) "identify three stages of digital transformation: digitisation, digitalisation, and digital transformation". Digitisation is "the encoding of analog information into a digital format", digitalisation describes how "digital technologies can be used to alter existing business processes", and digital transformation "describes a company-wide change that leads to the development of new business models". Fig. 13.1 depicts these three stages of digital transformation.

The digital-enabling infrastructure including the Internet of Things (IoT), e-commerce with all digitally-ordered, digitally-delivered, or platform-enabled transactions, and digital media and e-commerce are, thus, undoubtedly part of today's digital economy, comprising all economic effects of digitisation (see p. 6f in Barefoot, Curtis, Jolliff, Nicholson, & Omohundro, 2018).

There are estimates for the United States (U.S.), showing an annual average growth rate of the digital economy of 5.6% per year from 2006 to 2016 compared to 1.5% growth in the overall economy, with the digital economy, comprising all economic affects of digitisation, accounting for 6.5% of gross domestic product (GDP) in current U.S.-Dollar (USD) in 2016 (Barefoot, Curtis, Jolliff, Nicholson, & Omohundro, 2018).

It is not difficult to understand that the transformations of the economy to a digital economy on the one hand and to a circular economy on the other overlap, perhaps enhance each other, perhaps impede each other in a variety of ways. For sure, today's approaches to a circular economy would not be possible without the digital

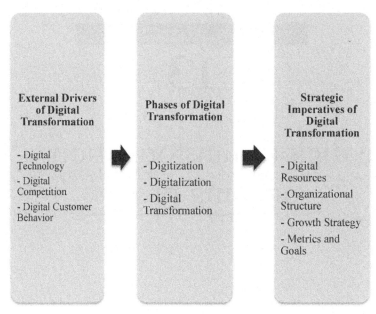

FIG. 13.1   Flow model of the digital transformation.Drivers, phases, and imperatives of digital transformation. © *From Verhoef, P. C., Broekhuizen, T., Bart, Y., Bhattacharya, A., Qi Dong, J., Fabian, N., & Haenlein, M. (2019). Digital transformation: A multidisciplinary reflection and research agenda.* Journal of Business Research. *https://doi.org/10.1016/j.jbusres.2019.09.022 (web archive link).*

transformation. The fast-changing digital environment provides a difficult framework for establishing a sustainable development – a critical issue for a circular economy. ElMassah and Mohieldin (2020) examine possible impacts of the digital transformation on achieving the sustainable development goals (SDG). In this context, Lobschat et al. (2019) come to the conclusion that "digital technologies that assist in human decision making or make decisions autonomously need to be subject to moral norms and ethical considerations similar to those that apply to humans", thus referring to "corporate digital responsibility".

In view of the continuous stream of often disruptive changes regarding digital technologies, the following remarks can only provide a brief and incomplete insight into the rapidly developing field of the digital transformation in its relation to implementing a circular economy.

13.1 Circular economy and digital economy – A close relationship

There is no doubt that, for example, the implementation of the waste hierarchy would not be possible without immense support form information and communication technologies (ICT). This refers, in particular to all dimensions of the sharing economy, which has fundamentally changed the way people share and conduct transactions by means of digital platforms and "platform mediation" Sutherland and Jarrahi (2018). Car sharing, or bicycle sharing and other business models in this context such as the "uberisation" would not be possible without ICT. The same is true for the various online platforms for buying and selling commodities, in particular also used commodities. Modern ICT tools thus allow the creation of virtual market places, enabling and coordinating decentralised buying and selling decisions. Reuse at the product level, an important part of the waste hierarchy, already got a new perspective with a great potential of further development (see also Section 13.3).

But also collecting, sorting, reuse at the material level, and recycling are more and more affected by the digital transformation, also by technologies based on artificial intelligence (AI): smart waste bins, smart trucks for waste collection, and robotics sorting waste can help to effectively manage different waste streams. Tsiliyannis (2019), adopting a systems perspective, quantifies waste monitoring that "can lead to efficient policies and implementation of proactive measures for waste reduction". Sarc et al. (2019) present and discuss the role of digitalisation and the use of robotic technologies in waste management, including business models, "to make treatment of waste more efficient". In this context, Jian, Xu, and Zhou (2019) explain and investigate with "Internet + recycling" a new and emerging collection mode for waste electric and electronic equipment (WEEE), which "is booming in conjunction with widespread internet use in China". This collaborative collection system is based on cost sharing mechanisms to promote recycling. Interestingly, Sujata, Khor, Ramayah, and Teoh (2019) outline the role of social media to affect recycling behaviour, thereby referring to aspects of behavioural economics. There is the goal to establish certain aspects of recycling as a "social norm" (see Section 8.2.3), also by means of the social media: "recycling behaviour among the wider community can be cultivated through the influence of social media and NGOs intervention".

Regarding the more technical parts of recycling, ICT can efficiently provide information on quantity and quality of secondary commodities, and can therefore help to improve material reusing, thereby perhaps also reducing the rebound effect (see Section 12.1.3). And, of course, the information about the materials used in certain commodities is of high relevance for a future design for environment (DfE) and also for further developing more efficient recycling methods.

Thus, there is a demand of further tools from the digital economy for implementing circular economy strategies, in particular for a sharing economy, for DfEs, for better processing secondary commodities as true substitutes for primary commodities, and for collecting, sorting and recycling waste, of course. In this sense, the implementation of a circular economy is not only dependent on tools from the digital economy, but provides also many opportunities for the digital economy.

A great economic potential of digitalisation is also seen in all matters regarding energy efficiency with a broad spectrum of innovative digital efficiency services. However, with respect to the transport sector, Noussan and Tagliapietra (2020) conclude "that in order to fully exploit the advantages of digitalisation, proper policies are needed to support an efficient and effective deployment of available technologies through an optimised and shared use of alternative transport options in the transport sector". Thus, the digital and the real economy have to adapt to each other – of special relevance for implementing a circular economy.

In view of the efforts to mitigate climate change as part of circular economy strategies, this topic will remain on the agenda of many countries, likely attracting increasing attention in the near future.

Looking at sustainable development as the other pillar of the circular economy, the "digital revolution" is considered a key enabler of sustainable development, "constituted by ongoing advances in artificial intelligence, connectivity, digitisation of information, additive manufacturing, virtual reality, machine learning, blockchains, robotics, quantum computing and synthetic biology", as the United Nations (UN) are arguing in their Global Sustainable Development Report. However, the UN is also pointing at some possible downsides of the digital transformation, "whether the digital revolution as a self-evolving evolutionary process that has generated huge global monopolies is even amenable to social steering" (see UN, 2019, p. 118). On the other hand, ElMassah and Mohieldin (2020) find that the digital transformation can "boost" sustainable development strategies.

This directs the attention to another area, gaining increasing attention and societal importance: smart cities and their role in a circular economy.

13.1.1 Smart cities for a circular economy

Although this is not the place to talk at length about smart cities, sustainability and digitalisation are nevertheless of relevance for the mega trend "Smart Cities" (see also the remarks in Section 9.2), which is closely related to "Circular Economy in Cities". Applications of new and advanced technologies, in particular ICT, are important for a smart city. They are meant to establish and enable fast and reliable connections between the inhabitants of a city, to strengthen existing and open additional channels for innovative activities, in order to reach the societal goals of a smart city, including a sustainable development and, as occasionally indicated, a development towards a circular economy (Wiesmeth, Häckl, & Schrey, 2020).

Applications of new and advanced technologies, in particular, ICT, to cities are important for a smart city, and business opportunities for the next years seem to be tremendous (Frost & Sullivan, 2019). The concept of a smart city itself, however, goes far beyond technical and technological issues. It is understood that a smart city is obviously dependent on its inhabitants, in particular on their willingness to accept these new framework conditions, on their willingness to bring themselves in with their creativity for the most important goal of a smart city, the sustainable development. This points then to the creativity of a city, to the "creative class", and to framework conditions enabling and enhancing this creativity, which is recognised as a key driver (Albino, Berardi, &

Dangelico, 2015; Komninos, 2011; Thuzar Moe, 2011). It is here that the concepts of a smart city and a circular economy meet in the common goal of establishing a sustainable development.

Similar to the implementation of a circular economy, the important question is to get this process, the development towards a smart city, started with a fundamental change in the mindset of the people – similar to the implementation of a circular economy. Without support from a tangible part of a city's inhabitants, the intended sustainable development risks to fail, both for a smart city and a circular economy. Whether and to what extent digital technologies can help to achieve this goal, to establish or modify social norms for sustainability, remains to be seen. In a recent ATG Access survey 68% of the respondents in the UK "do not know what a smart city is or how the concept can benefit urban residents". This people can hardly be reached by social media. Thus, the question remains, how to integrate individuals in all these efforts to establish a smart city – or to implement a circular economy?

Nevertheless, there could be the possibility to break habits, to establish new habits of relevance for implementing a circular economy by means of the digital economy. It will for sure expand its influence in the societies, perhaps with respect to enforcing traditional societal path dependencies, perhaps modifying them in view of the sustainable development goals. Both Jian et al. (2019) and Sujata et al. (2019) point to the potential of the digital transformation in this regard.

These considerations, in addition to others, show that digital technologies are essential for implementing a circular economy or a smart city, but whether they are amenable to "social steering", remains still somewhat unclear and open. The following subsection deals with some problematic aspects of the digital transformation in the context of implementing a circular economy.

13.2 Circular economy and digital economy – Various challenges

Among a variety of issues, which are signalling a potential controversy between the goals of a circular economy and aspects of the digital economy, is the rapid development of e-commerce in all its facets: online shopping, food delivery, the digital transformation of logistics, but also ride-sharing through transportation network companies such as Uber, Yandex and others (see also the case study in Section 13.3). These developments are made possible by a digital transformation of logistics with a "disruptive potential": lower entrance barriers allow startups to enter the market with new business models based on sophisticated platforms. Sucky and Asdecker (2019) point to the fact that these new business models are often not dependent on large investments in car pools, transport pools, or warehouse capacities.

E-commerce has grown tremendously in recent years. In Germany, for example, according to a survey of statista.de, sales from business to customers will soon reach 60 billion € per year, after some 20 billion € in 2010. The obvious conclusion is that online shopping, including express delivery, is simply convenient for many customers.

But what are the environmental issues associated with online shopping? There are indications that it, perhaps, reduces trips to the shopping districts, thereby reducing greenhouse gas emissions. The calculations of Smidfelt Rosqvist and Winslott Hiselius (2016), for example, indicate "that the predicted increase in online shopping behaviour together with the predicted increase of the Swedish population in 2030 would give a 22% decrease in CO_2 emissions related to shopping trips compared to 2012". There is, however, the increasing quantity of packaging waste coming along with this development, in particular plastic packaging. Duan, Song, Qu, Dong, and Xu (2019) provide insight into the associated challenges in China,

with plastic packaging materials possibly containing chemical residues from pesticide applications. Plastic packaging is also an issue in Germany, contributing to the high level of packaging consumption. Coffee to go and shrink-wrapped food, meanwhile also available for order and delivery on online platforms, are further adding to the packaging consumption – also some kind of a rebound effect of the digital transformation.

In general, beyond the perceived threats from cyberattacks or even cyber-warfare, it seems not yet to be clear, whether and to what extent digitalisation in the form of personalised advertisement, instant delivery, virtual shopping assistants and contactless payment lead or, rather, may lead to greater or less pollution. At least the potential to reduce damages to the environment is visible, although it may take a while, till these opportunities can be fully realised.

The position of the German Environment Agency (UBA) refers to some of the potential effects of digitalisation on consumption. According to the studies of the UBA, ordering online in rural areas often saves CO_2 emissions, because the collection of many consignments by the postal service replaces consumer car trips to go shopping. The situation is different regarding instant or same-day delivery service, because consignments can no longer be delivered together.

Also, using more digital technologies in collecting, sorting and recycling waste is necessary and important. But, on the other hand, the societal path dependencies addressed in the last chapter are waiting around the corner. The digital transformation – without having any intentions to this regard – may further delay waste prevention, which is often and in many environmental regulations on waste management considered the most important issue, leading the waste hierarchy (see also Section 9.3.2). Thus, one should carefully observe the development of the digital transformation and its effects on the implementation of a circular economy.

In general, the digital economy is in a state of flux, with the consequence that any interactions – and there are many – with the implementation of a circular economy, are a priori difficult to judge from an environmental point of view. For sure, and to repeat it: although circular economy activities were initiated some 30 years ago, long before the digital transformation, current circular economy strategies are unthinkable outside the digital framework. On the other hand, however, there are quite a few features of the digital economy, which seem to reverse or at least threaten various efforts to implement a circular economy.

These considerations make, once more, clear that implementing a circular economy is an ongoing process, which will always have to respect new scientific and technological developments, and which will always have to motivate further innovations. But some of these innovations need not directly be in favour of the circular economy strategies. The case study in the following section, concluding Part III, poses some questions on e-commerce in this regard.

13.3 Case study: E-commerce and circular economy

This is not just a traditional case study, exploring and analysing some development, which happened some time ago. It is rather an attempt to address various issues and ask some questions, which likely require some closer attention in near future. Right now it is, in view of the rapid development of the digital transformation, too early for a profound analysis of its persistent positive and/or negative impacts on the environment, and of its interaction with implementing a circular economy.

13.3.1 Critical aspects of the sharing economy

Innovative digital platforms for a variety of purposes, including the sharing economy, have been in existence for some time. How do these "sharing" activities interact with the implementation of a circular economy? Regarding the concept, Hossain (2020), reviewing the literature, mentions "extremely varied and constantly changing" practices of the sharing economy, and refers to a definition proposed by Muñoz and Cohen (2017): a sharing economy is "a socio-economic system enabling an intermediated set of exchanges of goods and services between individuals and organisations which aim to increase efficiency and optimisation of sub-utilised resources in society". This points to an important aspect of a circular economy: to save resources by employing otherwise underused facilities and technologies.

The sharing economy is of particular relevance in the accommodation and transportation sectors. Airbnb and Uber are prominent examples, which are, together with others, still under scrutiny in many countries. The discussion focuses on privacy and security issues, on consequences for traditional competitors, but yet not so much on environmental issues.

Regarding the accommodation and the transportation sectors, the following environmental issues and questions might be of relevance:

- To what extent do these new accommodation facilities induce increased vacancies in traditional hotels? Is one underused facility (private homes) just displacing another one (hotels)? If this displacement is significant, then the environmental effect of this part of the sharing economy on a circular economy seems to be small – with large economic effects, though.
- If, however, these new accommodation possibilities help to attract additional customers due to lower prices etc., then the question arises, whether increased travel activities cause additional environmental effects. In this case, the more efficient usage of facilities would be accompanied by a rebound effect (see Section 12.1).
- Similarly, to what extent do these additional transportation opportunities displace other

traffic? If this displacement is significant, then this might perhaps be a signal for a smaller number of private cars in the future. In the short-run, however, traditional taxi companies might suffer from this development.

- In case there is no displacement, the additional opportunities of transportation tend to produce additional traffic with additional cars. Thus, the positive environmental effects are limited in this case, and consumers seem to be trapped in the Tragedy of the Commons.

These questions represent somewhat extreme positions, but this helps to understand the underlying environmental issues. Preliminary results to some of these questions exist (see BMWi, 2018; Hossain, 2020), but as these new technologies are still developing, empirical results differ with no final answer yet. Of course, there are positive economic impacts: job opportunities and higher incomes. Weighing these rather clear economic effects in contrast to some fairly unclear environmental issues might not be an easy thing to do. Optimal framework conditions should, however, allow a positive development in both areas and, thus, modify obvious societal path dependencies.

13.3.2 Critical aspects of online shopping

The consistent growth of e-commerce has modified traditional shopping behaviours in a variety of ways – with environmental consequences. There are, first of all, the selling and buying platforms, eBay among them. These platforms certainly extend reusing activities and are thus in line with circular economy strategies. Questions might arise, however, to the volume of these sales.

Many e-commerce activities refer to online shopping, which increased tremendously in recent years. In this case some side-effects are of environmental relevance, pointing in particular to an increased volume of packaging material and, thus, packaging waste, to perhaps increased transportation activities, and to disposal of returned commodities. Jaller and Pahwa (2020) draw attention to the necessity of "managing the urban freight system, including delivery services and operations, to foster a more sustainable urban environment".

The following environmental issues and questions arise, again among others, in the context of online shopping:

- To what extent and in which ways is the shopping behaviour changing, given the new technologies? Kahlenborn, Keppner, Uhle, Richter, and Jetzke (2019) consider various aspects and point to the focus on all kinds of services in the context of travelling. But this refers to 2015 and meanwhile quite a few additional online services gained acceptance, in particular food deliveries etc. To what extent are these developments in accordance with circular economy strategies?
- Online shopping results in increased packaging waste. According to the German Environment Agency (UBA), plastic packaging increased in Germany by 79% between 2000 and 2017. To what extent are developments like this a consequence of online shopping? What is the role of the Tragedy of the Commons in this context (Section 7.1.1)?
- Part of the commodities bought online and returned thereafter are discarded. What are the economic reasons and what are the environmental issues associated with this observation? See also Sucky and Asdecker (2019) for a first environmental assessment of the returns of online ordered commodities.
- What are the probably positive environmental aspects of buying and selling platforms? Do increased deliveries or rush deliveries cause significant negative environmental effects?

Again, these are just a few questions in this context. Preliminary investigations exist, but due to the rapid development of online shopping activities, it is advisable to monitor them carefully (see, for example, Kahlenborn et al., 2019; Jaller & Pahwa, 2020). Of course, also in this case it seems necessary to weigh and compare economic and environmental effects. The optimal outcome would allow positive effects in both areas. Appropriate framework conditions are necessary to achieve such an outcome.

So far this "case study" focusing on possible future developments of e-commerce activities with relevance for implementing a circular economy.

References

Albino, V., Berardi, U., & Dangelico, R. M. (2015). Smart cities: Definitions, dimensions, performance, and initiatives. *Journal of Urban Technology*, 22(1), 3–21. https://doi.org/10.1080/10630732.2014.942092.

Barefoot, K., Curtis, D., Jolliff, W., Nicholson, J. R., & Omohundro, R. (2018). *Defining and measuring the digital economy. Working Paper: B ureau of Economic Analysis.* Vol. 3 (15). United Nations. Retrieved from (2018). https://www.unece.org/fileadmin/DAM/stats/documents/ece/ces/ge.20/2019/mtg1/Item_7_Defining_and_Measuring_the_Digital_Economy.pdf.

BMWi (2018). *Sharing Economy im Wirtschaftsraum Deutschland [in German]*. Bundesministerium für Wirtschaft und Energie (BMWi). Retrieved from (2018). https://www.bmwi.de/Redaktion/DE/Publikationen/Studien/sharing-economy-im-wirtschaftsraum-deutschland.pdf?__blob=publicationFile&v=3.

Duan, H., Song, G., Qu, S., Dong, X., & Xu, M. (2019). Post-consumer packaging waste from express delivery in China. *Resources, Conservation and Recycling*, 144, 137–143. https://doi.org/10.1016/j.resconrec.2019.01.037.

ElMassah, S., & Mohieldin, M. (2020). Digital transformation and localizing the sustainable development goals (SDGs). *Ecological Economics*, 169, 106490. https://doi.org/10.1016/j.ecolecon.2019.106490.

Frost & Sullivan (2019). *Smart cities. Retrieved from (2019)*. https://ww2.frost.com/wp-content/uploads/2019/01/SmartCities.pdf.

Hossain, M. (2020). Sharing economy: A comprehensive literature review. *International Journal of Hospitality Management*, 87, 102470. https://doi.org/10.1016/j.ijhm.2020.102470.

Jaller, M., & Pahwa, A. (2020). Evaluating the environmental impacts of online shopping: A behavioral and transportation approach. *Transportation Research Part D: Transport and Environment*, 80, 102223. https://doi.org/10.1016/j.trd.2020.102223.

Jian, H., Xu, M., & Zhou, L. (2019). Collaborative collection effort strategies based on the "internet + recycling" business model. *Journal of Cleaner Production*, 241, 118120. https://doi.org/10.1016/j.jclepro.2019.118120.

Kahlenborn, W., Keppner, B., Uhle, C., Richter, S., & Jetzke, T. (2019). *Konsum 4.0: Wie Digitalisierung den Konsum verändert [in German]*. Retrieved from: German Environment Agency (UBA).(2019). https://www.umweltbundesamt.de/sites/default/files/medien/1410/publikationen/fachbroschuere_konsum_4.0_barrierefrei_190322.pdf.

Komninos, N. (2011). Intelligent cities: Variable geometries of spatial intelligence. *Intelligent Buildings International*, 3(3), 172–188. https://doi.org/10.1080/17508975.2011.579339.

Lobschat, L., Mueller, B., Eggers, F., Brandimarte, L., Diefenbach, S., Kroschke, M., & Wirtz, J. (2019). Corporate digital responsibility. *Journal of Business Research*. https://doi.org/10.1016/j.jbusres.2019.10.006.

Mergel, I., Edelmann, N., & Haug, N. (2019). Defining digital transformation: Results from expert interviews. *Government Information Quarterly*, 36(4), 101385. https://doi.org/10.1016/j.giq.2019.06.002.

Muñoz, P., & Cohen, B. (2017). Mapping out the sharing economy: A configurational approach to sharing business modeling. *Technological Forecasting and Social Change*, 125, 21–37. https://doi.org/10.1016/j.techfore.2017.03.035.

Noussan, M., & Tagliapietra, S. (2020). The effect of digitalization in the energy consumption of passenger transport: An analysis of future scenarios for Europe. *Journal of Cleaner Production*, 258, 120926. https://doi.org/10.1016/j.jclepro.2020.120926.

Sarc, R., Curtis, A., Kandlbauer, L., Khodier, K., Lorber, K. E., & Pomberger, R. (2019). Digitalisation and intelligent robotics in value chain of circular economy oriented waste management – A review. *Waste Management*, 95, 476–492. https://doi.org/10.1016/j.wasman.2019.06.035.

Smidfelt Rosqvist, L., & Winslott Hiselius, L. (2016). Online shopping habits and the potential for reductions in carbon dioxide emissions from passenger transport. *Journal of Cleaner Production*, 131, 163–169. https://doi.org/10.1016/j.jclepro.2016.05.054.

Sucky, E., & Asdecker, B. (2019). *Digitale Transformation der Logistik – Wie verändern neue Geschäftsmodelle die Branche?*: (pp. 191–212) Wiesbaden: Springer Fachmedien Wiesbaden. https://doi.org/10.1007/978-3-658-22129-4_10.

Sujata, M., Khor, K.-S., Ramayah, T., & Teoh, A. P. (2019). The role of social media on recycling behaviour. *Sustainable Production and Consumption*, 20, 365–374. https://doi.org/10.1016/j.spc.2019.08.005.

Sutherland, W., & Jarrahi, M. H. (2018). The sharing economy and digital platforms: A review and research agenda. *International Journal of Information Management*, *43*, 328–341. https://doi.org/10.1016/j.ijinfomgt.2018.07.004.

Thuzar Moe (2011). Urbanization in Southeast Asia: Developing smart cities for the future? In *Regional Outlook Southeast Asia 2011–2012*. https://doi.org/10.1355/9789814311694-022.

Tsiliyannis, C. A. (2019). The cycle rate as the means for real-time monitoring of wastes in circular economy. *Journal of Cleaner Production*, *227*, 911–931. https://doi.org/10.1016/j.jclepro.2019.04.065.

UN. (2019). *Global sustainable development report 2019: The future is now – Science for achieving sustainable development.* New York: United Natiuons. Retrieved from (2019). https://sustainabledevelopment.un.org/content/documents/24797GSDR_report_2019.pdf.

Verhoef, P. C., Broekhuizen, T., Bart, Y., Bhattacharya, A., Qi Dong, J., Fabian, N., & Haenlein, M. (2019). Digital transformation: A multidisciplinary reflection and research agenda. *Journal of Business Research.* https://doi.org/10.1016/j.jbusres.2019.09.022.

Vial, G. (2019). Understanding digital transformation: A review and a research agenda. *SI: Review Issue*, *28*(2), 118–144. https://doi.org/10.1016/j.jsis.2019.01.003.

Wiesmeth, H., Häckl, D., & Schrey, C. (2020). *Smart institutions: Concept, index, and framework conditions*: (pp. 1–33). Cham: Springer International Publishing. https://doi.org/10.1007/978-3-030-15145-4_7-1.

PART IV

Features of environmental policies

The implementation of a circular economy is closely linked to the development of appropriate technologies for reasons, discussed in the chapters of Part II and Part III: the concept of the circular economy is embedded in its technological framework and there are various challenges in the allocation of environmental commodities. However, as the discussion has shown, even with a focus on technologies, guidelines, and various tools, policies are needed to address a wide range of problems arising from a lack of information and information asymmetries, to set and adapt environmental standards, to help consumers and producers to comply with these standards, but also to deal with issues such as rebound effects and technological and societal path dependencies.

This part of the book considers first once more aspects of allocating environmental commodities: laissez-faire, technology-guidance, command-and-control policies, and augmenting the market system are the different approaches, which are used here and there. It then formulates structural requirements of environmental policies that integrate the characteristics of a market system into the policies. In view of the information issues, these properties are important for implementing a circular economy and form the basis for a later examination of holistic policy approaches.

Environmental standards replace the generally unknown efficient levels of environmental commodities. Insufficient knowledge also highlights the challenges of adapting these standards for further scientific knowledge. All approaches – the application of the "best available technique", dynamic adaptations, and the use of public research institutions – have shortcomings regarding the incentives they provide. A case study on dynamic emission standards for CO_2 and NO_x for passenger vehicles in the European Union reveals details of this dilemma.

The then following chapter deals with market-oriented policy tools which have characteristics of the market mechanism, in particular decentralised decision-making. These tools include the pollution tax, the polluter pays principle, and tradable emission certificates. The broad outlines of these tools, the advantages and disadvantages for implementing a circular economy are discussed – with "avoidance possibilities" playing a special role. Of course, the "Emission Trading System" of the European Union is also presented with its main features. The chapter contains moreover a brief digression to the Coase Theorem, which deals with the possibility of direct negotiations on environmental issues when property rights have been assigned. Voluntary contributions and flexible, information-based environmental policies touch on aspects of behavioural environmental economics.

The basic tools can be combined to holistic policies, needed to achieve the goals of circular economy strategies which depend on the appropriate integration of various groups of stakeholders. This chapter begins with a brief overview of the concept of "Integrated Waste

Management", which has initiated the development of holistic approaches to waste management. This first step, however, was more concerned with the integration of various technical parts of waste management. The concept was then further developed to the now widely known and widely accepted principle of "Extended Producer Responsibility" (EPR), which is part of more than 400 EPR policies. Some critical aspects of the EPR principle and its implementation lead to the consideration of "Integrated Environmental Policies" (IEP) with the market-oriented structural requirements for an environmental policy which have been reformulated as "Constitutive Elements". Thereafter, the discussion returns to the implementation of the EPR principle in the context of an IEP. The focus is on waste management, more precisely on general aspects of collection systems and "Producer Responsibility Organisations" (PRO).

Due to its relation to the EPR principle and, in particular, its importance for the implementation of a circular economy, the waste hierarchy requires and deserves special attention. The last chapter therefore turns to the economics of the waste hierarchy, and its perception in practice. Particular attention is given to waste prevention, the "forgotten child". In waste management practice, this "priority goal" fades into the background. This chapter attempts to understand this observation and to provide possible explanations, also through societal path dependencies.

A similar observation applies to "reuse" as another important part of the waste hierarchy. As indicated in Section 12.2, path dependencies may lead to results related to the reuse of certain commodities which are ecologically dubious, at least to some extent. This chapter therefore examines also this topic – together with the "recycling" of waste commodities, which is linked to various interesting aspects.

In summary, the chapters of Part IV of the book describe and highlight components of more advanced policies, and explain relevant characteristics of holistic policies. These holistic approaches are then used in Part V to design concrete environmental policies, which combine technology-guidance with economy-guidance for the implementation of a circular economy.

14

Environmental policies for implementing a circular economy

"Environmental policy is primarily concerned with how best to govern the relationship between humans and the natural environment for the benefit of both" (Jordan, Smelser, & Baltes, 2001). This quote expresses quite well both the objectives of environmental policy and the challenges involved. Insufficient information characterises all these aspects of environmental policies and can temporarily lead to below-optimal results.

Indeed, the conclusions of Chapter 5 on the effectiveness of current and recent environmental policies in circular economy strategies are not overly optimistic: waste prevention or waste reduction is still a major issue, the reuse of old and the recycling of waste products is characterised by societal path dependencies, and the technological framework of a circular economy interacts with consumers and producers in many ways – critically influenced by uncertainties and information asymmetries.

This chapter therefore deals again – from a different point of view – with some aspects of the allocation of environmental commodities, which are addressed elsewhere in this book,

formulates certain requirements that are relevant for the implementation of a circular economy and compares these different approaches. The presentation includes "laissez-faire", "technology-guidance", "command-and-control", and "augmented market system", which can be part of a holistic policy. This discussion contributes to the use of environmental policies with appropriate structural properties as a tool for allocating environmental commodities, as an instrument for merging "economy-guidance" and "technology-guidance" in the context of the implementation of a circular economy.

14.1 How to allocate environmental commodities?

The following approaches to the allocation of environmental commodities, which complement the discussion in Chapter 7, played or continue to play a role in practical contexts. Each of them is characterised by certain properties that are important for the implementation of a circular economy. The challenge then is to combine these "tools" appropriately in order to achieve the objectives of circular economy strategies through a holistic policy.

14.1.1 Laissez-faire

The policy regime of laissez-faire largely postulates a separation of the government from the economic sector. A. Smith's metaphor of the "invisible hand", which leads decentralised economic decisions to a common social goal, refers to such a liberal attitude (see Section 6.3). Of course, laissez-faire can also be extended to the allocation of environmental commodities. It is understood that in industrialised, liberal societies a high level of environmental awareness prevents people from polluting the environment excessively – on the contrary, enlightened citizens voluntarily protect the environment (see also Section 16.4).

In this context, Immordino, Pagano, and Polo (2011) argue that laissez-faire can indeed be of advantage: if social harm from innovation activities is sufficiently unlikely, then penalties can reduce or even prevent these activities, which might otherwise lead to benefit for the society. Similarly, Bezin and Ponthière (2019) show that the introduction of a quota in a certain environmental context may induce the Tragedy of the Commons – leading to a situation worse than at laissez-faire. The development of the refillable quota in Germany (see the case study in Section 7.3) demonstrates such a situation: an attempt to keep the combined refillable quota above a certain threshold level triggered the Prisoners' Dilemma. It remains questionable whether a laissez-faire environmental regulation can be "guaranteed" by the threat of a potentially more restrictive regulation.

This implies that laissez-faire in an environmental context is likely characterised by the mechanisms of the Tragedy of the Commons and the Prisoners' Dilemma, at least as long as there is no perceived feedback – even if there

is a high level of environmental awareness. The situation in Saxony some 300 years ago (see Section 6.1) points to this fact: only when people personally experienced the consequences of the deforestation due to extensive mining activities did they start to look for alternatives. For many environmental contexts, including global warming, there are, however, tipping points which need to be considered before it is too late.

In consequence, laissez-faire regimes can be meaningful in environmental contexts with low and/or unlikely risks on the one hand, and likely high social gains on the other, thus stimulating the creative potential of a society. They can, as we shall see, play some role in environmental activities for which there is some guidance through a "social norm". This refers, for example, to the collection and segregation of waste in various countries.

14.1.2 Technology-guidance

Due to roots of the circular economy in industrial ecology (see Section 10.1) and due to the general challenges associated with allocating environmental commodities, the technological framework of a circular economy significantly affects the transformation of the economy into a circular economy. According to proponents of industrial ecology, this means to transfer the task of solving these allocation problems to the managers of the companies, who should act as "an important agent for accomplishing environmental goals", and who should therefore act as a policy-maker rather than as a policy-taker. Industrial ecology "is seen by many as a means to escape from the reductionist basis of historic command-and-control schemes" (Lifset & Graedel, 2002, p. 8).

There is no doubt that environmental technologies have a major impact on the details of the road to a circular economy. However, there are a few points that need to be taken into account (see also Chapter 11).

First and foremost, managers of the companies have clear and legitimate business goals. These objectives do not always have to be in line with circular economy strategies. There could be vested interests: profitable recycling could encourage managers to expand these activities while neglecting waste prevention. Managers may then be reluctant to introduce a design for environment (DfE) if the impact on corporate profits is negative or not clear. Unlike in the situation of a regular market economy, these vested interests can distract managers from the objectives of a circular economy.

Moreover, the supply of these environmental technologies is dependent on environmental regulations, which are enacted and enforced by governments (see Section 10.2). However, export subsidies often disrupt the circular economy strategies, leading to possibly suboptimal technological path dependencies. Informational issues, such as the economic reasonableness of implementing certain technologies, play a role in this direction. And finally, there are issues such as rebound effects and societal path dependencies that characterise the complicated relationship between the technological framework and decisions of consumers and producers, asking also for guidance in addition to the development of technologies (see Chapter 12).

In view of these comments, a kind of laissez-faire regime based on the development of environmental technologies is not recommended. There are too many additional interests that interfere with the achievement of the objectives of a circular economy. Nevertheless, this technological environment remains of considerable importance in this context.

14.1.3 Command-and-control policies

The implementation of a circular economy is often associated with achieving certain targets or "standards", not exceeding maximum levels of emissions of certain hazardous substances, for example. If these targets refer to actions of individual consumers and producers, it just remains to monitor and control these emissions. This is then an example of a simple command-and-control policy: observing the limits specified by the targets represents the "command", whereas the authorities control the observance. Theoretically, "environmental efficiency" is guaranteed if the economic agents comply with the regulations.

However, the question remains, how to achieve these standards? This applies in particular to desirable developments related to the implementation of a circular economy, such as designs for environment (DfEs) or a stricter observation of the waste hierarchy and many others. In general, these "soft" goals need to be supported by appropriate policies, and the simplest approach seems to be a command-and-control policy: the regulations of the government prescribe or proscribe certain actions, and the regulations should help to reach the targets or other environmental objectives. The authorities monitor and control the behaviour of the consumers and the producers, and violations of the rules are typically sanctioned.

Command-and-control policies dominate environmental policies in countries all over the world. Prominent examples are the Clean Air Act in the United States (U.S.) and the Federal Immission Control Act of Germany, both dating back to the early 1970s, to the early days of the environmental movement.

There are various issues with these command-and-control policies. First and foremost, because of their rather simple structure these policies are abundant: there is a command, there is control, and violations are sanctioned – a procedure, which is usually considered fair and which is socially accepted. Then, these policies are relatively easy to implement and they promise and enable a rapid response to some environmental degradation. Moreover, policy-makers, bureaucrats and politicians, can prove to be active in environmental protection – an aspect that should not be underestimated.

Nevertheless, the current implementation of such policies shows that their actual environmental efficiency is often low in terms of achieving the environmental objectives. This becomes obvious, when looking at waste prevention, often declared as "the first priority of waste management, … " (EU, 2015) in many environmental regulations, but with limited success so far (see Chapter 5). In addition to that, there are "avoidance possibilities" that weaken the environmental efficiency of these (and other) policies. For example, if landfilling of chemically or biologically active waste is prohibited, exporting this waste to other countries could be an option, which is probably not in the interests of policy-makers. In order to close these and other types of avoidance possibilities resulting from incomplete information, it is therefore necessary to adapt the policy accordingly, which requires time and effort.

What are possible reasons for these observations? Some of the examples of environmental objectives mentioned above involve or have to involve a large number of economic agents. This clearly applies to various aspects of waste management, in particular the prevention of waste and the collection and segregation of waste, among many others. However, due to mechanisms such as the Tragedy of the Commons and the Prisoners' Dilemma, classical command-and-control policies require the more or less continuous monitoring of the compliance of the economic agents with the environmental regulations. This is in general difficult, costly, and in most countries politically not feasible. Taylor (2000), considering incentives to minimise the generation of municipal solid waste, arrives at the conclusion that "the likelihood of command-and-control regulations being successfully implemented depends importantly on the social-psychological and economic incentives for waste minimisation provided in the regulations", and also Liao (2018) attests a limited effect of command-and-control instruments on the environmental performance of companies.

This implies that command-and-control policies can be used to control and monitor the behaviour of a small group of consumers or producers. For example, for large groups, all households or many companies, it is necessary to supplement the policies with "social-psychological and economic incentives" or to use different environmental policies. Nevertheless, command-and-control policies can, of course, be part of more complex, holistic approaches to environmental policies.

14.1.4 Augmented market system

In terms of performance, market systems often achieve widely accepted results. However, the unconstrained applicability of the market mechanism depends on certain framework conditions, which cannot be fully realised in a practical context. This refers to the requirements of perfect competition and to the existence of regular markets for each commodity. Especially the latter condition is not given for environmental commodities. Moreover, the situation is further complicated through issues such as perceived scarcity or perceived feedback, which may additionally interfere with establishing functional markets for environmental commodities (see Section 6.1).

Even if the market mechanism cannot be used directly as allocation mechanism for environmental commodities, there are features of the market system that are of relevance for the economic systems. Machlup (1974), referring to F. von Hayek's contribution to economics, points to "competition", but also to the integration of all individuals into the market system, allowing each one to profit from knowledge and the capabilities of others. For this reason, aspects of the market system should, if possible, be retained in the efforts to implement a circular economy.

Consequently, there is a certain necessity to augment the market system in order to take adequate care of environmental commodities. This issue of "completing the market system" (see

Section 7.1) is also the topic of Chapter 16, which refers to "market-oriented" environmental policy tools in more details.

Aspects of behavioural environmental economics are also coming into play (see Chapter 8), also with respect to augmenting the market system. It is necessary to "educate" environmental awareness and perceived feedback, to break old habits, to create new pro-environmental habits (Gneezy, 2019), to gain the participation of the individuals in all kinds of efforts to reduce environmental pollution. This refers in particular to the Tragedy of the Commons and the Prisoners' Dilemma, the mechanisms which interfere significantly with the allocation of environmental commodities.

More problematic in behavioural economics and in behavioural environmental economics, however, is the fact that generally valid results such as the "law of demand" in economics, are not available. This makes it more difficult to work with the results of behavioural economics in the context of the implementation of a circular economy (see also Shogren & Taylor, 2008). Nevertheless, these results, these "cases", could be helpful in gradually changing societal path dependencies.

An additional challenge for allocating environmental commodities in general and in an augmented market context in particular, arises from the increasingly international dimension of various environmental issues: the key challenge "lies in the ongoing search for a development which reflects the realities both in the industrialised and in the developing states" (Dolzer, Smelser, & Baltes, 2001). Indeed, differences in environmental awareness, in policy styles, and cultures must be observed to address international or even global environmental issues (Cocklin, Moon, & Kobayashi, 2020).

This international dimension invites behaviour known from the Tragedy of the Commons and the Prisoners' Dilemma. In particular, efforts to mitigate climate change are vulnerable to these mechanisms. The country, which first pursues a strict policy, could bear higher costs and other disadvantages for the competitiveness of its industry. On the other hand, it might have an advantage to be a first-mover if other countries follow suit. To sum up, the bottom line remains unclear to some extent.

In conclusion, augmenting the market system for allocating environmental commodities should be considered wherever this is possible. The reason is to integrate all individuals, all stakeholders with their knowledge, with their capabilities, into the solution of the allocation problems. This integration is indeed one of the important characteristics of the market system.

Given this review of possible basic allocation tools for environmental commodities including their advantages and their disadvantages, the following section formulates relevant structural requirements for environmental policies in the context of the implementation of a circular economy.

14.2 Structural requirements for environmental policies

The review of different approaches to the allocation of environmental commodities clearly shows that there is no a unique mechanism comparable to the market mechanism. Environmental policies for the implementation of a circular economy are therefore composed of various tools, which are presented and discussed in the following chapters. It is then necessary to combine these approaches appropriately to a holistic environmental policy. Fig. 14.1 illustrates this procedure. Concrete and detailed holistic policies will be drafted in the chapters of Part V.

This section covers fundamental structural requirements of these policies, which are relevant in all circumstances. These principles are derived from characteristics of the market

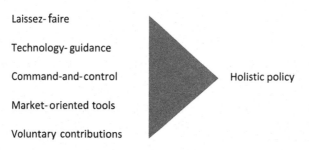

Allocating environmental commodities
for implementing a circular economy

Linking policy tools to a holistic policy with (market-affine) structural properties

FIG. 14.1 Allocating environmental commodities. Implementing a circular economy is dependent on special holistic policies. © *Source: Own Drawing.*

mechanism and transferred to the context of the allocation of environmental commodities. More details regarding these "Constitutive Elements" of an "Integrated Environmental Policy" are provided in Chapter 17 (see also Wiesmeth & Häckl, 2016).

14.2.1 Local framework conditions

The first aspect refers to the local situation that exists in a particular region or country. This "locality principle" has already been mentioned: there could be a different focus on the sustainable development goals (SDG), or a different focus on circular economy strategies – depending on the economic situation, geographic, climatic and demographic condition in a country.

For implementing a circular economy this means that the details of the various policies depend on these local conditions and have to be adapted to this conditions. Note that the local conditions are reflected in the characteristics of a market equilibrium.

14.2.2 Integrating the stakeholders

This second principle is also of the utmost importance. As we know already from a variety of contexts, implementing a circular economy is largely dependent on the knowledge, the information consumers and producers have. Allowing adequate use of this knowledge and information should therefore be one of the structural properties of an environmental policy. Note that this is also a characteristic of the market mechanism: each consumer, each producer is addressed to make decisions depending on the individual situation, depending on the individual knowledge.

In a market system, this task of integrating all stakeholders is achieved through the price system, which disseminates the required signals. As we know (see Chapter 7), this is not straightforwardly possible in the case of environmental commodities. This leaves the question, how to address the different groups of stakeholders. As indicated in the first section of this chapter, command-and-control policies may prove helpful in this respect, but only for small groups of stakeholders, such as producers of certain commodities. For large groups a command-and-control policy is, as experience shows (see Chapter 5), not the appropriate tool: it is simply not possible to monitor the behaviour of all individuals. And a laissez-faire approach is particularly vulnerable to mechanisms such as the Tragedy of the Commons.

Therefore, larger groups of stakeholders need to be addressed through appropriate framework conditions, which motivate most individuals to an environmentally friendly behaviour. The case study in Section 11.3 on the promotion of renewable energy sources in Germany provides an example, which, as it turned out, motivated too many economic agents – with further problematic consequences.

Of course, tools or approaches from behavioural environmental economics can also be helpful: in various situations, it may be necessary to raise environmental awareness, to increase perceived feedback from pro-environmental behaviour. This seems to be particularly important in the international environment with differences between participating countries.

14.2.3 Linking policy tools

This last basic structural property for an environmental policy for circular economy strategies is at the same time the most difficult one to implement in a practical context. Its origin can be found again in properties of the market mechanism. Due to the price signals available to all stakeholders involved, a change in one of the prices will have an impact with additional adjustments in other parts of the economy.

In order to mimic this characteristic feature of the market system through an environmental policy, it is, of course, necessary to link the various policy tools in an appropriate way. In more concrete terms, this means that certain consumer decisions must be linked to certain decisions made by producers in order to achieve a policy objective. This applies, for example, to the objective of waste prevention. Households must have incentives to prevent waste, but at the same time these incentives must be "transferred" to producers, for example to reduce packaging.

How these policy tools can be linked accordingly also depends in addition on the policy context. This is therefore the most complicated, but probably also the most interesting part of

designing environmental policies for the implementation of a circular economy. The various policies presented and explained in the chapters of Part V will show more aspects of the application of this principle in practical contexts.

In summary, environmental policies related to the implementation of a circular economy are complex constructs consisting of appropriate policy tools. The general objective is to transfer important features of a market system to these holistic policies.

The following chapters cover characteristics of other policy tools, which can be applied in these holistic environmental policies. Environmental targets or standards are often used in circular economy strategies.

References

Bezin, E., & Ponthière, G. (2019). The tragedy of the commons and socialization: Theory and policy. *Journal of Environmental Economics and Management*, 98, 102260. https://doi.org/10.1016/j.jeem.2019.102260.

Cocklin, C., Moon, K., & Kobayashi, A. (2020). *Environmental policy*: (pp. 227–233). Oxford: Elsevier. https://doi.org/10.1016/B978-0-08-102295-5.10788-7.

Dolzer, R., Smelser, N. J., & Baltes, P. B. (2001). *Environmental policy*: (pp. 4638–4644). Oxford: Pergamon. https://doi.org/10.1016/B0-08-043076-7/04496-X.

EU (2015). *Closing the loop – an EU action plan for the circular economy*. Retrieved from (2015). https://eur-lex.europa.eu/resource.html?uri=cellar:8a8ef5e8-99a0-11e5-b3b7-01aa75ed71a1.0012.02/DOC_1&format=PDF.

Gneezy, U. (2019). *Introduction. Incentives and behaviour change. In A. Samson (Ed.), The behavioural economics guide 2019* (p. VI). Retrieved from (2019). https://www.behavioraleconomics.com/the-be-guide/the-behavioral-economics-guide-2019/.

Immordino, G., Pagano, M., & Polo, M. (2011). Incentives to innovate and social harm: Laissez-faire, authorization or penalties? *Journal of Public Economics*, 95(7), 864–876. https://doi.org/10.1016/j.jpubeco.2011.01.011.

Jordan, A., Smelser, N. J., & Baltes, P. B. (2001). *Environmental policy: Protection and regulation*: (pp. 4644–4651). Oxford: Pergamon. https://doi.org/10.1016/B0-08-043076-7/04176-0.

Liao, Z. (2018). Environmental policy instruments, environmental innovation and the reputation of enterprises. *Journal of Cleaner Production*, 171, 1111–1117. https://doi.org/10.1016/j.jclepro.2017.10.126.

Lifset, R., & Graedel, T. E. (2002). Industrial ecology: Goals and definitions. In R. U. Ayres, & L. W. Ayres (Eds.), *A handbook of industrial ecology*. Cheltenham: Edward Elgar Publishing. https://doi.org/10.4337/9781843765479.00009.

Machlup, F. (1974). Friedrich Von Hayek's contribution to economics. *The Swedish Journal of Economics*, 76(4), 498–531. https://doi.org/10.2307/3439255.

Shogren, J. F., & Taylor, L. O. (2008). On behavioral-environmental economics. *Review of Environmental Economics and Policy*, 2(1), 26–44. https://doi.org/10.1093/reep/rem027.

Taylor, D. C. (2000). Policy incentives to minimize generation of municipal solid waste. *Waste Management & Research*, 18(5), 406–419. https://doi.org/10.1177/0734242X0001800502.

Wiesmeth, H., & Häckl, D. (2016). Integrated environmental policy: A review of economic analysis. *Waste Management & Research*, 35(4), 332–345. https://doi.org/10.1177/0734242X16672319.

15

Environmental standards

Environmental standards play an important role in all possible environmental contexts and policies. In principle, they replace unknown efficient levels of environmental commodities. After explaining the concept and the challenges of setting appropriate standards, the following section turns to the international dimension. This is particularly important because it concerns a possible "race to the bottom" in relation to environmental standards, which is sometimes referred to as a potential conflict between international trade and environmental protection.

In the light of new scientific knowledge, new technologies and/or increased environmental awareness, it is necessary from time to time to raise these standards. The following section introduces and examines various procedures for adapting standards. Due to information asymmetries, this is not easy to achieve, especially in the international context. A case study in the last section analyses emission standards for CO_2 and NO_x in the automotive sector of the European Union (EU).

15.1 Economic background

In the context and with the assumptions of a pure market economy, the "invisible hand" of the market mechanism leads to an equilibrium, which represents a Pareto-efficient allocation (see Section 6.3). The efficient level of the various commodities is the result of the decentralised decisions of consumers and producers, which are coordinated by the price system and lead to a market equilibrium.

With environmental commodities the situation is changing and a universal allocation mechanism is not available. Due to a lack of information, the efficient values of the environmental commodities directly linked to an appropriate level of environmental protection remain largely unknown.

Nevertheless, the allocation of environmental commodities, the solution of the basic allocation problems, requires precisely this kind of information. Environmental standards can be used as an approximation to the unknown efficient levels. Typical environmental standards therefore represent, among other things, the maximum permissible emissions of air, soil and water pollutants over a certain period of time or the highest possible concentrations of these pollutants. They provide for limit values for immission levels for certain pollutants, lay down minimum quotas for recovery and recycling activities, establish shares of renewable energy sources in the generation of electrical energy, quotas for

173

refillable drinks containers and maximum levels of average fuel consumption of the fleets of car manufacturers. The fact that environmental goods and bads are just two sides of the same medal justifies these formulations.

However, whether these standards are too high or too low compared to efficient levels remains unclear, although this is important for the economic system. The level of standards clearly affects production costs and other economic activities, such as transport. The question of the economic reasonableness of the standards (Section 11.1) must therefore be raised. A well-founded answer is not easy: the answer depends not only on technological issues, but also on the specific situation in one country, and a universal answer that applies to the situations in other countries is unfortunately, not available (see, for example, Ragossnig & Schneider, 2017 for the "right level of recycling plastic waste"). Nevertheless, the requirement of economic feasibility, or the economic reasonableness of a particular technology or standard is a common postulate in many environmental regulations (see Section 11.1), but without providing a precise definition. The concept developed in Wiesmeth (2011) results from a formal model and cannot be immediately translated into a practical context. Only a rather complex cost-benefit analysis enables empirical evaluations (see, for example, WHO, 2013; Holland, 2014 for cost-benefit analyses regarding air pollution).

Another issue deserves to be mentioned, despite the challenges of setting appropriate standards: the need to adopt appropriate standards makes a significant contribution to developing environmental policy into a separate policy area, with some implications for the relationship between environmental policy and other competing policy areas. In addition, choosing the "right" levels of the standards and adapting them to new scientific findings poses additional challenges. This also applies to the implementation of a circular economy, as Flynn and Hacking (2019) emphasise. They are convinced that "policy instruments like standards need to challenge existing neoliberal market relations rather than simply follow them", and Goulden, Negev, Reicher, and Berman (2019) mention that "standard-setting is affected by resources, pragmatism and economic considerations".

Environmental policy based on environmental standards is therefore structurally comparable to economic policy based on adequate growth rates, or monetary policy based on target inflation rates, or labour policy based on target unemployment rates, etc. There is not much left of the "harmony" between the environment and the economy, which manifests itself in the Pareto criterion, which provides a benchmark for the allocation of all commodities – including environmental commodities. In case of a critical decision, it need not always be the environment, which is on the winning side: an economic recession with rising unemployment usually supports opponents of stricter environmental policies, because of the direct feedback from loosing your job. The current Corona crisis is proof of this (see Section 25.4). Thus, the intrinsic relationship between environmental and economic issues, known from the theory of environmental economics (see Chapter 6), breaks down into a dichotomy between the environment and the economy – ultimately a consequence of information deficits (see also Wiesmeth, 2011, Section 4.2).

And countries integrated into the global trading system could extend their competition to environmental standards if there are serious risks to their economies. This, as already indicated, feeds fears of a race to the bottom (see, for example, Bhagwati, 1996; Weber & Wiesmeth, 2003).

15.2 Environmental standards in an international context

In a global context, the setting of environmental standards affects international trade in general and countries' trade policies. In view of the Porter hypothesis, high standards could stimulate technological innovation in the industrialised countries in particular (Rubashkina, Galeotti, & Verdolini, 2015). Moreover, the export of environmental technologies depends on sufficiently high standards in importing countries (Section 10.2), and developing countries could also benefit from higher environmental standards "as a strategy of international technology transfer": international manufacturers may be more likely to be encouraged to supply new technologies if countries thus keep promises as future markets for high technologies (Saikawa & Urpelainen, 2014). And also Bhattacharya and Pal (2010), analysing the "strategic nature of choice of environmental standards", conclude that equilibrium values of environmental standards tend to be higher in an open economy.

Although these arguments clearly support high standards, there have always been fears among groups of environmental activists that trade policies and the accompanying efforts to promote corporate competitiveness – in the sense of the Prisoners' Dilemma – could trigger a "race to the bottom" in terms of environmental standards.

In one of his many contributions to international trade, Bhagwati (1996) points to some potentially serious conflicts between free trade and the environment. Different environmental regulations, in particular different environmental standards, may affect the international competitiveness of local industry and therefore constitute one of the main objections to free trade. In 2018, there were 663 environment-related notifications and 1363 environment-related measures by members of the World Trade Organisation (WTO), a significant increase over the last years. Kohn (2003), for example, discusses the possibility that environmental standards for gasoline, as set out in the Clean Air Act (CAA) of the United States (U.S.), could be considered as barriers to trade by

countries that want to export to the U.S. below-standard qualities – and thus put a pressure on standards.

A "race to the bottom" in environmental standards is, according to Bhagwati (1996) one of the "genuine problems" pointing to a potentially serious conflict between free trade and the environment. It had been the topic of intense debate in the 1990s (see Weber & Wiesmeth, 2003 for a formal approach), now anew gaining interest and attention in the context of global efforts to mitigate climate change. Polanco, de Sépibus, and Holzer (2017), for example, analyse the impact the Transatlantic Trade and Investment Partnership (TTIP) could have on existing and future climate policies and laws.

The story behind such a race to the bottom is simple: in the face of global competition governments could try to improve competitiveness of their industry by alternating reductions in environmental standards. It is the concern of various environmental and social movements, such as the "Our World is not for Sale" network, that international trade could actually hinder efforts to reduce environmental degradation, including climate change. Kim and Wilson (1997) explore "the possibility of a race to the bottom, under which intergovernmental competition for mobile capital leads to inefficiently lax environmental standards". On the other hand, both Bhagwati (1996) and Weber and Wiesmeth (2003) in a formal model, point out the possibility that profits from trade could be used for additional measures to protect the environment. Levy and Dinopoulos (2016) analyse the impact of global environmental standards and trade-liberalisation policies. They achieve ambiguous results for global pollution and economic wellbeing.

Therefore, for the implementation of a circular economy, it is still important to take into account this international dimension of setting environmental standards with further potential consequences for environmental policies. There are aspects of a Prisoners' Dilemma: let other countries maintain their higher standards, our lower standards will help our industry. Here, too, behavioural environmental economics could help to increase the perceived feedback from environmental protection. This seems to characterise part of the negotiations on climate change.

15.3 Adjusting environmental standards

In many cases, the concrete values of environmental standards are justified by health concerns, scientific analyses or precautionary considerations, as well as by aspects of the international competitiveness of national industry. From time to time, standards are adjusted or need to be adjusted to reflect new insights into the risk potential of a product or specific production process. For similar reasons, more and more regulations are introducing additional standards, for example with regard to emissions of some air pollutants such as nitrous oxides or particulate matter, as in the Clean Air Programme of the EU.

But, how do you choose the right time and level to adjust a standard, such as making a standard dynamic? As these adjustments are usually associated with higher costs, firms will not always voluntarily contribute with their expertise, which in turn is generally not fully available to the government. Moreover, in export-oriented countries, adverse effects on the international competitiveness of the companies could also play a role. Some recent developments at EU level, mainly affecting car manufacturers, point to the problematic context of dynamic standards. The case study in the last section will point to the challenges associated with this at first glance simple issue.

In the following subsections, various practical attempts are presented and analysed to make environmental standards dynamic.

15.3.1 Best available techniques

One prominent method is to define in appropriate environmental regulations, in command-and-control policies, for example, the application of the "best available techniques" or the "best available control technologies" (see also Section 11.2). This refers to the Clean Air Act (CAA) of the U.S., and the Federal Immission Control Act (BImSchG) of Germany, so to speak the role models of command-and-control policy. These environmental regulations date back to the beginnings of the environmental movements in the 1970s, and provide a reference point for many of the regulations implemented thereafter.

The Clean Air Act: The CAA is the comprehensive federal law in the U.S. that regulates all kinds of air emissions. It authorises the Environmental Protection Agency (EPA) to protect public health and public welfare and to regulate emissions of hazardous air pollutants. Section 169 "Definitions" of the CAA provides a definition of the term "best available control technology":

> "The term best available control technology means an emission limitation based on the maximum degree of reduction of each pollutant subject to regulation under this Act emitted from or which results from any major emitting facility, which the permitting authority, on a case-by-case basis, taking into account energy, environmental, and economic impacts and other costs, determines is achievable for such facility through application of production processes and available methods, systems, and techniques, including fuel cleaning, clean fuels, or treatment or innovative fuel combustion techniques for control of each such pollutant ...".

It is important to note that in the case of the CAA the "permitting authority" is responsible for determining the relevant parameters for the best available control technology based on the "maximum degree of reduction" of the pollutants. Moreover, the focus is on emissions of hazardous air pollutants, somewhat different from the broader Federal Immission Control Act of Germany, which also has a stronger focus on precautionary measures.

The Federal Immission Control Act: The purpose of this Act is to ensure integrated prevention and control of any harmful environmental damage caused by emissions to air, water and soil by ensuring the involvement of the waste management sector, in order to achieve a high level of protection for the environment as a whole and the taking of precautions against any hazards, significant adverse effects and significant nuisances caused in any other way. Section 3 "Definitions" provides in Subsection (6) a definition of the term "best available technique":

> "Best available techniques as used herein shall mean the state of development of advanced processes, establishments or modes of operation which is deemed to indicate the practical suitability of a particular technique for restricting emission levels to air, water or soil, for guaranteeing installations safety or for preventing or reducing any effects on the environment with a view to achieving a high level of protection for the environment as a whole …".

A list of criteria for determining best available techniques is specified in the Annex to the Federal Immission Control Act. In determining best available techniques particular account must be taken of the cost and benefit of any measures considered and the principles of precaution and prevention, therefore also pointing to the question of "economic reasonableness" (see Section 11.1).

Of course, there is an issue with determining the "best available technique": these regulations could, at least occasionally, "motivate" engineers and scientists to reduce their efforts to develop innovative environmental technologies. For this reason, these regulations lead to the development and the application of new and environmentally more efficient technologies, but the process of updating techniques and technologies may be slower than necessary if the expected consequences for businesses are not unambiguously positive. Therefore, the implementation of a circular economy requires the appropriate framework conditions to improve innovation activities and successfully raise environmental standards. This will be further explored in Part V.

Interestingly, Nishitani and Itoh (2016) in their study on refrigerators on the Japanese retail market, find that environmentally friendly attributes of refrigerators, such as energy efficiency and the absence of chlorofluorocarbons allow a price premium. The authors refer to a reflection of environmental standards. In this case, the market situation in Japan with high environmental awareness and a high perceived feedback from the purchase of environmentally friendly electrical appliances obviously enables this design for environment (DfE). But this situation need not hold for other countries.

The requirement of a DfE, philosophically close to "best available techniques", is found in regulations for Extended Producer Responsibility (EPR) and is therefore quite common for many environmental regulations (see Chapter 17 for further details on this aspect).

15.3.2 Dynamic standards

Another method of adapting environmental standards is to raise them at the intervals specified in the regulations. This method, which is becoming increasingly important, usually applies, among other things, to the collection and recycling rate of various categories of household waste, to the share of electricity generated from renewable resources or to emissions of vehicles. Table 15.1 shows the dynamic increase in recycling targets in Germany.

A particular example shows some problematic issues associated with this procedure: the average CO_2 emissions of the fleets of car manufacturers. From 2020, new passenger cars will be allowed to emit an average of only 95 g CO_2

TABLE 15.1 Dynamic recycling targets in Germany.

Material Category	2019	2022
Glass	80%	90%
Paper, board and cartons	85%	90%
Aluminium	80%	90%
Beverage carton packages	75%	80%
Plastics	90%	90%

Legal minimum recycling targets.

per km in the EU, with some adjustments for heavier vehicles and for the quota of cars to be taken into account. As already mentioned, e-vehicles are considered with 0 grams emissions and are counted twice, at least in 2020 (see Section 10.2.4). However, these "super credits" will be phased out in 2022.

Whether the standard of $95\,g\,CO_2$ per km can be achieved in Germany in 2020 and 2021 therefore depends decisively on the sales of sufficiently many e-vehicles, which are not yet on a satisfactory level. And whether the market for e-vehicles will provide some relief remains to be seen, in addition to related environmental issues: the production of large quantities of batteries with natural resources produced under somewhat dubious environmental conditions, and later the recycling of the same quantities of used batteries. Moreover, the next challenge is already waiting: by 2030, average fleet emissions have to be reduced by a further 37.5% compared to 2021 levels, at least under the latest EU regulations.

This example illustrates the problematic points of such "dynamic" adjustments: on the one hand, raising the standards at certain intervals provides clarity on the efforts required to develop new and innovative technologies. Rubashkina et al. (2015) "find evidence of a positive impact of environmental regulation on the output of innovation activity ..." in the sense of the Porter hypothesis. On the other hand, alternative strategies in the context of implementing

a circular economy need not always be the optimal ones if, for whatever reason, difficulties arise in achieving the standards. Insufficient information and information asymmetries lead to this result with possibly serious consequences, as documented in the case study on passenger car emission standards in the EU (Section 15.4).

15.3.3 Public research activities

Another way to raise standards is through publicly funded research activities in universities and other appropriate research institutions. In this case, the results of the research would have to be disclosed and made available to the interested public, such as companies, car manufacturers, or manufacturers of electronic equipment. Of course, these publicly funded research projects would relate to the further development of the techniques and technologies used and would then have to be declared the best available techniques and technologies. Consequently, the companies would have to use these or similar technologies to comply with the regulations.

The problem with this recommendation is that most universities and research institutions around the world currently receive basic funding from their governments, while they need to attract third-party funding in a competitive way for their large-scale research activities. The use of this competitive framework is certainly a way of stimulating research activities in these research institutions. However, when private companies pay for research, and this is indeed seen as an indicator of the quality of research at this institution, the results belong exclusively to them.

This dilemma could be solved, for example, by pooling certain research questions from related producers and presenting them jointly to public research institutions, which would then have to compete for the grant. In order to not interfere too much with competition between car manufacturers, this procedure has to be limited to a few really fundamental issues.

In this context of externalities and characteristics of public goods related to environmental commodities, it should not be forgotten that things are different from the situation of a regular market economy and should therefore allow a slightly different approach – deviating from certain societal path dependencies.

15.4 Case study: Emission standards for vehicles

There is clearly no need to discuss the importance of clean air for a healthy life. Indeed, much has changed in this regard in recent decades. Filter technologies in power plants, catalytic converters and unleaded gasoline have helped to significantly reduce air pollution in many countries. Additional measures to limit emissions of particulate matter, ozone, and nitrous oxides continue to improve the situation in metropolitan areas and efforts to limit greenhouse gas emissions contribute to mitigating climate change.

Nevertheless, in recent years, two issues in particular have been under rigorous investigation in various industrialised countries: emissions of nitrous oxides (NO_x) and carbon dioxide (CO_2) – obviously for different reasons. Gases such as NO_x contribute "to the formation of greater quantities of secondary particulate matter than the amounts emitted directly" and long-term exposure can cause respiratory problems (see Leopoldina, 2019, p. 8f). While NO_x emissions are local pollution, CO_2 emissions relate to global warming – with road transport linking the emissions.

Both CO_2 and NO_x emissions have declined in Germany in recent decades. However, the German Environmental Agency (UBA) refers to greenhouse gas emissions from the transport sector, which have been on more or less at the same level for 20 years. As transport activity has increased since 2000, this means at least decoupling emissions from the growth of transport. Nevertheless, there are these ambitious reduction targets, which will reduce emissions to 55% below 1990 levels by 2030. With regard to NO_x, the goal is simply to reduce emissions as much as possible due to the documented health risks.

This case study aims to analyse the impact of dynamic emission standards on vehicle fleets. Since there seem to be "socially and historically related interchanges amongst people", the research design could be "social construction of reality" according to the design portfolio in Ridder (2017) – this case "represents a good opportunity to understand a theoretical issue". The "theoretical issue" is the expected reaction of car manufacturers to dynamic emission standards.

15.4.1 Dynamic emission standards for vehicles

In order to control and gradually reduce both NO_x and CO_2 emissions from vehicles, various countries have introduced standards. Table 15.2 shows the standards and their adaptations for passenger cars in the EU. The values for CO_2 emissions refer to the average fuel economy of the car manufacturers.

With these dynamic emission standards the EU is trying to stimulate research activities for increasingly efficient engines. For technical reasons, diesel engines contribute to reducing greenhouse gas emissions more than petrol engines. On the other hand, NO_x emissions from road transport come mainly from diesel vehicles. This was not much of a problem, as long as it was possible to comply with emission standards. With regard to the Kyoto Protocol of 1997, sales of diesel cars were encouraged and sales soared, particularly in the U.S.

However, it soon became apparent that there was a significant difference between the

TABLE 15.2 EU emission standards for passenger cars.

Year	NO$_x$ in g/km (gasoline)	NOx in g/km (diesel)	CO$_2$ in g/km (gasoline + diesel)
2000	0.15	0.50	140
2005	0.08	0.25	
2009	0.06	0.18	130
2014	0.06	0.08	
2020	0.06	0.08	95

The adjustments of the emission standards for passenger cars in the EU.

"theoretical" NO$_x$ emission standards in the EU and the "empirically" measured emissions. Fig. 15.1 shows these differences, which are particularly evident for cars with diesel engines.

This situation, which also arose from technical constraints, created a dilemma for the car manufacturers: on the hand, they had to pay attention to the rising standards for greenhouse gas emissions, which were also threatened by the increasing demand for larger passenger cars. On the other hand, there were the standards for NO$_x$ emissions, which also attracted increasing attention due to health concerns.

This was probably the case when car manufacturers, especially Volkswagen, resorted to so-called "defeat devices" which activated emission control mainly during laboratory emissions testing. Under Art. 5(2) of the EU

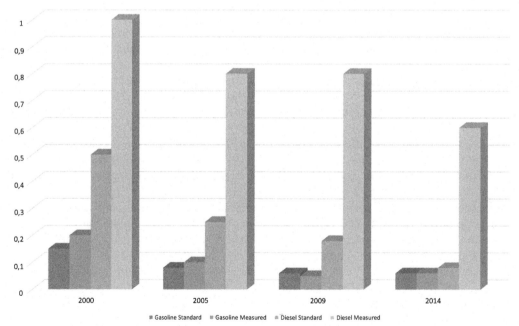

FIG. 15.1 NOX emissions of petrol cars and diesel cars.Comparison of NOX emission limits and measured emissions. © *Own Drawing with data taken from https://www.eea.europa.eu/media/infographics/comparison-of-nox-emission-standards/view.*

Regulations No 715/2007 the use of defeat devices is generally prohibited, exceptions apply. However, these exceptions left room for interpretation – not in terms of sales in the U.S., but in the EU.

15.4.2 Social construction of reality

The interesting question, of course, is what went wrong in this case. The aim of these environmental regulations, the emission standards, was to motivate car manufacturers to additional efforts to reduce these emissions – to mitigate climate change while reducing the health risks of air pollution with nitrous oxides. It was therefore an attempt to overcome the issue of asymmetric information regarding DfEs, perhaps also the effects of the Prisoners' Dilemma (see Section 7.1).

Technically, it was possible to keep the emissions within the legal limits – regarding NO_x with additional quantities of AdBlue, for example. But for whatever reason, that was not Volkswagen's choice of strategy. Perhaps the additional need to refill AdBlue regularly at comparatively short intervals has prevented the company from acting in full compliance with the regulations. Or it seemed difficult for the company to convincingly communicate an "end-of-the-pipe" technology to clean up the exhaustion fumes, without which diesel cars could not pass emission tests in the U.S.

For sure, this strategy had and still has consequences, which resemble path dependencies. First of all, new registrations of cars with diesel engines in Germany in 2018 and 2019 were more than 30% below the previous years. The market share of new cars with diesel engines has thus fallen from around 50% in 2016 to 32% in 2018, increasing the fleet's CO_2 emissions – thus closing one of the backdoors for a feasible reduction of CO_2 emissions from transport activities. In addition, the increase in sales of heavier cars with petrol engines is contributing to this problematic situation.

The dynamic adjustments of these emission standards combined with some technical constraints and particular consumer preferences, led to this result. It is indeed a kind of "social construction of reality": different parts of society have been included in order to "achieve" this not necessarily expected result with further path dependencies.

The case study shows once again the relevance of information: dynamic standards were supposed to help to overcome information problems related to DfEs, but in this case this approach failed as car manufacturers used their knowledge and tried to avoid some of the consequences of fully complying with the higher standards in terms of business interests.

References

Bhagwati, J. (1996). *Trade and the environment: The false conflict? In R. C. Feenstra, G. M. Grossman, & D. A. Irwin (Eds.), The political economy of trade policy: Papers in honor of J. Bhagwati.* The MIT Press Retrieved from (1996). https://econpapers.repec.org/RePEc:mtp:titles:0262061864.

Bhattacharya, R. N., & Pal, R. (2010). Environmental standards as strategic outcomes: A simple model. *Resource and Energy Economics, 32*(3), 408–420. https://doi.org/10.1016/j.reseneeco.2009.11.001.

Flynn, A., & Hacking, N. (2019). Setting standards for a circular economy: A challenge too far for neoliberal environmental governance? *Journal of Cleaner Production, 212*, 1256–1267. https://doi.org/10.1016/j.jclepro.2018.11.257.

Goulden, S., Negev, M., Reicher, S., & Berman, T. (2019). Implications of standards in setting environmental policy. *Environmental Science & Policy, 98*, 39–46. https://doi.org/10.1016/j.envsci.2019.05.002.

Holland, M. (2014). *Cost-benefit analysis of final policy scenarios for the EU clean air package.* Retrieved from (2014). https://ec.europa.eu/environment/air/pdf/TSAP%20CBA.pdf.

Kim, J., & Wilson, J. D. (1997). Capital mobility and environmental standards: Racing to the bottom with multiple tax instruments. *Japan and the World Economy, 9*(4), 537–551. https://doi.org/10.1016/S0922-1425(97)00014-5.

Kohn, R. E. (2003). Environmental standards as barriers to trade. *Socio-Economic Planning Sciences, 37*(3), 203–214. https://doi.org/10.1016/S0038-0121(02)00041-1.

Leopoldina (2019). *Clean air—Nitrogen oxides and particulate matter in ambient air: Basic principles and recommendations.* Halle (Saale): German National Academy of Sciences

Leopoldina. Retrieved from (2019). https://www.leopoldina.org/uploads/tx_leopublication/2019_Leo_Stellungnahme_Saubere_Luft_en_web_05.pdf.

Levy, T., & Dinopoulos, E. (2016). Global environmental standards with heterogeneous polluters. *International Review of Economics & Finance, 43*, 482–498. https://doi.org/10.1016/j.iref.2016.01.009.

Nishitani, K., & Itoh, M. (2016). Product innovation in response to environmental standards and competitive advantage: A hedonic analysis of refrigerators in the Japanese retail market. *Journal of Cleaner Production, 113*, 873–883. https://doi.org/10.1016/j.jclepro.2015.11.032.

Polanco, R., de Sépibus, J., & Holzer, K. (2017). TTIP and climate change. *Carbon & Climate Law Review, 11*(3), 206. https://doi.org/10.21552/cclr/2017/3/14.

Ragossnig, A. M., & Schneider, D. R. (2017). What is the right level of recycling of plastic waste? *Waste Management & Research, 35*(2), 129–131. https://doi.org/10.1177/0734242X16687928.

Ridder, H.-G. (2017). The theory contribution of case study research designs. *Business Research, 10*(2), 281–305. https://doi.org/10.1007/s40685-017-0045-z.

Rubashkina, Y., Galeotti, M., & Verdolini, E. (2015). Environmental regulation and competitiveness: Empirical evidence on the porter hypothesis from European manufacturing sectors. *Energy Policy, 83*, 288–300. https://doi.org/10.1016/j.enpol.2015.02.014.

Saikawa, E., & Urpelainen, J. (2014). Environmental standards as a strategy of international technology transfer. *Environmental Science & Policy, 38*, 192–206. https://doi.org/10.1016/j.envsci.2013.11.010.

Weber, S., & Wiesmeth, H. (2003). From autarky to free trade: The impact on environment and welfare. *Jahrbuch für Regionalwissenschaft, 23*, 91.

WHO (2013). *Health risks of air pollution in Europe—HRAPIE project recommendations for* concentration–response functions for cost–benefit analysis of particulate matter, ozone and nitrogen dioxide. Bonn: World Health Organization, Regional Office for Europe. Retrieved from (2013). http://www.euro.who.int/en/health-topics/environment-and-health/air-.

Wiesmeth, H. (2011). *Environmental economics: Theory and policy in equilibrium. Springer texts in business and economics.* Berlin: Springer Nature. https://doi.org/10.1007/978-3-642-24514-5.

Market-oriented policy tools

Similarly to command-and-control policies (see Section 14.1), market-oriented environmental policies, which include the price-standard approach, focus on achieving environmental standards. In contrast to the command-and-control policies, these tools seek to internalise the environmental effects by completing the market system (see Section 7.1): characteristics of the market or price mechanism are therefore applied in order to achieve the objectives of the policy in general, and to meet the standards in particular. The policies covered in this chapter are sometimes referred to as "policy tools", as they can be part of complex and holistic environmental policies for implementing a circular economy. This discussion will continue in the next chapter.

As explained in the last chapter, environmental standards play the role of the generally unknown efficient level of production and consumption of environmental commodities (see also Chapter 6 and Chapter 7). Unlike a "simple" command-and-control policy, in which policy-makers set standards and enforce them with appropriate measures, market-oriented policies, similar to the market mechanism, use the information, which is available on a decentralised basis. It is therefore up to consumers and producers to decide to what extent they want to continue an environmentally harmful activity and thus pay for transfers to other parts of the economy, or whether they prefer to bear the economic costs of investing in environmentally friendly production technologies. Such transfer payments, such as pollution taxes or expenses for emission certificates, mean higher prices for certain environmental commodities whose supply is limited by an environmental standard. Economic costs arise when economic resources are used, for example, to install better filters to clean wastewater, or catalytic converters to reduce certain emissions from vehicles.

The following sections explain the main features of various examples of market-based environmental policy – the implementation of a circular economy depends also on this type of policy. The first section deals with the pollution tax with a view on the polluter pays principle and avoidance possibilities for paying the tax. Thereafter markets for tradable emission certificates are introduced including some applications. The chapter concludes with a brief examination of the Coase Theorem and some behavioural aspects relating to voluntary contributions and flexible information-based policies.

Implementing the Circular Economy for Sustainable Development
https://doi.org/10.1016/B978-0-12-821798-6.00016-8

16.1 The pollution tax

The theory of environmental economics shows that environmental or external effects are due to incomplete market systems (see Section 7.1). There are commodities such as clean air or some kind of waste, for which there are no regular markets. Completing the market system therefore means to "internalise" the environmental effects, to integrate those commodities, for which there is initially no regular market, into the market system. This implies the introduction of artificial markets: producers or consumers must then buy "clean air", in addition to other factors or "sell" waste at a negative price in addition to their regular products. These additional prices can be interpreted as a "tax", as an environmental tax, or the "Pigou Tax", named after A. C. Pigou, who introduced this concept some 100 years ago (Pigou, 1920). Of course, negative prices for waste are already common in the form of charges for the collection of household waste.

While the Pigou Tax is used in the theory of environmental economics to internalise an environmental effect in order to achieve an efficient allocation, a pollution tax is the counterpart in an empirical context: a tax is levied in order to achieve an environmental standard. In a formal, theoretical context, it is of course possible, to choose these taxes in such a way that an efficient allocation including the environmental commodities results (see, for example, Wiesmeth, 2011, Chapter 6). In a practical context, this is not possible due to information deficits. The recent negotiations in Germany on a German carbon tax highlight these difficulties. Also, King, Tarbush, and Teytelboym (2019) point to issues related to carbon tax reform, which may lead to results, which are not expected and are not at all clear from the start.

There are a number of issues relating to the introduction of a pollution tax: the polluter pays principle, which is often based on a pollution tax, requires some explanations.

16.1.1 The polluter pays principle

The polluter pays principle is laid down in many environmental regulations. Also the 2008 Waste Directive of the European Union (EU) refers to this principle as "a requirement that the costs of disposing of waste must be borne by the holder of waste, by previous holders or by the producers of the product from which the waste came".

This principle is appealing: why should the polluter not pay for the environmental damage he or she has caused – fair enough. However, things are not always so simple and straightforward. Even the above quote from the EU Directive allows for a variety of potential "polluters". Is the polluter the manufacturer of the product from which the waste originates, or rather the consumer who bought the product and now "holds" the waste? Hansmeyer (1980) points out precisely "that in a varied economy environmentally damaging effects are mostly not retraceable to the activities of a single economic unit".

Looking at global climate policies in this context: it is obvious that today's industrialised countries have contributed to the higher concentrations of greenhouse gases in the last decades, or even centuries. In accordance with the historical polluter pays principle, therefore, they should undoubtedly be identified as the main polluters, and they should contribute to the fund to help developing countries adapt to the consequences of climate change. Nevertheless, these are sunk costs that, for reasons of efficiency and fairness, should not be important for future-oriented economic decisions: bygones are bygones. Therefore, the polluter pays principle should integrate all greenhouse gas emitters, wherever they are, in addition to the fact that developing countries should receive support (Tilton, 2016).

Similarly, Glazyrina, Glazyrin, and Vinnichenko (2006) point to the problem of pollutants accumulating in the air, in the soil, in the water. Here, too, the question arises as to how the polluter pays principle can be appropriately adapted in order to "avoid conflict between the interests of society as a whole and the interests of private business".

There are other issues relating to the implementation of the polluter pays principle and, therefore, to the appropriate level of a pollution tax or other environmental levy. For example, given the incomplete information of the policy-makers, what exactly should be taken into account? What is the environmental damage caused by the polluter? Should victims of environmental harm be directly compensated? What should happen if the perpetrators cannot be identified or are insolvent? Luppi, Parisi, and Rajagopalan (2012) deal with such questions about the situation in developing countries.

These considerations show that it is often easy to blame somebody else for a negative development, for an environmental pollution in this case, and it is a common observation in our societies and perhaps immanent in humans to shift responsibility to someone else. Apart from the cases where a household, a producer deliberately or carelessly pollutes the environment, the application of this principle should be carried out with great care.

Nevertheless, it is clear that there should be a visible and understandable link between "polluters", whoever they may be and without calling them as such, and their payments in relation to a pollution tax or other environmental levy. This is certainly mandatory when it comes to the need for a broad acceptance of these fees in society, because in the end they will have to be borne by many people. This relates somewhat to the question of a "perceived feedback" raised in Section 6.1.3.

16.1.2 Avoidance possibilities

Taxes and fees related to environmental issues are plentiful. They are disguised as "levies" on some polluting activity such as the discharge of waste water, they appear as "fees", for example for waste collection; they act as "fiscal taxes" with the primary objective to raise revenue, as is the case with the gasoline tax, which is mainly used, at least in Germany, for the construction and maintenance of the transport infrastructure; and they should have some characteristics of a pollution tax, such as the ecological tax in general and the highway toll in particular.

There is a crucial difference between fiscal taxes and environmental taxes. As already indicated, the primary purpose of a fiscal tax is to generate revenue for more or less specific government expenditure. However, the primary purpose of an environmental or pollution tax is to encourage economic agents to reduce pollution. As a result, a pollution tax works best when tax revenues are low, and a fiscal tax works best when revenues are large. Therefore, the difference between a fiscal and a pollution tax could be associated with the "avoidance strategies" regarding the tax, which are available to the economic agents. In the case of a pollution tax, economic agents should have legal possibilities to avoid paying the tax, as opposed to a fiscal tax. Of course, these opportunities should favour the protection of the environment.

In a practical context, there are usually many ways to reduce the tax burden, and not all of them meet the objective of the pollution tax. So when municipalities charge a fee on household waste that depends on the weight of the waste, then households could – and experience shows that some will – dispose of their garbage in the forest or, somewhat better, in the garbage bins of the rest areas along the highways. These are avoidance strategies which are clearly not in line with the objective of this pollution tax and which result from the Tragedy of the Commons (Section 7.1).

If there are no or almost no avoidance strategies for a particular tax, or if the market conditions are such that the avoidance strategies are not applied, then this is the case with a fiscal tax. This relates, for example, to the situation of a gasoline tax, which is levied on a largely price-inelastic demand. Experience shows that after a few days consumers are (almost) no longer responding to the price increases resulting from the tax. Typical fiscal taxes are income taxes, or sales taxes.

Energy taxes are of particular interest in this case. Although they are considered to be environmental taxes, as the generation and consumption of energy are associated with a variety of environmental problems and issues, the tax base "energy" is so broad that avoidance strategies remain limited in the short run. Saving energy is just as much part of this as is the installation of energy-saving devices. The extent to which they are used, depends on a variety of other issues, including price and income elasticities, the extent of the rebound effect (Section 12.1), but also on a society's acceptance of a certain tax, a carbon tax, for example (McLaughlin, Elamer, Glen, AlHares, & Gaber, 2019).

The "eco-taxes", the ecological taxes introduced in various countries in recent decades, deserve a special remark, referring also to the sustainability goal of a circular economy: tax revenues are intended to support social security systems. However, in order to internalise the environmental effects of energy consumption, the revenue from a tax that supports this objective should remain low; in order to make a substantial contribution to the social security system, the revenue from a tax that supports this objective must be large. It is therefore conceptionally difficult to achieve both objectives simultaneously with an eco-tax, thereby proving the "double-dividend hypothesis" often attributed to it. Empirical investigations lead to mixed results. Parry and Bento (2000), incorporating tax-favoured consumption goods into their model, arrive at a double dividend, while Carraro, Galeotti, and Gallo (1996) show "that recycling carbon tax revenues may provide an 'employment double dividend' only in the short run". Also Nerudová and Dobranschi (2014) question "the conditional occurrence of the hypothesis", and propose alternative measures "to boost carbon taxation efficiency".

The introduction of a pollution tax should therefore always be accompanied by a careful analysis of the available avoidance strategies. Those opportunities, which can be used by economic agents to reduce the tax burden, should support a clearly defined environmental objective.

16.1.3 Cost efficiency of a pollution tax

The economic advantage of a pollution tax is a consequence of the decentralisation of the relevant decisions. Every company, every consumer, decides on the individually available information, thereby reducing the necessity for the policy-makers to collect additional information. As simple examples show, this can help to reduce the economic costs, originating from behavioural changes associated with environmental policies and drawing on economic resources (cf. Wiesmeth, 2011, Chapter 11).

Despite this positive effect on economic costs, there are also some problematic issues, which dampen the willingness to introduce pollution taxes. One of these issues is the tax burden to be borne by individual companies or households. Even if the aim of a pollution tax is to reduce the polluting activity in the broadest sense, taxes usually remain to be paid. Although these are transfer payments which do not burden the economy as a whole, they can place a significant additional burden on individual consumers and producers.

Moreover, the ecological efficiency attributed to a (functioning) command-and-control policy is not necessarily given for a pollution tax. In fact, it is in general a priori unknown whether a particular tax level will lead to the desired result, will help to meet environmental standard. Then a usually complicated and time-consuming procedure for adjusting the tax level is necessary. In the case of Germany's carbon pricing system, there is a regular update to take account of unforeseen technical and economic developments. This includes the planned prices for carbon, the "carbon tax".

In consequence, pollution taxes gain political interest when certain tax revenues are to be collected, which can be used for any kind of social measures: to support the social security system, to reduce income inequality, to support adaptation to climate change and others. However, this does not really correspond to the original philosophy of these taxes.

16.2 Tradable emission certificates

This alternative example of a market-oriented environmental policy is also based on the internalisation of environmental effects by completing the market system. In this case through a market for artificial commodities, certificates linked to emissions of certain pollutants in production. The environmental standard is indicated by the number of certificates issued in a given period of time. The individual demand for certificates depends on the production possibilities of the polluter and the situation on the relevant markets. Here, too, decentralised information is used in decision-making.

In theory, the Pigou Tax and a market for certificates constitute just two sides of the same medal: they are different ways to complete the market system, but they lead to the same efficient market equilibria. From a practical point of view, however, there are significant differences in the perception of these instruments. While a pollution tax tries to meet a standard with a predetermined price, a market for certificates specifies the standard, while the question is asked about a reasonable price. In this context, Hu, Yang, Sun, and Zhang (2020) analyse the tradeoffs between a carbon tax and a cap and trade model for the Chinese remanufacturing industry.

In addition to this structural difference, there is another issue with certificates, which are often considered a "right" to pollution. Discussions in environmental organisations sometimes focus on this issue, making it difficult for policymakers to propose and establish a market for certificates.

A further remark relates to the "distance" of this type of policy tools to policy-makers, to governments. Once such a market for certificates is established, there is not much need to interfere with its operations. But it is precisely this fact that could be a problem, because policy-makers sometimes prefer to be seen as active on environmental issues. The fact that the German carbon tax is only gradually transferred to and merged with the EU ETS, the Emission Trading System of the EU, seems to be an indicator of this assumption.

16.2.1 Features of markets for certificates

An important aspect of a market for tradable certificates is the limitation of the number of the certificates by environmental standards. In contrast to a pollution tax, for which the appropriate value for the tax rate is to be found in a trial and error process, it is therefore possible to use a standard to put an effective "cap" on pollution. Trading activities result then in a market price for the certificates. For this reason, this policy is well-known as a "cap and trade policy".

Setting up such a policy requires, first of all, a decision on what and who will be priced. Germany's Climate Action Plan (Germany, 2016), a cap and trade system, refers to fuels for transport and heating fuels, it covers heating emissions from buildings, transport emissions except of air transport. The participants are not emitters themselves, but companies that put fuels into circulation or suppliers of the fuels. This means that around 4000 companies are integrated into the package.

Another question refers to socially or economically motivated exceptions for certain industries or sectors of the economy. In the EU ETS, for example, small sources of greenhouse gases are not required to participate in the trading scheme, and certain energy-intense production activities, such as steel production, are given special conditions, similar to the aviation sector. The German approach, on the other hand, does not include non-fuel emissions such as methane from agriculture.

A particularly critical issue in the context of markets for tradable certificates is the initial allocation of the certificates to the relevant economic agents. The possibility of auctioning certificates is becoming increasingly important in the EU ETS. Grandfathering formed a common allocation mechanism for the initial endowments: the reference value for the endowment with certificates is a recent or historical value of the pollutant emissions or a combination thereof. However, in order not to incentivise the increase in polluting activities before the introduction of a market for tradable certificates, it is generally preferable to base the initial endowment on an average of historical emission levels.

Again the German approach is different: in the first few years, there are fixed prices per tonne of CO_2 emissions with fixed annual increases. Then auctions, initially with price corridors, will replace the "carbon tax" with the prices originally set.

The integration of such a system into an international framework is important in the case of international environmental commodities. This is a critical issue in the EU, where currently different systems affect each other. There is, for example, the German effort to increase the share of electricity from renewable sources (see Section 11.3). The greenhouse gas emissions avoided in Germany mean lower demand and possibly a lower price for certificates in the EU ETS, which could motivate other countries to increase their purchases of certificates. Thus, the double system in Germany does not contribute to a significant reduction in greenhouse gas emissions in the EU (Rathmann, 2007). On the

other hand, del Río (2017) shows that this negative interaction between targets for electricity from renewable sources and the carbon price from the EU ETS can be mitigated by appropriate measures.

Moreover in the international context, when the systems of certificates do not only include countries with comparable economic framework conditions, additional issues need to be addressed. A price of, say, 50 USD (United States Dollar) per tonne of carbon means different things in developed and developing countries. Emerging economies in their process of economic growth must at least temporarily increase their greenhouse gas emissions, and the integration into a global emission trading system could be seen as expensive and – probably even worse – unfair. Support measures are therefore needed to take into account the creation of global markets for emission certificates.

Separated, different systems for different countries or parts of the world are of course possible, but due to the intrinsic nature of environmental commodities with their public good characteristics, these systems cannot be completely independent. Certain spillover effects can affect their operations. The international carbon market currently includes emission trading systems in Canada, China, Japan, New Zealand, South Korea, Switzerland and the United States. So far, the EU ETS has closer links with the systems in China and South Korea. Jiang, Xie, Ye, Shen, and Chen (2016) provide further insight into China's emission trading scheme, and Li, Weng, and Duan (2019) examine the impact of linking China's national ETS to the EU ETS. One of their conclusions is "to limit the permits that can be traded to reduce the negative impacts of linking on the two systems", pointing out the other challenges in this context.

The following subsection provides some additional information on the EU ETS, which plays an important and growing role in the climate policy of the EU.

16.2.2 The emission trading system of the European Union (EU ETS)

The EU ETS, operating in all EU countries plus Iceland, Liechtenstein and Norway, was established in 2005 and is the world's first and largest trading system, accounting for more than 75% of international emissions trading. It covers around 45% of the EU's greenhouse gas emissions, and the target set for the emissions is 21% lower in 2020 than in 2005 and will be 43% lower in 2030 than in 2005. The system limits emissions from some 11,000 heavy energy-using installations (power plants, industrial plants) and airlines operating between these countries.

Currently, sectors such as housing, agriculture, waste, and transport (excluding aviation) are not included in the ETS. However, they must reduce emissions till 2030 by 30% compared to 2005. Moreover, the share of renewable energy should increase by at least 32% and energy efficiency by at least 32.5% by 2030.

The benefits expected from these measures, according to the EU, include affordable energy, security of the EU's energy supply, reduced dependence on energy imports, new opportunities for economic growth and employment, and environmental and health benefits.

Issues addressed in the literature refer, among others, to the combination of the ETS with targets for renewable energies (del Río, 2017; Rathmann, 2007), to carbon leakage, i.e., the increase of CO_2 emissions outside the countries participating in the EU ETS as a consequence of the ETS (Naegele & Zaklan, 2019), and the extension of the ETS to aviation, touching both issues of carbon leakage and potential economic effects (Efthymiou & Papatheodorou, 2019; Scheelhaase, 2019). Whether the EU ETS is "fit for purpose", is investigated by Marcu et al. (2019) in their "State of the EU ETS Report". Among other things, they discuss emerging issues and explore areas that require further examination, possible consequences of Brexit, for example.

Thus, although the concept of a market for tradable certificates is attractive: the participating economic agents can decide upon the information available to them in order to buy additional or to sell a surplus of certificates, there are a variety of issues to be taken into account. For example, the monitoring aspect must not be forgotten: similar to a command-and-control policy, the compliance with the regulations must be controlled. Then, the public perception of this policy, together with occasionally more problematic reports (Tapia Granados & Spash, 2019), contribute to a somewhat negative attitude towards markets for tradable certificates. However, as Fig. 16.1 shows, EU policy has been successful in terms of decoupling economic growth and emissions of greenhouse gases and various air pollutants.

Nevertheless, markets for tradable certificates are and will remain an important tool for implementing a circular economy. They must be given an optimal area with careful configuration, perhaps in the context of mitigating climate change at international level.

16.3 The Coase theorem

There often is a reciprocal nature of environmental effects: noise, for example, is pollution when someone feels disturbed. An environmental issue therefore arises only from the presence of both the issuer of noise and people who feel uncomfortable.

The formal problem, in turn, is the absence of a regular market for noise. However, instead of creating an artificial market through a Pigou Tax or emission certificates, it may also be possible to assign property rights to either of the two sides: either the noise issuer has the right to pollute the environment, or the potentially affected individuals have the right to a calmer environment. Once these property rights have been assigned, the parties may negotiate and agree on appropriate financial compensation.

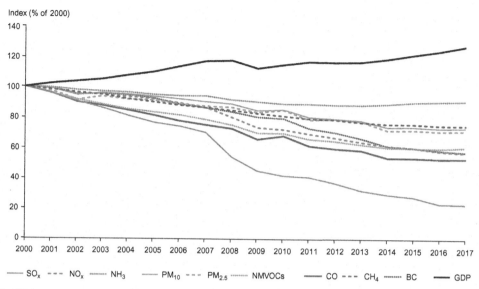

FIG. 16.1 EU-28 air pollutant emissions and economic growth.Evolution of main air pollutant emissions and of the gross domestic product (GDP) as percentages of 2000 levels. © *From EEA. (2019). EU-28 emissions, 2000–2016. European Environment Agency (EEA). Retrieved from https://www.eea.europa.eu/data-and-maps/figures/eu-28-emissions.*

With this approach, reducing the government to an institution that assigns and guarantees property rights, Coase (1992) relativised the need for government measures, such as a tax proposed by Pigou (1920)), for example, to internalise environmental effects.

The statement of the Coase Theorem (Coase, 1960) refers to a situation without transaction costs. Negotiations between the parties lead to an efficient allocation, irrespective of the initial assignment of the property rights. Consequently, a discretionary policy by the government to internalise the externality, such as imposing a Pigou Tax, does not always have to be necessary. Coase thus opened the discussion for a more "market-oriented" or "negotiation-oriented" environmental policy.

As far as applications in the environmental context are concerned, to name but two of them, there is an analysis of the negotiations between waterworks and farmers in Denmark (Abildtrup, Jensen, & Dubgaard, 2012), and there is an attempt "to connect various versions

of the Coase Theorem to carbon trading …" (Lai, Lorne, & Davies, 2018). However, the ubiquitous existence of transaction cost, violating the assumption of the theorem, makes direct applications difficult.

16.4 Voluntary contributions

Behavioural economics allows for non-standard behaviour of economic agents, also in the context of the environment, of contributing towards meeting the standards. Nevertheless, the researchers are trying to understand this behaviour, which is related to "laissez-faire" as a means to allocate environmental commodities (see Section 14.1.1) in somewhat more general contexts.

There is, for example, the case of Japan, where the government is negotiating with polluters (Arimura et al., 2019). The "light-handed" regulations used in these negotiations are meant and understood as a threat of "heavy-handed" and certainly more expensive regulations. Similar to the Coase Theorem, these negotiations become difficult in terms of transaction costs once a large number of stakeholders need to be involved.

Nosenzo and Tufano (2017) arrive at comparable conclusions. In their analysis, voluntary participation in collective action problems, such as environmental issues, may result from the threat of costly penalties for non-participation. They also examine the impact that the "company on board" can have, i.e. the impact on potential followers of those stakeholders who are already contributing. However, at least in their model, this effect plays only a minor role.

The model of Chih (2016) deals more directly with aspects of behavioural economics. Individuals, facing higher standards of social norms tend to contribute more to a public good, an environmental commodity, for example. Therefore, this approach reiterates the need to establish appropriate social norms, to break old habits, to create new, environmentally friendly habits as postulated in behavioural environmental economics (see Chapter 8).

Another interesting project is the Guangzhou pilot project (Tan, Wang, & Zaidi, 2019), which aims at developing "voluntary personal carbon trading" in Guangdong province in China. The aim is to reduce greenhouse gas emissions from the domestic sector, thereby supporting the pilot emission trading scheme. The empirical results show that one of the strong drivers is the "perceived usefulness" of the project, relating this kind of experiment to the discussion on perceived scarcity of environmental commodities and a perceived feedback from one's own actions (see Section 6.1).

Thus, so-called "voluntary" contributions are often less voluntary than they appear at first glance. The conclusion is that economic decisions are based on opportunity costs, i.e. the (perceived) consequences of alternative decisions. This basic economic principle naturally also extends to the protection of the environment, although of course "costs" do not have to be just monetary costs.

As far as other instruments of the market-oriented approach are concerned, voluntary contributions therefore seem to lie somewhere between the pollution tax and the Coase Theorem – enriched by aspects of behavioural environmental economics.

16.5 Flexible, information-based environmental policies

Flexible, information-based standards are important for both command-and-control policies and market-based policies. In order to have some flexibility, a certain possibility of reacting quickly to new information with a similarly flexible policy is generally certainly reasonable and can help to deliver better results for the environment.

Flexibility reflects the need to design environmental policies for a concrete framework – with uncertainties, insufficient information and informational asymmetries. If, for whatever reason, there are significant changes to this framework, then it might be useful to adapt the policy. The German policy for promoting renewable energy sources, established in the 2000s, currently presents some financial challenges (Section 11.3). This, and in particular the relatively high renewables surcharge of more than 20% of the electricity price, could have been avoided with a flexible policy that would make necessary changes quicker and easier.

On the other hand, the problem remains to use the appropriate indicators and to adapt the policy optimally. This in turn is influenced by the informational issues mentioned above. Moreover, and perhaps more importantly, the economic agents affected by this policy can learn from these changes and adapt their own behaviour. This "behavioural rebound effect" (see Section 12.1) could therefore weaken the effects of the intended changes.

In the literature, flexible policies are generally seen as positive. Tsiliyannis (2007) promotes flexible recycling rates that are adjusted to the growth rate. Therefore, rapid economic expansion should provide for higher recycling rates to meet the growing demand for packaging, etc., and Yuan and Zhang (2020) consider the flexibility to better stimulate environmental innovation in the sense of the Porter Hypothesis: appropriate regulation can stimulate innovation, for example, to reduce costs. Flexible policies make it easier to find the right values for environmental standards.

Despite the obvious need to keep environmental policy flexible, this opportunity should be used carefully. Information issues concern the details of environmental policies and also affect the possible directions of change. Moreover, there could be counter-effects resulting from changes in the behaviour of economic agents which, like the rebound effects, could reduce the effectiveness of the adjustments of the policies.

References

Abildtrup, J., Jensen, F., & Dubgaard, A. (2012). Does the Coase theorem hold in real markets? An application to the negotiations between waterworks and farmers in Denmark. *Journal of Environmental Management, 93*(1), 169–176. https://doi.org/10.1016/j.jenvman.2011.09.004.

Arimura, T. H., Kaneko, S., Managi, S., Shinkuma, T., Yamamoto, M., & Yoshida, Y. (2019). Political economy of voluntary approaches: A lesson from environmental policies in Japan. *Economic Analysis and Policy, 64*, 41–53. https://doi.org/10.1016/j.eap.2019.07.003.

Carraro, C., Galeotti, M., & Gallo, M. (1996). Environmental taxation and unemployment: Some evidence on the 'double dividend hypothesis' in Europe. *Proceedings of the Trans-Atlantic Public Economic Seminar on Market Failures and Public Policy, 62*(1), 141–181. https://doi.org/10.1016/0047-2727(96)01577-0.

Chih, Y.-Y. (2016). Social network structure and government provision crowding-out on voluntary contributions. *Journal of Behavioral and Experimental Economics, 63*, 83–90. https://doi.org/10.1016/j.socec.2016.05.008.

Coase, R. H. (1960). The problem of social cost. *Journal of Law and Economics, III*, 1–44. Retrieved from https://www.law.uchicago.edu/files/file/coase-problem.pdf.

Coase, R. H. (1992). The institutional structure of production. *The American Economic Review, 82*(4), 713–719. Retrieved from(1992). http://www.jstor.org/stable/2117340.

del Río, P. (2017). Why does the combination of the European Union emissions trading scheme and a renewable energy target makes economic sense? *Renewable and Sustainable Energy Reviews, 74*, 824–834. https://doi.org/10.1016/j.rser.2017.01.122.

Efthymiou, M., & Papatheodorou, A. (2019). EU emissions trading scheme in aviation: Policy analysis and suggestions. *Journal of Cleaner Production, 237*, 117734. https://doi.org/10.1016/j.jclepro.2019.117734.

Germany (2016). *Climate action plan 2050: principles and goals of the German government's climate policy.* Federal Ministry for the Environment, Nature Conservation, Building and Nuclear Safety (BMUB). Retrieved from https://www.bmu.de/fileadmin/Daten_BMU/Pools/Broschueren/klimaschutzplan_2050_en_bf.pdf.

Glazyrina, I., Glazyrin, V., & Vinnichenko, S. (2006). The polluter pays principle and potential conflicts in society. *Ecological Economics, 59*(3), 324–330. https://doi.org/10.1016/j.ecolecon.2005.10.020.

Hansmeyer, K.-H. (1980). "Polluter pays" v. "public responsibility". *Environmental Policy and Law, 6*(1), 23–24. https://doi.org/10.1016/S0378-777X(80)80013-8.

Hu, X., Yang, Z., Sun, J., & Zhang, Y. (2020). Carbon tax or cap-and-trade: Which is more viable for Chinese remanufacturing industry? *Journal of Cleaner Production. 243*, https://doi.org/10.1016/j.jclepro.2019.118606.

Jiang, J., Xie, D., Ye, B., Shen, B., & Chen, Z. (2016). Research on China's cap-and-trade carbon emission trading scheme: Overview and outlook. *Applied Energy, 178*, 902–917. https://doi.org/10.1016/j.apenergy.2016.06.100.

King, M., Tarbush, B., & Teytelboym, A. (2019). Targeted carbon tax reforms. *European Economic Review, 119*, 526–547. https://doi.org/10.1016/j.euroecorev.2019.08.001.

Lai, L. W. C., Lorne, F., & Davies, S. N. G. (2018). A reflection on the trading of pollution rights via land use exchanges and controls: Coase theorems, Coase's land use parable, and Schumpeterian innovations. *Progress in Planning.* https://doi.org/10.1016/j.progress.2018.10.001.

Li, M., Weng, Y., & Duan, M. (2019). Emissions, energy and economic impacts of linking China's national ETS with the EU ETS. *Applied Energy, 235*, 1235–1244. https://doi.org/10.1016/j.apenergy.2018.11.047.

Luppi, B., Parisi, F., & Rajagopalan, S. (2012). The rise and fall of the polluter-pays principle in developing countries. *The Economics of Efficiency and the Judicial System, 32*(1), 135–144. https://doi.org/10.1016/j.irle.2011.10.002.

Marcu, A., Alberola, E., Caneill, J.-Y., Mazzoni, M., Schleicher, S., Vailles, C., … Cecchetti, F. (2019). *2019 state of the EU ETS report.* Retrieved from (2019). https://ercst.org/wp-content/uploads/2019/05/2019-State-of-the-EU-ETS-Report.pdf.

McLaughlin, C., Elamer, A. A., Glen, T., AlHares, A., & Gaber, H. R. (2019). Accounting society's acceptability of carbon taxes: Expectations and reality. *Energy Policy*, *131*, 302–311. https://doi.org/10.1016/j.enpol.2019.05.008.

Naegele, H., & Zaklan, A. (2019). Does the EU ETS cause carbon leakage in European manufacturing? *Journal of Environmental Economics and Management*, *93*, 125–147. https://doi.org/10.1016/j.jeem.2018.11.004.

Nerudová, D., & Dobranschi, M. (2014). Double dividend hypothesis: Can it occur when tackling carbon emissions? In: *17th international conference enterprise and competitive environment 2014*, Vol. 12, (pp. 472–479). https://doi.org/10.1016/S2212-5671(14)00369-4.

Nosenzo, D., & Tufano, F. (2017). The effect of voluntary participation on cooperation. *Journal of Economic Behavior & Organization*, *142*, 307–319. https://doi.org/10.1016/j.jebo.2017.07.009.

Parry, I. W. H., & Bento, A. M. (2000). Tax deductions, environmental policy, and the "double dividend" hypothesis. *Journal of Environmental Economics and Management*, *39*(1), 67–96. https://doi.org/10.1006/jeem.1999.1093.

Pigou, A. C. (1920). *The economics of welfare*. London: Macmillan and Co. Retrieved from(1920). https://www.econlib.org/library/NPDBooks/Pigou/pgEW.html?chapter_num=15#book-reader.

Rathmann, M. (2007). Do support systems for RES-E reduce EU-ETS-driven electricity prices? *Energy Policy*, *35*(1), 342–349. https://doi.org/10.1016/j.enpol.2005.11.029.

Scheelhaase, J. D. (2019). How to regulate aviation's full climate impact as intended by the EU council from 2020 onwards. *Journal of Air Transport Management*, *75*, 68–74. https://doi.org/10.1016/j.jairtraman.2018.11.007.

Tan, X., Wang, X., & Zaidi, S. H. A. (2019). What drives public willingness to participate in the voluntary personal carbon-trading scheme? A case study of Guangzhou Pilot, China. *Ecological Economics*, *165*, 106389. https://doi.org/10.1016/j.ecolecon.2019.106389.

Tapia Granados, J. A., & Spash, C. L. (2019). Policies to reduce CO_2 emissions: Fallacies and evidence from the United States and California. *Environmental Science & Policy*, *94*, 262–266. https://doi.org/10.1016/j.envsci.2019.01.007.

Tilton, J. E. (2016). Global climate policy and the polluter pays principle: A different perspective. *Resources Policy*, *50*, 117–118. https://doi.org/10.1016/j.resourpol.2016.08.010.

Tsiliyannis, C. A. (2007). A flexible environmental reuse/recycle policy based on economic strength. *Waste Management*, *27*(1), 3–12. https://doi.org/10.1016/j.wasman.2006.06.015.

Wiesmeth, H. (2011). Environmental economics: Theory and policy in equilibrium. In *Springer texts in business and economics*. Springer Nature. https://doi.org/10.1007/978-3-642-24514-5.

Yuan, B., & Zhang, Y. (2020). Flexible environmental policy, technological innovation and sustainable development of China's industry: The moderating effect of environment regulatory enforcement. *Journal of Cleaner Production*, *243*, 118543. https://doi.org/10.1016/j.jclepro.2019.118543.

17

Holistic policy approaches

Although the implementation of a circular economy depends on the development of appropriate technologies (Chapter 9), orientation through complex, holistic policies is required. Holistic policies are needed to properly integrate relevant stakeholders, for example in waste prevention; holistic policies are needed to address all types of information issues (Chapter 11); and holistic policies are needed to provide tools to achieve the targets or standards of the circular economy strategies (Chapter 15). It is not realistic to hope or even expect that a large proportion of business companies and households will voluntarily and enthusiastically support the implementation of a circular economy – with their own contributions, their own pro-environmental actions. It is not only the meagre outcome of COP 25, the United Nations Climate Change Conference in Madrid in December 2019, that is proof of this.

Policies to support the implementation of a circular economy often have to address different groups, the large group of households, and smaller groups of certain business companies, for example. Moreover, these groups need to be addressed in the context of implementing a circular economy in relation to different obligations. A policy for all purposes will not do it, policy tools have to be assembled into a policy mix or even a holistic policy.

There is some literature on the need for a policy mix. In their review, Aldieri, Carlucci, Vinci, and Yigitcanlar (2019) point to the necessity to implement "a coherent policy mix" to achieve sustainable economic growth. Similarly, to promote the transition to sustainable development Kern, Rogge, and Howlett (2019) argue that policy mixes should focus both on policy goals and on policy strategies, while Edmondson, Kern, and Rogge (2019) study the co-evolutionary dynamics of policy mixes and socio-technical systems in this sustainability context.

This chapter examines holistic policy approaches, their origins, their economic background, and practical experiences gained. Moreover, various tools of holistic approaches are introduced and their relevance for implementing a circular economy is analysed.

In the first section "Integrated Waste Management" (IWM) is seen as a first step into the world of holistic policies, followed by a more detailed study of "Extended Producer Responsibility" (EPR), or rather the "EPR Principle". The concept of an "Integrated Environmental Policy" (IEP) makes full use of the EPR principle and prepares the ground for the design of policies for implementing circular economy strategies in Part V. Since the IEPs are intended in particular to support the implementation of the waste hierarchy, this chapter also analyses the incentive-compatibility of various collection systems and "Producer Responsibility Organisations" (PRO).

17.1 Integrated waste management

The concept of IWM originated in 1975 from the mission statement of the Solid Waste Authority of Palm Beach County, Florida. The Authority would "develop and implement programs in accordance with its Comprehensive Plan by integrating solid waste transportation, processing, recycling, resource recovery and disposal technologies" (McDougall, White, Franke, & Hindle, 2001). Since then, the concept has evolved into a holistic approach to waste management, based on these more technical aspects.

Bilitewski, Härdtle, and Marek (1997) initiated this development with a comprehensive discussion of all aspects of IWM, and Wilson (1996) examined "the development of integrated sets of policy measures by countries around the world..." (p. 389). These changes in the understanding of waste management were followed by an adaptation of the technical infrastructure. For Greater Manchester in the United Kingdom (UK), Uyarra and Gee (2013) characterise this process of a "sustainable transformation of urban infrastructure" as moving "away from landfill and towards more sustainable waste management" (p. 102).

IWM refers to life cycle assessment, which continues to play an important role in the analysis of all types of environmental impacts and other issues (Boustead, 1996). While Winkler and Bilitewski (2007) compare and evaluate different life cycle models for solid waste management, more recent approaches use this method to review detailed aspects of buildings, green roofs, recycling polymers and polyesters – to name but a few.

The European Union (EU) also promoted "Life Cycle Thinking and Assessment" in order "to move to a more resource-efficient and sustainable future". The EU believes that "the waste hierarchy will generally lead to the most resource-efficient and environmentally sound choice" and that life cycle thinking and assessment "provide a scientifically sound approach to ensure that the best outcome for the environment can be identified and put in place". By looking at the overall environmental impacts, life cycle thinking is a holistic support tool for environmental decision-making, especially in the context of the waste hierarchy.

According to the opinion of the EU, the following typical questions arising from the concept of life cycle thinking require an answer:

- Is it better to recycle waste or to recover energy from it? What are the trade-offs for particular waste streams?
- Is it better to replace appliances with new, more energy-efficient models or to use the old ones and to avoid waste?
- Are the greenhouse gas emissions generated by the collection of waste justified by the expected benefits?

As a result, IWM is one of the first holistic approaches to waste management, with "holistic" referring to waste as an environmental good (or bad), for which the allocation problems should be solved in an environmentally and socially responsible manner.

What does this holistic approach mean for environmental policy and the implementation of a circular economy? After the mechanical integration of various technical activities in waste management, the waste hierarchy is already the result of a more integrated approach, of embedding waste management into environmental economics, of shifting traditional linear waste management to "circular" integrated waste management, thus establishing the link to implementing a circular economy (Cobo, Dominguez-Ramos, & Irabien, 2018). The EPR principle is an important cornerstone for achieving this integration.

17.2 The EPR principle

The EPR principle is the common basis for EPR policies – the most important and widespread holistic environmental policies. This section discusses some general aspects and provides experiences with applications of this principle. The implementation of the EPR principle in some basic practical contexts with regard to circular economy strategies is discussed in the following sections.

17.2.1 General aspects of the EPR principle

It is a key feature of EPR policies that they transfer some responsibility for a product's end-of-life environmental impact to the original producer and seller of that product. This corresponds to the OECD's definition of EPR as an "an environmental policy approach in which a producer's responsibility for a product is extended to the post-consumer stage of a product's life cycle" (see OECD, 2001; Walls, 2006), thereby preventing waste, improving reusability of old items and recycling of waste.

In a more detailed way an EPR policy, or rather the EPR principle, is according to OECD (2001) "characterised by: (1) the shifting of responsibility (physically and/or economically; fully or partially) upstream toward the producer and away from municipalities; and (2) the provision of incentives to producers to take into account environmental considerations when designing their products". Moreover, "EPR seeks to integrate signals related to the environmental characteristics of products and production processes throughout the product chain", while most other policy tools focus on a single target. Lindhqvist (2000) and Ghisellini, Cialani, and Ulgiati (2016) provide some more details on the origin and first applications of this important concept.

In view of the discussion in Section 16.1.1 of the polluter pays principle with its attempt to single out a "polluter" for a certain environmental issue, this introductory characterisation of the EPR principle raises a few questions, which are of relevance for the performance of EPR policies (see also Wiesmeth & Häckl, 2011 for further details):

1. If the EPR principle is meant to provide incentives for producers for a design for environment (DfE) (see Lindhqvist, 2000; Walls, 2006, for example), then the EPR concept seems to blame, in the first place, the producers for an environmental problem, for an environmental degradation. The role of the consumers and other stakeholders is neglected, although demand for specific commodities and designs may lead to the environmental problem in question.

2. If the EPR principle seeks to integrate signals related to a product throughout the product chain, why should 'municipalities', in particular, and consumers and other stakeholders, in general, be excluded from the consideration? After all, municipalities can play a decisive role in promoting a DfE – by organising waste management in a way, which supports or alleviates recycling activities, for example. Similarly, consumers can substantially support efforts in waste management.

According to the definition, the EPR principle is intended to stimulate DfEs. However, as has already been shown by the practical examples in Chapter 5, without appropriate policy guidelines, producers are likely to pay attention mainly to the market situation and other relevant conditions. After all, they have the knowledge required and relevant to a DfE for their products. Policy-makers are largely dependent on the cooperation of the producers, and this "cooperation" must be stimulated through appropriate policy tools as part of a holistic policy.

This is clearly stated, for example, in the regulations for waste electrical and electronic equipment (WEEE). Art. 4 "Product Design" of the EU Directive on WEEE of 2012 stipulates: "Member states shall take appropriate measures so that the eco-design requirements facilitating re-use and treatment of WEEE ... are applied ...". And the further reference in Art. 4 "... that producers do not prevent, through specific design features or manufacturing processes, WEEE from being re-used ..." points out that if EPR, or DfE in this context, is only declared as a policy principle, then a substantial DfE will only happen, if the market situation is − coincidentally − in favour of specific environmentally friendly designs. Strategic considerations such as a first-mover advantage or a late-mover advantage (see also Chapter 4) can of course play a role. Consequently, more needs to be done to properly integrate the EPR principle into environmental policies.

It is likely that the reasons for the often not optimal performance of EPR policies (see Chapter 5) are due to the generally incomplete specification of the EPR policies or the incomplete integration of the EPR principle into environmental policies. Consequently, the EPR principle, or EPR policies as a holistic approach, requires a more careful analysis in the context of the implementation of a circular economy.

What are main concrete applications of the EPR principle in environmental policies? Since there are a variety of EPR policies − OECD (2016) refer to a survey identifying about 400 EPR systems in operation in 2016 − it is indeed better to refer to the EPR principle, which is then translated into environmental policies. While the polluter pays principle establishes or seeks to establish a link between a polluting activity and environmental taxes or charges in order to reduce pollution, the EPR principle is a holistic approach: by focusing on the end-of-life environmental impact of a product, manufacturers

should be motivated for a DfE and various other aspects of their products that are important for environmental protection.

Therefore, and due to its appealing characteristics as a holistic concept, the implementation of various aspects of a circular economy, in particular the waste hierarchy, need to be based on the EPR principle. The important question will be how best to use this principle, how to integrate it optimally into environmental policies.

17.2.2 Experiences with EPR policies

EPR policies have played and continue to play a prominent role in environmental regulations in developing and developed countries, in emerging economies and countries in transition, around the world. As a concept, not yet under the label "EPR", it already appeared in the German Packaging Ordinance of 1992. It is of relevance for the European Strategy for Plastics in a Circular Economy of 2018, with many references to EPR − interestingly also as a means of raising funds to promote efforts to raise environmental awareness. And it is important in the various versions of the EU Directive on Waste, with a focus on strengthening "the re-use and the prevention, recycling and other recovery of waste". Various aspects of the following literature review are taken from Wiesmeth, Shavgulidze, and Tevzadze (2018).

Gupt and Sahay (2015), referring to implementations of the EPR principle in relation to the end-of-life management of recyclable goods, conclude that "regulatory provisions, take-back responsibility and financial flow" are most important for EPR, and that "changes in the design of the products to reduce environmental impacts (DfE) ... do not seem to have a major role in the success of EPR ..." (p. 609). According to Massarutto (2014), "EPR has become a cornerstone of solid waste management policies throughout the world", yet "policies inspired by EPR have been indeed successful, but probably for different purposes and for different reasons than initially believed" (p. 11), and "the impact of EPR on green design and product innovation has been much lower than expected" (p. 14).

The review report BIO Intelligence Service (2014) covers four main issues that are important for implementing the EPR principle, and therefore relevant for implementing a circular economy (p. 21):

- Allocation of responsibilities among stakeholders: financial and/or organisational responsibility?
- Costs coverage: What are the costs covered by EPR and in what proportions?
- Fair competition: how is economic competition organised within the framework of EPR policies?
- Transparency and control: who monitors the different aspects of an EPR policy and how?

Similarly, the review report OECD (2016) highlights problematic aspects and key challenges of a wide range of EPR policies in different environmental areas and in different countries, and points to difficulties assessing these policies. OECD (2016) complains in particular "a considerable lack of data, analytical difficulties in distinguishing the impact of EPR systems from other factors, and the wide variety of EPR systems which limits comparison among them" (p. 13). Nevertheless, according to their findings, there is "evidence that in some countries, EPRs have helped to shift some of the financial burden for waste management from municipalities and taxpayers to producers, and to reduce the public costs of waste management" (p. 13).

In the context of a market system, the final outcome of this shifting of the financial burden remains unclear. The extent to which consumers and producers have to shoulder this burden, depends on the market situation: on price and income elasticities, and on the availability of alternative products – with and without DfEs.

As far as the environmental context is concerned, this is probably not the most important issue. OECD (2016) lists various specific recommendations, which are considered crucial for the performance of an EPR policy: the issues relate to the governance of EPR, to competition policy, to stronger incentives for DfE, and to the appropriate involvement of informal workers in emerging economies and developing countries.

Tencati, Pogutz, Moda, Brambilla, and Cacia (2016) deal specifically with packaging waste and consider "the promotion of collaborative efforts along the packaging supply chains" as potential positive effects on packaging prevention (p. 43). Such "collaborative efforts" do play a role in some EPR policies. Cahill, Grimes, and Wilson (2010) look at WEEE in addition to packaging waste and also point to the different EPR approaches in the EU member states. In their view, these systems "differ primarily due to the contrasting opinion on the legitimacy of local authorities and stakeholders and, in some cases, a fear on the part of the industry of associated costs". Gu, Wu, Xu, Wang, and Zuo (2017) consider a redesign of the WEEE fund system in China to better stimulate DfEs, and Wang et al. (2017) analyse details of the EPR systems for WEEE in Japan, Germany, Switzerland and China, and "look ahead" to the development directions of EPR systems. Salhofer, Steuer, Ramusch, and Beigl (2016) compare WEEE management in Europe and China and emphasise the still dominating role of the informal sector in China, and Corsini, Rizzi, and Frey (2017) examine the impact of organisational dimensions on WEEE collection, "the need for stronger coordination of EPR and waste policies…". With regard to end-of-life vehicles (ELV), Xiang and Ming (2011) characterise China's approach on the basis of an EPR system for an emerging economy.

Packaging waste, a major concern of municipal solid waste management, is, of course, the object of a wide range of EPR policies. As indicated in Chapters 5 and 20, the prevention of packaging waste, which is a top priority under many environmental regulations, has not yet been fully achieved through EPR policies, although there are some positive results (Tencati et al., 2016). For Portugal and Spain, however, decoupling packaging waste from economic growth still requires attention and further efforts (Rubio, Ramos, Leitão, & Barbosa-Povoa, 2019).

This brief review of the literature on different EPR policies shows that the performance of these systems, in particular environmental performance, is limited – despite various positive cases. Whether the reasons for these observation are due to structural deficiencies in the implementation of the EPR principle, or whether economic growth and other factors affect EPR policies, remains to be seen.

As mentioned, there were around 400 EPR policies in 2016. This large number of mostly different approaches makes it difficult to find some guiding principles: what are the most important applications, where can we identify relevant implementations of the EPR principle with regard to the circular economy? These questions will be addressed in the sections after a brief discussion on "Integrated Environmental Policies" (IEP).

17.3 Constitutive elements of an IEP

The EPR principle is important for holistic policies in the context of implementing a circular economy. Nevertheless, there are some questions regarding the perception of the principle, and practical experience shows ambiguous results (see the review in the last section).

In order to address these remarks from an economic point of view, the next step following IWM and the EPR principle is to "integrate" environmental commodities into the economic allocation problems in the sense that holistic policies with market-oriented structural properties

solve the allocation problems. This brings us back to the structural requirements arising from the market context and set out in Section 14.2 for environmental policies. Wiesmeth and Häckl (2016) develop the IEP concept and draw attention to so-called "Constitutive Elements" (CE) of these policies, derived from their intrinsic nature. An IEP is therefore a holistic policy based on the EPR principle and satisfies the constitutive elements listed in the following subsections.

17.3.1 Constitutive element I – The locality principle

CE I – Dependence of the policy on local conditions: the background is that feasible solutions to the allocation problems in the market context depend on local conditions. This should be reflected in the allocation of environmental commodities in IEPs.

This first constitutive element is now recognised in many holistic environmental policies. It justifies, for example, that IWM policies vary from region to region, as market solutions depend on local conditions. These include wealth and preferences of the consumers, production possibilities, availability of natural resources, climatic and geographical conditions, the economic situation in general, trade relations and, in the context considered here, the technical infrastructure.

Interestingly, the waste policies of the EU, like many initiatives of the EU, are still characterised by harmonisation efforts, with unifying and standardising attempts, which contradict CE I and which occasionally lead to criticism (Bartl, 2014). In this context, Broitman, Ayalon, and Kan (2012) draw attention to schemes of waste management, which are often designed at a national level with local conditions neglected, and Agamuthu and Victor (2011) refer to "the pitfall of … the simplistic adoption of policies and legislations from other countries' 'purportedly' successful waste management system without taking into context the local, cultural and socio-economic waste management issues" (p. 952).

Consequently, the challenge is to carefully examine the appropriate framework conditions for implementing a circular economy. Local conditions need to be respected as far as possible, particularly with regard to the export of environmental technologies. Unfortunately, this requirement makes it difficult to draft IEP policies for international environmental commodities (see also Chapters 22 and 23).

17.3.2 Constitutive element II – Integrating affected agents

CE II – Integrating affected economic agents into the policy and addressing a potentially large number of involved agents through appropriate framework conditions: the background is that the allocation problems in a market system are interdependent. Therefore, any interference with the allocation problems may affect a large number of economic agents. This should be taken into account in IEPs. Agents who are not adequately integrated into the policy may undermine policy objectives (see the case study in Section 7.3).

The design of an IEP therefore requires the identification of the agents affected by this policy. Any change, any interference with the technical and organisational infrastructure, can affect a wide range of economic agents. In this context, Heidrich, Harvey, and Tollin (2009) propose "a template and matrix model for identification of stakeholder roles and influences by rating the stakeholders".

The integration of relevant stakeholders is linked in particular to the question of how to adequately address different groups of economic agents, how they can be effectively integrated without too much monitoring and high control costs for society. The common need to target a potentially large number of agents has so far attracted little attention in the literature. Command-and-control policies often dominate.

However, due to the control costs, they are better suited to target smaller group of agents.

This issue is of great importance for waste management, which has to address all households for an efficient operation. The cooperation of the households is necessary for the collection and separation of waste, and in particular for waste reduction and waste prevention (see Part V). Guerrero, Maas, and Hogland (2013) argue that "waste management involves a large number of different stakeholders, with different fields of interest. ... Detailed understandings on who the stakeholders are and the responsibilities they have in the structure are important steps in order to establish an efficient and effective system" (p. 227).

17.3.3 Constitutive element III – Linking policy tools

CE III – Identifying appropriate signals and linking them appropriately through policy tools: here the background is that in a market economy individual agents, consumers and producers, use the price signals to formulate their demand and supply decisions. Due to the existence of external effects, policy tools have to be made dependent on other, additional signals, and appropriately linked in order to achieve the policy objectives.

Relevant signals refer to different phases of a product's lifecycle, including the design and the post-consumer phase, collection and recycling rate. This requirement complies with the EPR principle: "EPR seeks to integrate signals related to the environmental characteristics of products and production processes throughout the product chain" (OECD, 2001). McKerlie, Knight, and Thorpe (2006) highlight EPR programs "with clear legislation that encourages sustainable product design by delivering a full range of signals to producers". The difficult task of an IEP is then to select appropriate signals for all groups of stakeholders, not only producers, and to link them adequately.

As experience shows, this is not always easy and straightforward. A large number of EPR policies and implementations of the EPR principle in the management of WEEE, of drinks packaging and other areas of waste management reveal more or less serious shortcomings: environmental goals, which are not achieved and not really successful efforts to reduce waste. Part V of the book will have to deal particularly closely with this issue, which is, of course, highly relevant for implementing a circular economy.

So far the main holistic approaches of environmental policy, its development from technical integration to a rather sophisticated policy with different parts, which must be carefully linked in order to achieve optimal results. In the next sections, the EPR principle and EPR policies will be reexamined with regard to these constitutive elements of an IEP. Regarding the overarching objectives of a circular economy, some general aspects of the waste hierarchy – collection systems and producer responsibility organisations – will be in the focus of the following considerations. The chapters in Part V will then provide concrete IEPs based on the EPR principle.

17.4 Implementation of the EPR principle

The implementation of the EPR principle as part of an IEP requires some consideration. One of the main roads that motivates producers to a DfE is to "push" them to make proper use of their knowledge. This could perhaps be achieved by making them pay for the collection and recycling of the waste generated from their products, thus putting some pressure on them to consider a DfE for their products in order to reduce costs and thus prevent waste.

This section refers to issues of relevance mainly in waste management, although IEPs can also be designed for other circular economy strategies (see Part V). The following subsections

examine collection systems and so-called EPR systems for managing waste under the umbrella of an IEP. Wiesmeth et al. (2018) investigate these aspects for an IEP for drinks packaging in Georgia, a country in transition.

17.4.1 Collection systems

This straightforward "holistic" approach of making producers pay for the collection and recycling of waste from their product contains some pitfalls which need to be examined carefully. First of all, collecting the waste products is aimed at all consumers, thereby referring to the organisation of the collection and recycling system, not to forget financing these activities. Note that also the "response" of households to the details of a collection system must also be taken into account (see Chapters 11 and 12). Ideally, technical details should help households to participate appropriately in the collection and segregation activities.

With regard to recyclable waste, there are two major collection systems with different characteristics: the separate collection system and the take back system. Due to their role in the implementation of a circular economy, they are briefly checked for incentive-compatibility and feasibility. For a comprehensive review of a variety of cases, consider Gupt and Sahay (2015), who identify separate collection as one of the success factors of EPR systems. Corsini et al. (2017) analyse certain organisational framework conditions with regard to the collection of WEEE and point to "the need for stronger coordination of EPR and waste policies in order to achieve adequate levels of WEEE collection, …".

Separate Collection System: A simple version of a separate collection system requires separate waste bins for the recyclables to be collected: packaging waste, in particular all types of drinks packaging with separate collection for plastics, paper, glass, and aluminium, but also small WEEE, which are then prepared and consigned for recycling. Collection is then organised as a bring system with containers within walking distance from the house, or as a door-to-door collection system.

Seyring, Dollhofer, Weißenbacher, Bakas, and McKinnon (2016) and also OECD (2016) provide comprehensive assessments of the various collection systems in operation. They find that certain collection systems can help to raise the capture rates: bring sites, in particular, seem to be preferred, at least for glass packaging, while door-to-door collection leads to higher capture rates for paper and bio-waste. Hahladakis, Purnell, Iacovidou, Velis, and Atseyinku (2018), focusing on kerbside collection of plastics packaging, refer to the ambitious recycling targets of the EU for plastic waste to "explain how different collection modalities affect the quantity and quality of recycling …".

These observable differences in connection with this laissez-faire approach (Section 14.1.1) provide an insight into the "relationship" between households and the technical details of the organisation of waste collection. In addition, proposals from behavioural environmental economics should be respected: depending on environmental awareness, on economic well-being, new habits need to be established that develop adequate waste separation into a social norm (see Section 8.2). Then laissez-faire could prove beneficial due to the low cost of waste collection.

Take Back System with Deposit: A take back system with a deposit fee focuses more on individual incentives to return waste items, such as empty drinks packaging or WEEE. This "individual responsibility", an important feature of a market system, raises incentive-compatibility and return rates.

Take back systems for drinks packaging have been investigated by Dace and Pakere (2013) for Latvia, Numata (2016) for Finland and Norway, Groth (2008) for Germany, and by Wiesmeth et al. (2018) as a proposal for Georgia. These systems obviously need a more sophisticated infrastructure: charging and returning the deposit

fee, a clearing house, logistics to collect the returned bottles (see, for example, the DPG Deutsche Pfandsystem GmbH for details of the German deposit system). Fig. 17.1 presents some details of a deposit system for drinks packaging.

One of the findings of Dace and Pakere (2013) for the deposit system for beverage containers in Latvia is that collection costs range between 1.5 Euro Cent (manual collection) and 3.6 Euro Cent (automatic collection) per unit of packaging (see p. 323, Fig. 9). The OECD report refers to a deposit system for beverage producers in the Republic of Korea, with costs amounting to 40% of costs of manufacturing a bottle (see OECD, 2016, p. 279f). Wiesmeth et al. (2018) arrive at costs of 2.24 Euro Cent per bottle including licensing for participation in a collective PRO (see the next section).

One aspect regarding mandatory deposit fees needs a closer consideration. What is surprising is that rather low deposit fees for one-way drinks packaging in Germany, 25 Euro Cents for a regular one-way bottle, motivate consumers to return 98% of the empty drinks containers. After all, for most households, these fees do not really represent an amount of money that causes financial difficulties. However, the difference between Germany and Austria, with Austria not having a mandatory deposit system for drinks packaging, shows that the deposit fee matters (see Section 8.2.2).

There are two ways to explain this observation, both coming from behavioural environmental economics: first, there could be a behavioural aspect associated with such a fee, which influences the mindset in an environmentally friendly way. Alternatively, returning the empty bottles dutifully could be a "cheap excuse" for not behaving environmentally friendly in other areas (Engel & Szech, 2017). Support from behavioural environmental economics is obviously needed to fully understand this observation (see Chapter 8).

Advanced Disposal Fees: These "advanced" fees are levied on certain recyclable products

FIG. 17.1 Deposit system for drinks packaging. A basic deposit system with the manufacturers as Account Managers and the retailers as Account Claimants. © *Source: Own Drawing.*

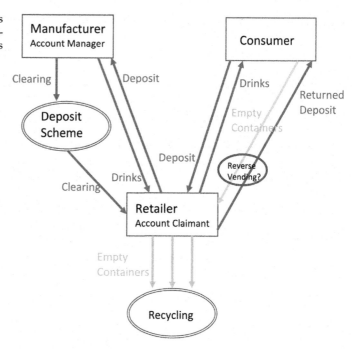

based on the estimated costs of collection and recycling (OECD, 2016, p. 22). However, compared to a refundable deposit, they do not generate individual responsibility and there are fewer incentives for returning the waste products. Nevertheless, according to OECD (2016) these advanced disposal fees constitute a frequently used instrument (see p. 24).

Not to be misunderstood: these advanced disposal fees help to cover collection and recycling costs of certain waste items. They do not provide specific incentives to return waste products to collection points and, since these fees also relate to recycling costs, the incentives for producers for a DfE are limited as long as these fees are not sufficiently differentiated and cannot much or not at all be avoided by producers (see Section 16.1.2). Advanced disposal fees can be levied on fireworks, for example, to financially support street cleaning after New Year's Eve celebrations.

A take back system with a deposit, establishing individual responsibility, integrates waste best according to the constitutive elements of an IEP. However, there is the cost of the necessary infrastructure and the organisation of such a system. Therefore, local conditions need to be observed when deciding on a collection system.

17.4.2 Producer responsibility organisation

After the decision on appropriate collection and segregation systems for recyclable waste products, the question remains as to how to organise the technical and financial responsibilities. These activities can be outsourced to a third-party organisation, a producer responsibility organisation (PRO), as business companies need not be burdened with these issues, which are often characterised by economies of scale. Vested interests to be discussed in the course of this subsection provide an additional reason to take these decisions from producers. This aspect again refers to the constitutive elements of an IEP.

With the help of a PRO, producers can collectively manage the collection and treatment of the waste products and thus comply with the legal regulations. Consequently, PROs are a widely used concept in the context of EPR policies. Both BIO Intelligence Service (2014) and OECD (2016) provide surveys and reviews of this concept and its different manifestations, some of which are covered in this subsection.

The reports of Cahill et al. (2010), BIO Intelligence Service (2014), Gupt and Sahay (2015), OECD (2016), Tencati et al. (2016), and Wang et al. (2017) point to a huge variance regarding the structure and the performance (technical, economic) of the systems in use. One of the important issues refers to the question of for-profit PROs. BIO Intelligence Service (2014), for example, referring to certain performance indicators regarding "fair competition", comes to the conclusion that "…there is no evidence that a centralised organisation is preferable to the introduction of competition among PROs or vice-versa" (p. 24f). However, since it is not possible to create "perfect competition" (see Section 6.3) among PROs anyway, none of these systems or organisations can be expected to function properly. For example, comparisons of different EPR systems allow for many results, depending on the selection of the performance indicators, among other things.

The following review focuses on some basic PRO systems, which play a role in practical EPR policies, are of relevance for industrialised and developing countries, for emerging economies and transition countries. Even more importantly, they allow an examination of the constitutive elements of an IEP. Other systems often contain features of these basic systems or combinations thereof: Walther, Steinborn, Spengler, Luger, and Herrmann (2010) look at the WEEE system in Germany, Gupt and

Sahay (2015) and (OECD, 2016, Chapter 3) consider EPR systems in various developing and developed countries, and Wang et al. (2017) analyse the Chinese fund system for WEEE. Again, further aspects regarding drinks packaging in Georgia are provided in Wiesmeth et al. (2018).

Individual Implementation Systems: An individual system transfers the obligations to an individual company, which establishes its individual PRO or which transfers its collection and recycling obligations to an individual PRO. The polluter pays principle (see Section 16.1) seems to be perfectly integrated – at least at first glance (see OECD, 2016, p. 164f). After all, the "polluter", the company under consideration, is in charge of paying for these basic environmental services. Occasionally, such systems are associated with "Individual Producer Responsibility" (IPR). Rotter, Chancerel, and Schill (2011) argue that IPR creates incentives for "Design for Recycling" (DfR) (see also Chapter 20), but that "in practice, implementing IPR is challenging", in particular if applied to WEEE. Atasu and Subramanian (2012) investigate implications of "collective" and "individual" producer responsibility for EPR policies regarding WEEE. According to their findings, IPR offers superior incentives for DfR, and Wang et al. (2017) argue that "IPR makes an enterprise itself be responsible for funding the recovery and disposal of the products it produces, …" and a DfE "by the enterprise will no longer be 'making a wedding dress for others' but rather is in effect related to self-interest".

However, there are some aspects, which reduce the attractiveness of IPR, both for individual companies, but also from an environmental point of view. First of all, costs of such an individual system can be high and prohibitive (Huisman, 2013), and, as already argued before, the question about the "polluter" may have been posed incompletely, because, in a competitive environment, producers also have to observe the demand for their products (see Section 16.1.1, or Wiesmeth & Häckl, 2011).

Moreover, the ultimate goal of EPR policies is to provide incentives for a DfE, for waste prevention. It needs to be investigated carefully to what extent these incentives are not "disrupted" by certain PRO systems, in particular systems based on IPR. This in turn refers to the constitutive elements of an IEP.

Regarding costs, such a system could be useful in the following situation: chain stores, with shops all over a country, may consider setting up their own system, as there is already the logistic system for deliveries, which can also be used for collecting the waste products, empty drinks containers, for example. This is happening in Germany, where various discounters offer beverages in specially designed one-way containers (see, for example, the statement of Lidl Deutschland). The German Packaging Act allows in §8 certain individual producers or distributors to take back their packaging waste and consign it to treatment and recycling.

What are the incentives for reducing packaging waste and for a DfE in such an individual system? For example, if recycling plastic waste, let's say PET bottles, is profitable, then there is no apparent reason for a DfE. This extends to the case, where recycling, including the collection of waste, is not profitable but is the least unprofitable alternative in the given circumstances. In these and related cases, the proportion of one-way PET bottles could be increased, if this is in line with demand for beverages in one-way bottles.

If there is a mandatory deposit fee for the collection of certain waste products, e.g. PET bottles, then the "deposit leakage" (see Section 20.4.2) provides additional incentives to extend the share of drinks in single-use plastic bottles. This indicates that the interactions of the different parts of an EPR policy need to be observed and analysed with regard to the constitutive elements of an IEP.

Such an individual system can therefore be characterised by "vested interests": companies have legitimate business interests. However, these business interests may conflict with environmental issues, with the waste hierarchy.

This lack of incentive-compatibility in individual systems has not been adequately addressed so far (see, however, Wang et al., 2017 for highlighting some incentive issues in this context). One reason may be that individual systems do not play much of a role in practice, probably due to too excessive costs for most producers (see Huisman, 2013; OECD, 2016, p. 175). However, the issue of vested interests is also of relevance for collective systems to be discussed next.

Collective Implementation Systems: The review reports BIO Intelligence Service (2014), Gupt and Sahay (2015), and OECD (2016) refer to a collective system, "when several producers decide to collaborate and thus transfer their responsibility to a specific organisation", a PRO, which is a for-profit or non-profit organisation (BIO Intelligence Service, 2014, p. 29). Moreover, companies may choose to collectively fulfil their environmental obligation through an "association" or rather join an independent PRO in a competitive environment, a "compliance scheme". The availability of these alternatives depends, of course, on the detailed legal regulations in a country, and fostering competition is the driving force behind the postulation of several PROs (see BIO Intelligence Service, 2014, p. 133f; OECD, 2016, p. 113f).

In both reports, but also in the reviews of Cahill et al. (2010) and Tencati et al. (2016), there are no detailed comparisons of collective systems based on associations on the one hand, and on compliance schemes on the other in terms of their incentive structure. This could be problematic due to the possibility of vested interests of producers collaborating in an association.

Collective Implementation System – Association: The idea to form an association among producers and distributors to implement the obligations of the EPR policy collectively seems, again at first glance, to be an optimal way to implement the polluter pays principle. Economies of scale reduce costs and the association can transfer the relevant obligations to a PRO controlled by the association. Gupt and Sahay (2015) and Wang et al. (2017) refer to such systems and provide additional information. Fig. 17.2 shows the basic structure, with the companies as members of the association potentially benefiting from the operations.

There are some obvious issues with such associations, and there are some more subtle aspects, relating to the environmental performance of EPR systems based on associations. First of all, problems could arise in sharing the costs of the system or perhaps also in the profits or revenue from recycling. In this context, Gui, Atasu, Ergun, and Toktay (2016) refer to weight-based proportional cost allocations, which do not take into account the heterogeneity of the products of different manufacturers, and thus jeopardise the stability of the association.

Another problem refers to the fact that, due to competition, the association or some members could have a critical view of potential new entrants: such an association could function as an instrument to prevent market entry. In addition, why should the association care much about the quality of its services, about the environment? The public authorities have only incomplete information on the operations, and limited possibilities to influence these operations. In particular, the export of comparatively large quantities of WEEE to developing countries to reduce recycling costs (see Ongondo, Williams, & Cherrett, 2011; Sovacool, 2019) could also be related to such unfavourable organisational structures.

Further aspects refer to the obvious difficulties to extend such an association to other areas of waste management, from drinks packaging to packaging waste or WEEE, for example. Additional associations would have to be established with further challenges and probably higher costs, although some organisational issues could remain simpler. OECD (2016) provide a survey on the advantages and disadvantages of different EPR systems in France (p. 249f).

As in individual systems, the incentives for reducing waste or a DfE are therefore threatened by vested interests: reasonable prices for recycled plastics, for example, or the deposit leakage, can provide incentives to increase the

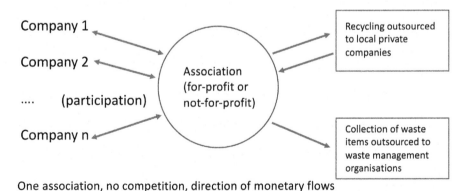

Association

Companies are members of the Association

Company 1

Company 2

....   (participation)

Company n

Association (for-profit or not-for-profit)

Recycling outsourced to local private companies

Collection of waste items outsourced to waste management organisations

One association, no competition, direction of monetary flows

FIG. 17.2   Collective implementation – Association. The producers, as members of the Association, influence decisions with vested interests. © *Source: Own Drawing.*

share of plastic bottles, while non-profitable recycling can provide incentives to ship plastic waste to other countries, such as from the EU to various Asian countries.

Collective systems based on associations thus lose their initial attractiveness, they do not fully meet the constitutive elements of an IEP.

Collective Implementation – Compliance Schemes: As already indicated, a compliance scheme is (usually) a private company, largely independent from the producers, certified and accredited by the public authorities. For handling waste products, packaging, WEEE, and others, it receives revenue from licence fees. Each producer or distributor has to join such a scheme for licensing the products to which the regulations refer. This can be, for example, packaging material in general, drinks packaging in particular, or also electronic equipment. The fees to be paid result from competition among the compliance schemes. The German Green Dot is one such compliance system – in competition with others (see also Section 20.5 for some information on these licence fees).

The main difference between a compliance scheme and a general PRO is the aspect of independence from the producers. By assumption, a compliance scheme is a PRO, whose decisions cannot be directly influenced by the producers. As outlined above, this is usually not the case for a PRO operating a collective system based on an association. Producers as members of the association always have some possibilities to influence the decisions of the PRO, which may arise from vested interests. One of the features of a collective system based on compliance schemes is therefore the weakening or even elimination of the vested interests of producers, which are not always in accordance with the environmental goals of the EPR policy. Fig. 17.3 shows the basic structure of a collective system with compliance schemes. In their activities the compliance schemes are independent of the companies.

In such a system there is a clear incentive for a DfE, which also contributes to waste reduction: less packaging material, lighter bottles or refillable bottles lower the licensing fees to be paid to the compliance scheme – regardless of the situation in the volatile recycling markets. Moreover, producers have to be aware that, if consumers are properly integrated, most old

Compliance Schemes

Companies cannot influence the operations of the compliance schemes

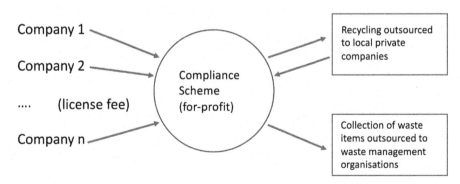

Several independent schemes, competition, direction of monetary flows

FIG. 17.3 Collective implementation – Compliance schemes. The companies joining a compliance scheme pay the licence fees, but cannot affect its decisions. © *Source: Own Drawing.*

equipment and packaging will be collected and consigned to recycling, which will drive up costs and then licensing fees. This provides further incentives for a DfE, and vested interests are no longer apparent, as producers can only reduce their overall costs by reducing the environmental impact of their products.

What about not-for profit compliance schemes? In the report on EPR systems in the EU, OECD (2016) list only 13 for-profit PROs in their sample of 36 (p. 28). The EPR Alliance, an association specifically dedicated to not-for profit PROs, outlined the typical arguments against for-profit PROs (OECD, 2016, p. 77): the question of an unclear distribution of profits, of a preferential treatment of profitable participants, of entry barriers for small producers and of non-profit interests in a PROs operations such as education and prevention. However, these questions relate mainly to a PRO for an association of producers, characterised by vested interests. Therefore, these arguments are not relevant for independent compliance schemes.

Experience shows that the quality of services, which can only be incompletely controlled by the public authorities, could decrease without for-profit compliance schemes in competition (OECD, 2016, p. 47f). Competition forces compliance schemes to offer the best services at reasonable prices, and competition acts as a disseminator of information. Companies can learn and do learn from each other, even when they are in competition. A monopoly, even a temporarily awarded monopoly, or not-for profit PROs, clearly disable or weaken this function.

These considerations show that an important aspect of implementing the EPR principle in the context of an IEP refers to vested interests, which need to be controlled. The design of concrete IEPs for different areas of relevance for implementing a circular economy in the chapters of Part V will provide additional insight.

Since the implementation of the waste hierarchy is closely linked to the EPR principle, the following final chapter of Part IV examines the "economics of the waste hierarchy".

References

Agamuthu, P., & Victor, D. (2011). Policy trends of extended producer responsibility in Malaysia. *Waste Management & Research, 29*(9), 945–953. https://doi.org/10.1177/073424 2X11413332.

Aldieri, L., Carlucci, F., Vinci, C. P., & Yigitcanlar, T. (2019). Environmental innovation, knowledge spillovers and policy implications: A systematic review of the economic effects literature. *Journal of Cleaner Production, 239*, 118051. https://doi.org/10.1016/j.jclepro.2019.118051.

Atasu, A., & Subramanian, R. (2012). Extended producer responsibility for E-waste: Individual or collective producer responsibility? *Production and Operations Management, 21*(6), 1042–1059. https://doi.org/10.1111/j.1937-5956.2012.01327.x.

Bartl, A. (2014). Ways and entanglements of the waste hierarchy. *Waste Management, 34*(1), 1–2. https://doi.org/10.1016/j.wasman.2013.10.016.

Bilitewski, B., Härdtle, G., & Marek, K. (1997). *Waste management*. Berlin: Springer-Verlag. https://doi.org/10.1007/978-3-662-03382-1.

BIO Intelligence Service (2014). *Development of guidance on extended producer responsibility (EPR). Report for the European Commission DG ENV*. Retrieved from (2014). https://ec.europa.eu/environment/archives/waste/eu_guidance/pdf/Guidance%20on%20EPR%20-%20Final%20Report.pdf.

Boustead, I. (1996). LCA — How it came about. *The International Journal of Life Cycle Assessment, 1*(3), 147–150. https://doi.org/10.1007/BF02978943.

Broitman, D., Ayalon, O., & Kan, I. (2012). One size fits all? An assessment tool for solid waste management at local and national levels. *Waste Management, 32*(10), 1979–1988. https://doi.org/10.1016/j.wasman.2012.05.023.

Cahill, R., Grimes, S. M., & Wilson, D. C. (2010). Review article: Extended producer responsibility for packaging wastes and WEEE—A comparison of implementation and the role of local authorities across Europe. *Waste Management & Research, 29*(5), 455–479. https://doi.org/10.1177/0734242X10379455.

Cobo, S., Dominguez-Ramos, A., & Irabien, A. (2018). From linear to circular integrated waste management systems: A review of methodological approaches. *Sustainable Resource Management and the Circular Economy, 135*, 279–295. https://doi.org/10.1016/j.resconrec.2017.08.003.

Corsini, F., Rizzi, F., & Frey, M. (2017). Extended producer responsibility: The impact of organizational dimensions on WEEE collection from households. *Waste Management,* *59,* 23–29. https://doi.org/10.1016/j.wasman.2016.10.046.

Dace, E., & Pakere, I. (2013). Evaluation of economic aspects of the deposit-refund system for packaging in Latvia. *Management of Environmental Quality: An International Journal,* *24*(3), 311–329. https://doi.org/ 10.1108/14777831311322631.

Edmondson, D. L., Kern, F., & Rogge, K. S. (2019). The co-evolution of policy mixes and socio-technical systems: Towards a conceptual framework of policy mix feedback in sustainability transitions. *Policy Mixes for Sustainability Transitions: New Approaches and Insights through Bridging Innovation and Policy Studies,* *48*(10), 103555. https://doi.org/10.1016/j.respol.2018.03.010.

Engel, J., & Szech, N. (2017). *Little good is good enough: Ethical consumption, cheap excuses, and moral self-licensing. Discussion Paper No. 17-28 GEABA.* Retrieved from (2017). http://www.geaba.de/wp-content/uploads/2017/07/DP_17-28.pdf.

Ghisellini, P., Cialani, C., & Ulgiati, S. (2016). A review on circular economy: The expected transition to a balanced interplay of environmental and economic systems. *Towards Post Fossil Carbon Societies: Regenerative and Preventive Eco-Industrial Development,* *114,* 11–32. https://doi.org/10.1016/j.jclepro.2015.09.007.

Groth, M. (2008). *A review of the German mandatory deposit for one-way drinks packaging and drinks packaging taxes in Europe. 87.* University of Lüneburg Working Paper Series in Economics. Retrieved from (2008). https://core.ac.uk/download/pdf/6781128.pdf.

Gu, Y., Wu, Y., Xu, M., Wang, H., & Zuo, T. (2017). To realize better extended producer responsibility: Redesign of WEEE fund mode in China. *Journal of Cleaner Production,* *164,* 347–356. https://doi.org/10.1016/j.jclepro.2017.06.168.

Guerrero, L. A., Maas, G., & Hogland, W. (2013). Solid waste management challenges for cities in developing countries. *Waste Management,* *33*(1), 220–232. https://doi.org/10.1016/j.wasman.2012.09.008.

Gui, L., Atasu, A., Ergun, Ö., & Toktay, L. B. (2016). Efficient implementation of collective extended producer responsibility legislation. *Management Science,* *62*(4), 1098–1123. Retrieved from (2016). https://ideas.repec.org/a/inm/ormnsc/v62y2016i4p1098-1123.html.

Gupt, Y., & Sahay, S. (2015). Review of extended producer responsibility: A case study approach. *Waste Management & Research,* *33*(7), 595–611. https://doi.org/10.1177/0734242X15592275.

Hahladakis, J. N., Purnell, P., Iacovidou, E., Velis, C. A., & Atseyinku, M. (2018). Post-consumer plastic packaging waste in England: Assessing the yield of multiple collection-recycling schemes. *Waste Management,* *75,* 149–159. https://doi.org/10.1016/j.wasman.2018.02.009.

Heidrich, O., Harvey, J., & Tollin, N. (2009). Stakeholder analysis for industrial waste management systems. *Waste Management,* *29*(2), 965–973. https://doi.org/10.1016/j.wasman.2008.04.013.

Huisman, J. (2013). Too big to fail, too academic to function. *Journal of Industrial Ecology,* *17*(2), 172–174. https://doi.org/10.1111/jiec.12012.

Kern, F., Rogge, K. S., & Howlett, M. (2019). Policy mixes for sustainability transitions: New approaches and insights through bridging innovation and policy studies. *Policy Mixes for Sustainability Transitions: New Approaches and Insights through Bridging Innovation and Policy Studies,* *48*(10), 103832. https://doi.org/10.1016/j.respol.2019.103832.

Lindhqvist, T. (2000). *Extended producer responsibility in cleaner production: Policy principle to promote environmental improvements of product systems.* Doctoral Thesis-IIIEE, Lund University. Retrieved from (2000). https://portal.research.lu.se/portal/files/4433708/1002025.pdf.

Massarutto, A. (2014). The long and winding road to resource efficiency – An interdisciplinary perspective on extended producer responsibility. *SI: Packaging Waste Recycling,* *85,* 11–21. https://doi.org/10.1016/j.resconrec.2013.11.005.

McDougall, F. R., White, P. R., Franke, M., & Hindle, P. (2001). *Integrated solid waste management: A life cycle inventory* (2nd ed.). Wiley-Blackwell.

McKerlie, K., Knight, N., & Thorpe, B. (2006). Advancing extended producer responsibility in Canada. *Advancing Pollution Prevention and Cleaner Production – Canada's Contribution,* *14*(6), 616–628. https://doi.org/10.1016/j.jclepro.2005.08.001.

Numata, D. (2016). Policy mix in deposit-refund systems – From schemes in Finland and Norway. *Waste Management,* *52,* 1–2. https://doi.org/10.1016/j.wasman.2016.05.003.

OECD (2001). *Extended producer responsibility: A guidance manual for governments.* Paris: OECD Publishing. https://doi.org/10.1787/9789264189867-en.

OECD (2016). *Extended producer responsibility: Updated guidance for efficient waste management.* Paris: OECD Publishing. https://doi.org/10.1787/9789264256385-en.

Ongondo, F. O., Williams, I. D., & Cherrett, T. J. (2011). How are WEEE doing? A global review of the management of electrical and electronic wastes. *Waste Management,* *31*(4), 714–730. https://doi.org/10.1016/j.wasman.2010.10.023.

Rotter, V. S., Chancerel, P., & Schill, W.-P. (2011). Practicalities of individual producer responsibility under the WEEE directive: Experiences in Germany. *Waste Management & Research,* *29*(9), 931–944. https://doi.org/10.1177/0734242X11415753.

Rubio, S., Ramos, T. R. P., Leitão, M. M. R., & Barbosa-Povoa, A. P. (2019). Effectiveness of extended producer responsibility policies implementation: The case of Portuguese and Spanish packaging waste systems. *Journal of Cleaner Production*, 210, 217–230. https://doi.org/10.1016/j.jclepro.2018.10.299.

Salhofer, S., Steuer, B., Ramusch, R., & Beigl, P. (2016). WEEE management in Europe and China – A comparison. *WEEE: Booming for Sustainable Recycling*, 57, 27–35. https://doi.org/10.1016/j.wasman.2015.11.014.

Seyring, N., Dollhofer, M., Weißenbacher, J., Bakas, I., & McKinnon, D. (2016). Assessment of collection schemes for packaging and other recyclable waste in European Union-28 member states and capital cities. *Waste Management & Research*, 34(9), 947–956. https://doi.org/10.1177/0734242X16650516.

Sovacool, B. K. (2019). Toxic transitions in the lifecycle externalities of a digital society: The complex afterlives of electronic waste in Ghana. *Resources Policy*, 64, 101459. https://doi.org/10.1016/j.resourpol.2019.101459.

Tencati, A., Pogutz, S., Moda, B., Brambilla, M., & Cacia, C. (2016). Prevention policies addressing packaging and packaging waste: Some emerging trends. *Waste Management*, 56, 35–45. https://doi.org/10.1016/j.wasman.2016.06.025.

Uyarra, E., & Gee, S. (2013). Transforming urban waste into sustainable material and energy usage: The case of Greater Manchester (UK). *Special Issue: Advancing Sustainable Urban Transformation*, 50, 101–110. https://doi.org/10.1016/j.jclepro.2012.11.046.

Walls, M. (2006). *Extended producer responsibility and product design: Economic theory and selected case studies. Discussion Papers, Resources for the Future, (dp-06-08).* Retrieved from (2006). https://ideas.repec.org/p/rff/dpaper/dp-06-08.html.

Walther, G., Steinborn, J., Spengler, T. S., Luger, T., & Herrmann, C. (2010). Implementation of the WEEE-directive—Economic effects and improvement potentials for reuse and recycling in Germany. *The International Journal of Advanced Manufacturing Technology*, 47(5), 461–474. https://doi.org/10.1007/s00170-009-2243-0.

Wang, H., Gu, Y., Li, L., Liu, T., Wu, Y., & Zuo, T. (2017). Operating models and development trends in the extended producer responsibility system for waste electrical and electronic equipment. *Resources, Conservation and Recycling*, 127, 159–167. https://doi.org/10.1016/j.resconrec.2017.09.002.

Wiesmeth, H., & Häckl, D. (2011). How to successfully implement extended producer responsibility: Considerations from an economic point of view. *Waste Management & Research*, 29(9), 891–901. https://doi.org/10.1177/0734242X11413333.

Wiesmeth, H., & Häckl, D. (2016). Integrated environmental policy: A review of economic analysis. *Waste Management & Research*, 35(4), 332–345. https://doi.org/10.1177/0734242X16672319.

Wiesmeth, H., Shavgulidze, N., & Tevzadze, N. (2018). Environmental policies for drinks packaging in Georgia: A mini-review of EPR policies with a focus on incentive compatibility. *Waste Management & Research*, 36(11), 1004–1015. https://doi.org/10.1177/0734242X18792606.

Wilson, D. C. (1996). Stick or carrot? The use of policy measures to move waste management up the hierarchy. *Waste Management & Research*, 14(4), 385–398. https://doi.org/10.1006/wmre.1996.0039.

Winkler, J., & Bilitewski, B. (2007). Comparative evaluation of life cycle assessment models for solid waste management. *Life Cycle Assessment in Waste Management*, 27(8), 1021–1031. https://doi.org/10.1016/j.wasman.2007.02.023.

Xiang, W., & Ming, C. (2011). Implementing extended producer responsibility: Vehicle remanufacturing in China. *Journal of Cleaner Production*, 19(6), 680–686. https://doi.org/10.1016/j.jclepro.2010.11.016.

18

The economics of the waste hierarchy

Due to its importance, this chapter reconsiders the waste hierarchy (see Section 9.3) with a particular focus on the economic context. First of all, there is a review of the perception of this concept in the practice-oriented literature, which is of special relevance for implementing a circular economy. The following sections examine economic aspects of waste prevention, reuse and recycling, again with a view to literature and practice. Societal path dependencies (Section 12.2) illustrate some additional empirical observations. To sum up, this review prepares the ground for the development of appropriate "Integrated Environmental Policies" (IEP) in the chapters of Part V.

18.1 The waste hierarchy — Revisited

According to Van Ewijk and Stegemann (2016), the concept of the waste hierarchy most likely originated as early as the 1980s, with the reduction, recycling and reuse of hazardous waste taking precedence over treatment or disposal (p. 123). The Waste Directive of the European Union (EU) of 2008 encourages member states "to apply the waste hierarchy and, in accordance with the polluter-pays principle, a requirement that the costs of disposing of waste must be borne by the holder of waste, by previous holders or by the producers of the product from which the waste came". The Directive further formulates: "The following waste hierarchy shall apply as a priority order in waste prevention and management legislation and policy: (a) prevention; (b) preparing for re-use; (c) recycling; (d) other recovery, e.g. energy recovery; and (e) disposal" (see Art. 4).

Section 9.3 sets out scientific reasons for compliance with the waste hierarchy. In practice, a lack of suitable sites for landfilling is one of them. Meanwhile, however, "scarcity" is no longer just a matter of the size of a country. The potential hazards posed by more or less uncontrolled and unregulated landfills are causing serious concern to more and more residents in a wide range of countries. Obviously, the assimilative capacity of the environment to receive additional and increasing amounts of waste is "perceived" as exhausted. This perception is supported by various reports on pollution from waste electrical and electronic equipment (WEEE) and chemicals, plastic waste and micro-plastics, and other problematic and toxic substances (see EEA, 2019, Part I).

Implementing the Circular Economy for Sustainable Development
https://doi.org/10.1016/B978-0-12-821798-6.00018-1

213

The other important reason for adhering to the waste hierarchy is the saving of natural resources – by reducing waste through innovative products, through a "Design for Environment" (DfE), but also through reuse and recycling. In particular, recycling activities are meanwhile widespread, and the collecting and separation of waste for recycling is a recognised behavioural routine for households in many countries, especially but not only in developed countries.

18.1.1 Waste hierarchy in practice – A literature review

There is some literature on the waste hierarchy and its implementation in practice. Gharfalkar, Court, Campbell, Ali, and Hillier (2015) analyse the waste hierarchy in the EU Waste Directive, complaining a lack of clarity on the main concepts. In their view, there is "an overlap between measures such as 'prevention' and 'reduction', 'preparing for reuse' and 'reuse' and lack of clarity on why the measure of 'reuse' is included in the ... definition of 'prevention'". They then develop an alternative "hierarchy of resource use", which "includes the measures of 'reuse' and 'recovery' in the hierarchy and considers 'waste' as a 'resource'. Interestingly, they also propose replacing the measure of 'prevention' with a new measure of 'replacement', with the primary objective of 'replacement' to prevent the consumption of virgin natural resources" (p. 310).

Also Pires and Martinho (2019) propose a waste hierarchy index "to measure the waste hierarchy within a circular economy context, applied to municipal solid waste". This index focuses on "recycling and preparing for reuse ... as positive contributors to the circular economy, and incineration and landfill as negative contributors". Thus, this index provides a benchmark for the implementation of the waste hierarchy, with a focus on various forms of reuse and recycling (p. 299).

Hultman and Corvellec (2012) examine how the waste hierarchy, or more precisely the European waste hierarchy, which ranks "the desirability of different waste-management approaches according to their environmental impact", has been interpreted in different organisational contexts. They point to some "paradoxical relationships between economy and society on the one hand, and environment and nature on the other" (see Abstract), which leaves room for further discussion.

Finally, Van Ewijk and Stegemann (2016) point out that the aims of the waste hierarchy "coincide largely with those of dematerialisation", which "simplifies the complex interactions between nature and the economy by linking environmental degradation to material throughput" (p. 124). In this sense, the waste hierarchy certainly helps to divert waste from landfills, but in their opinion the relationship between the waste hierarchy and environmental impact is not so clear: landfilling waste could become more attractive than incineration with increasing transport distances, or "when the environmental impacts from landfill are assumed to occur over a relatively short time frame". In addition, food waste, garden waste and other kinds of waste are not necessarily covered by the waste hierarchy (p. 124), and there appear to be shortcomings in a priority order: for example, including an option such as landfilling "legitimises its existence" (p. 126).

18.1.2 Economic integration of the waste hierarchy

With the above considerations, the waste hierarchy suddenly seems to be transformed into a complex concept: the different parts cannot be defined unambiguously and without overlapping. There are obviously different ways of interpreting the waste hierarchy, and it does not seem to cover all types of waste. What to do with such a construct, which, on the other hand, is of utmost importance for implementing a circular economy?

Traditional economics provides an explanation for this seemingly problematic situation: consider waste as an environmental good, or, rather, an environmental bad, for which the allocation problems have to be solved, as discussed in Chapter 6. These basic problems will be solved, even if no measures are taken to protect the environment. They will also be solved, with more environmentally friendly outcomes, if there are appropriate environmental policies. What should be the result, what should be relevant characteristics of a feasible allocation with environmental commodities?

If waste is seen as a "bad", which is undesirable for everyone, or at least for most people, then any proposal that seeks an efficient allocation must attempt to reduce waste to some minimum, without exactly defining this minimum, which may vary for different types of waste and depend on local conditions.

Even if waste is considered a "design flaw" (see Section 18.2.3), not all waste can be prevented, as this would likely imply a more or less complete dematerialisation of the economy. So what to do with the remaining waste commodities? There is the possibility of reuse, and there is the possibility of recycling. Again, it is difficult, if not outrightly impossible, to find the optimal extent of the reuse of certain commodities or to identify optimal targets for recycling and the appropriate recycling method, such as material or thermal recycling. Of course, this lack of information leads to the question of the economic reasonableness of targets etc. (Section 11.1).

Thus, the integration of the waste hierarchy into economics provides an explanation for these vaguenesses and inaccuracies in the attempts to outline the waste hierarchy. It is an explanation, which helps to understand this situation, but does not lead to an immediate solution. It will remain, and it must remain the responsibility of each country, of each region,

to decide on an appropriate interpretation of the waste hierarchy, on possible differences between waste prevention and waste reduction, and on the targets and methods to be associated with it.

In this sense, the waste hierarchy offers some guidelines from economics, which must be brought to life for the implementation of a circular economy. Interestingly, the EU Waste Directive also calls on decision-makers to apply the waste hierarchy carefully: "when applying the waste hierarchy ... member states shall take measures to encourage the options that deliver the best overall environmental outcome. This may require specific waste streams departing from the hierarchy where this is justified by life-cycle thinking on the overall impacts of the generation and management of such waste" (see Art. 4).

18.2 Prevention of waste – The forgotten child

As there are, as indicated above, different concepts and perceptions of "waste prevention" of use in practical contexts, we are first referring to the EU Waste Directive of 2008, which defines "prevention" as "measures taken before a substance, material or product has become waste, that reduce: (a) the quantity of waste, including through the re-use of products or the extension of the life span of products; (b) the adverse impacts of the generated waste on the environment and human health; (c) the content of harmful substances in materials and products" (see Art. 3).

By comparing various definitions, also Gharfalkar et al. (2015) point to some conceptual issues regarding waste prevention, arguing, in particular, that the definition of waste prevention given in the Waste Directive is primarily aimed at reducing waste and not at prevention (p. 308). And Van Ewijk and Stegemann (2016) already draw attention to a deeper topic, "a conceptual problem of the waste hierarchy that may hamper prevention efforts through policy": the inclusion of prevention in a tool that is essentially intended for waste managers (p. 125). In fact, reducing "the adverse impacts of the generated waste on the environment and human health" could be applied to recycling, if one overlooks the fact that this refers only to "measures taken before a substance, material or product has become waste".

Wilts (2012) considers preventing waste "nothing less than a fundamental change of the socio-technical system of waste infrastructures and requires a transition from end-of-pipe technologies towards an integrated management of resources" (p. 29). There are also the questions of how to measure waste prevention and how waste prevention can be linked to the waste management sector.

In order to find a practical definition of waste prevention, Corvellec (2016) presents and discusses a sample of Swedish waste prevention initiatives. His analysis shows that "the initiatives in the sample boil down to three main types of actions: raising awareness about the need to prevent waste, increasing material efficiency, and developing sustainable consumption" (see Abstract).

18.2.1 Various aspects of waste prevention

These introductory remarks on the perception of waste prevention in the practice-oriented literature reveal aspects similar to those addressed in the context of the waste hierarchy: there are questions as to the precise definition of the concept and there are differences in practical implementation. These questions, too, result, at least to some extent, from the economic background of waste prevention, as discussed together with the economic integration of the waste hierarchy: due to a lack of information, it is simply not possible to describe all the characteristics of an efficient allocation, including waste prevention, not to mention the fact that there are typically many

efficient allocations with probably very different characteristics – also with respect to the level of preventing waste.

This helps to explain the conceptual questions of Gharfalkar et al. (2015) and the observations of Corvellec (2016) mentioned above, and is probably also responsible for the difficulties to measure prevention of waste: according to Wilts (2012), how to measure something that does not exist because it was prevented? In the context of the economic allocation problems, it is similarly difficult to obtain information on prevention efforts. Wilts (2012) presents various indicators for waste prevention and points to the need for more process-oriented indicators instead of waste-based indicators (p. 31).

There are some more questions, which need a closer examination. The conceptual problem of the waste hierarchy addressed by Van Ewijk and Stegemann (2016) is "vested interest among waste collectors to collect as much as possible" (p. 125), thereby reducing the prevention efforts. "Vested interests" have already been discussed (see Section 17.4) and they need to be taken into account in environmental policies for implementing a circular economy.

Another aspect relates to the remark of Gharfalkar et al. (2015) that waste prevention in the EU Waste Directive is actually focused on waste reduction. Assuming that too much waste is generated in the current situation, the policy objective of reducing waste is obvious, and measuring waste reduction seems to be easier, as it is possible to compare waste generation in different situations, either per unit of product, per capita, or per unit of gross domestic product (GDP), thus making again use of traditional waste-based indicators (see Wilts, 2012, p. 31). This is also the approach taken in Germany (see the case study in Section 19.4).

Fig. 18.1 shows this situation for Germany with significant reductions in net waste volume. However, the reductions result mainly from smaller quantities of construction and

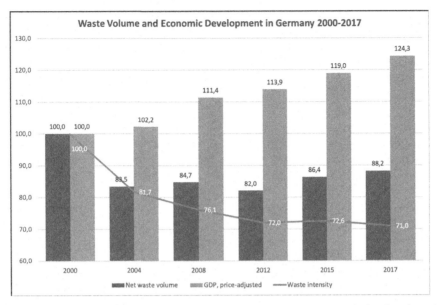

FIG. 18.1 Waste volume and economic growth in Germany 2000–2017. Waste intensity declined in the period 2000–2017. © *Own Drawing with Data from the German Federal Statistical Office 2019 (https://www.destatis.de/DE/Themen/Gesellschaft-Umwelt/Umwelt/Abfallwirtschaft/Publikationen/Downloads-Abfallwirtschaft/abfallbilanz-xlsx-5321001.xlsx?_blob=publicationFile).*

demolition waste, which comprises about 60% of total waste volume. Household waste has been more or less on the same level since 2000.

This focus on waste reduction in the EU Waste Directive leads immediately to the important issue of design changes, the DfEs. Reducing waste is therefore closely linked to DfEs, and appropriate policies related to the implementation of a circular economy need to respect this high-priority goal. Of course, this also applies in a similar way to the prevention of waste: there should be motivation to design new commodities or develop new technologies such that the input of raw materials is kept to a minimum, so that producers make use of the "best available techniques and technologies" (see Section 15.3).

In summary, these results show that policies that encourage prevention and reduction of waste should therefore provide incentives for DfEs, should promote the application of the latest techniques and technologies, and should be free of vested interests, which tend to undermine the environmental objectives of the policies. These policies should therefore be based on the EPR principle (see Section 17.2). What is the current situation regarding waste reduction and prevention?

18.2.2 Some empirical evidence

According to the empirical results gathered from current environmental policies in different countries and referring to different types of waste streams, the evidence regarding waste prevention is not very convincing, to say the least. The review reports of BIO Intelligence Service (2014) and OECD (2016) and others certainly point to some success, but if we have to accept that packaging waste in Germany has increased by 19% and plastics packaging has increased by 74% since 2000, according to a study of the German Environment Agency (UBA), it seems clear that this development is not in line with successful circular economy strategies. Moreover, the decrease in the volume of waste shown in Fig. 18.1 is mainly due to declining quantities of demolition and construction waste (see also BMU, 2018, Fig. 18.1).

It is true, of course, that it is difficult to measure and document evidence of waste prevention or waste reduction (Wilts, 2012). It is, moreover, also true that other effects, such as economic growth, lead to an increasing demand for commodities resulting in more waste products and, in particular, more packaging waste (Tencati, Pogutz, Moda, Brambilla, & Cacia, 2016). However, this does not really explain some observations and developments regarding waste in different countries in recent decades (see also Chapter 5).

Once again, waste prevention, although considered a priority issue in most waste management regulations, seems to disappear from the agenda as soon as we have arrived at the practice of waste management. The Green Dot of Germany, for example, one of the major producer responsibility organisations (PRO) or compliance schemes mainly for packaging waste, thus promotes a "Design for Recycling". Certainly, this is a promising attempt for a DfE, a "sustainable package is not only characterised by the use of the materials concerned, but is also crucially influenced by the design and the processing involved". But, the primary focus is clearly on recycling, not on prevention of waste.

Moreover, if recycling is established, because it promises to be a profitable business, recycling as much waste as possible could be one of the business strategies – not necessarily in agreement with the strategies for implementing a circular economy. Not to be misunderstood: profitable recycling activities can be a good thing, especially for developing countries. However, problems may arise once recycling activities are set up with the primary objective of generating economic profits, of creating additional jobs. Waste can then be regarded as a "renewable resource" of which as much as possible should be produced. This distortion of

what needs to be done indicates the influence of societal path dependencies.

These observations and these remarks seem to reflect the opinion of Van Ewijk and Stegemann (2016) that the inclusion of waste prevention into a policy intended for waste management is probably not very helpful, as practical waste management wants to prove and has to prove its success by showing the figures of high collection and recycling rates.

This again points to societal path dependencies that are likely to have arisen from the traditional consideration and handling of waste.

18.2.3 Societal path dependencies

The review of the literature on the waste hierarchy in general and waste prevention and reduction in particular reveals a certain discrepancy between the priority goal of waste prevention and the remaining obligations of the waste hierarchy. Thus, waste prevention is difficult to measure and, even more importantly, seems to be at odds with various other business interests under the umbrella of waste management.

In addition, there seems to be a broad, more or less tacit agreement that waste collected and recycled, materially or thermally, is no longer considered waste, as it has lost its harmful nature for the environment. As indicated in Section 12.2, this could be associated with societal path dependencies (Wilts, 2012): waste is something we want to get rid of. And one way to achieve this is to collect waste and "eliminate" it by incineration, by recycling.

Not so long ago, waste that was simply land-filled, was no longer considered waste: it was no longer visible and did not bother anyone. In this respect, at least in most industrialised countries, some progress has been made, justified by the different types of waste streams with a higher damaging potential for both humans and the environment compared to the situation a few decades ago, when household waste consisted to a large extent of biodegradable components, such as bio-waste or paper. Plastic waste in particular made the difference in addition to the ever-increasing quantities of waste, requiring more and more landfill area. Developing countries, as well as emerging economies, were, and some of them are still "surprised" by the quantities of new types of waste clogging their landfills.

These new and different waste streams, most of which are non-biodegradable and potentially pose long-term environmental and health risks, are also a priority objective for taking into account waste prevention: as discussed in Section 9.3, it is simply not possible to collect all the waste commodities. Thus, with each additional unit of waste generated, part of it remains in the environment. Plastic waste in rivers and oceans, including microplastics (Shahul Hamid et al., 2018), is proof of this.

The question, of course, is how to break these societal path dependencies, how to make waste prevention and reduction, so far not too successful (see again the examples in Chapter 5), a genuine priority objective of waste management? In this context, Wilts (2012) points to the path dependencies arising from "waste management as an end-of-pipe approach", which has been optimised at all steps for decades. Probably, these technical path dependencies resulted from or were connected to the societal path dependencies mentioned above. Frondel, Horbach, and Rennings (2007) argue that Germany had the lowest share of investment in cleaner technologies among OECD countries due to the dominance of end-of-pipe technologies and related regulations (see also Wilts, 2012, p. 34). A change is therefore necessary towards more integrated measures to waste prevention. But this implies: more DfEs, more innovative and environmentally friendly technologies and commodities. Here, too, the role of appropriate environmental policies to motivate DfE becomes apparent.

18.3 Reuse of commodities

According to the EU Waste Directive, "reuse means any operation by which products or components that are not waste are used again for the same purpose for which they were conceived". Observe that "the quantity of waste, including through the re-use of products or the extension of the life span of products" is also registered under prevention of waste (see Art. 3).

What are the economic and environmental issues related to reuse? How can the environmentally sound reuse of commodities be stimulated? What about the economic background?

18.3.1 Economic context of reusing commodities

The reuse of commodities is, first of all, an economic activity, motivated by a variety of factors: the price, the uniqueness of the product, its quality, but in combination with the price also the better affordability of used products such as cars or electronic equipment. Some commodities, such as the fine arts, paintings, can even gain value over time, also in connection with reuse.

Thus, from an economic point of view, there is nothing strange about reusing commodities, and in fact, second-hand shops, which now include luxury consignment shops (Orlean, 2019) and a growing variety of online platforms, are helping to extend reuse activities and generate many business opportunities. This also applies to the context of a "sharing economy" with its many characteristics, ranging from transport to food and grocery deliveries to hotel services. Reuse of all kinds of commodities and equipment is undoubtedly flourishing, fuelled by digital transformation (Chapter 13).

If it were only a matter of economic issues, no one would have to worry much about the reuse of commodities. It would not even be necessary to measure the economic extent of reuse in one way or the other, the market mechanism could be trusted to find the balance, the equilibrium. However, the potential environmental benefits make reuse interesting for the context of the implementation of a circular economy.

18.3.2 Environmental benefits of reusing commodities

In principle, reusing commodities can help to prevent and reduce waste by delaying the time till a new product needs to be produced, requiring additional resources, possibly affecting sustainability. Thus, there seems to be no discrepancy between the economic and the environmental contexts of reuse.

However, there is one issue, which immediately comes to mind with a more detailed look: if commodities are reused on a large scale, then business companies will obviously sell less of these products. Consequently, expanding reuse is not necessarily in favour of business activities, which usually focus on all kinds of growth opportunities.

There is an exception to this rule: if used commodities can be sold to customers, who, for whatever reason, would not buy the new products, then reuse will indeed become an additional business opportunity. This applies in particular to used commodities, which can be sold to customers in other countries, perhaps even in countries outside the EU, in order to avoid, for example, special take back requirements for end-of-life vehicles (ELV). In this case, the raw materials used for producing these commodities, these cars, probably remain in these countries, having recycled them under whatever conditions. In addition, there is an incomplete displacement of new commodities through the reused.

The reuse of old products is not always environmentally advantageous. Boldoczki, Thorenz, and Tuma (2020), for example, by pointing out that "reuse is beneficial, if the impacts that arise during a certain usage duration of a reused product are smaller than those of a new product", show that especially so-called "white goods" (washing machines, refrigerators, etc.) have a significant impact on the categories global warming, water consumption, and cumulative energy demand during their use phase. Reuse of inefficient devices should therefore be avoided. Note that the reuse of these commodities as a second washing machine or as a second refrigerator is only a special form of rebound effect (Section 12.1), which is therefore also associated with the reuse of certain commodities. Fig. 18.2 shows this environmental impact of an older, reused washing machine in comparison to a new one, which is more efficient in terms of energy and water consumption.

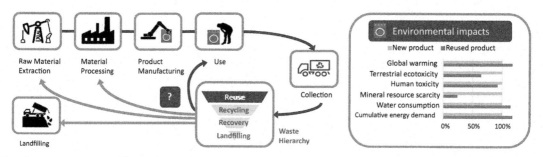

FIG. 18.2   Comparison new and reused product. Environmental impacts of new and reused product. © *From Boldoczki, S., Thorenz, A., & Tuma, A. (2020). The environmental impacts of preparation for reuse: A case study of WEEE reuse in Germany. Journal of Cleaner Production, 119, 736. https://doi.org/10.1016/j.jclepro.2019.119736 (web archive link).*

18.3.3 Societal path dependencies

Zink and Geyer (2017) look at reuse of commodities from a different perspective, from the perspective of rebound effects (Section 12.1). They first point out the widely accepted benefits of avoiding primary production through reuse at the product level. However, it may result in an incomplete replacement of new products with reuse. This is exactly the case, when, for example, used cars are sold to customers in other countries who would not, for whatever reason, buy a new car. Thus, there is incomplete displacement with limited environmental benefits. In addition, the lack of availability or insufficient maintenance of these cars in combination with low quality gasoline can contribute to high levels of air pollution in densely populated cities. Therefore, in such a case, reuse is unlikely to be beneficial from an environmental point of view.

On the other hand, it is ethically and politically impossible to deny people in poorer regions of the world access to cars, computers and other electrical and electronic equipment. So, what to do?

In order to avoid this critical "circular economy rebound", Zink and Geyer (2017) recommend to "market secondary goods in the same way as primary goods", although this "may include educating the public to overcome stigmas", about the quality of secondary goods (p. 599). Where both countries produce the commodities in question, trade activities may also lead to import and export of used commodities and to a certain displacement of new products.

In summary, these observations again refer to societal path dependencies. As explained above, the economic background to reuse does not really correspond to the daily needs of market economies tied to a more or less global network of countries, most of which promote international trade. It will be difficult to convince these export-oriented countries, or rather their export-oriented companies, that they should fundamentally change their business strategies for reusing their products.

Societal path dependencies, in this case the traditional export-orientation of the countries to support their economic growth, are likely to create some obstacles to the reuse of certain commodities, thus helping to more effectively displace the production of new commodities. It remains to be seen whether and how these challenges can be managed.

18.4 Recycling of waste commodities

Referring again to the EU Waste Directive, "recycling" means any recovery operation by which waste materials are reprocessed into products, materials or substances whether for the original or other purposes. It includes the reprocessing of organic material but does not include energy recovery and the reprocessing into materials that are to be used as fuels or for backfilling operations" (see Art. 3). For the sake of completeness, "energy recovery", which is often counted among "recycling" or "thermal recycling", is included in "recovery", following recycling in the EU waste hierarchy.

Of course, there is much to be said about recycling, because it brings together business opportunities worth an incredible amount of money, helps to recover valuable resources for sustainable development, and the good feeling of living in an environmentally conscious country. It is, moreover, in line with societal path dependencies, although there could be hazardous and environmentally dubious recycling processes that affect the implementation of a circular economy.

Not all of these questions, many of which are of a more technical nature, can be examined here. Instead, the focus will be on some economic and, of course, environmental aspects.

18.4.1 Economic context of recycling

Compared to waste prevention and reuse of certain commodities, recycling is attracting much more attention from the general public.

The environmental regulations set targets for collection and recycling of various waste streams, and official reports such as "Waste Management in Germany 2018" or the "2016 Recycling Economic Information Report" of the United States (U.S.), for example, indicate high recycling and recovery rates, thereby pointing to the goals of a circular economy in general and to sustainable development in particular.

What does "recycling" mean from an economic point of view, what can we learn from integrating recycling into the fundamental allocation problems?

In an economic context with the market mechanism solving the allocation problems, recycling plays a role, whenever it is a business case, a business opportunity: the market value of recycled materials exceeds the costs of work and equipment needed for the collection and recycling of waste. This depends, of course, crucially on local costs, wage rates, especially on the recycling technologies, but also on global markets for recycled materials. For this reason, this type of "unregulated" recycling is often seen in developing countries and emerging economies with their still low pay rates and simple technologies. Dias, Bernardes, and Huda (2019) are investigating this issue with respect to the recycling of WEEE in Australia with labour costs accounting for more than 90% of the costs.

This observation immediately points to an environmental problem that may be associated with certain recycling activities: recycling with outdated technologies and perhaps inexperienced workers can lead to environmental degradations, which could outweigh the positive effects of recovering valuable resources. This observation naturally applies to WEEE recycling in some developing countries, such as Ghana (Sovacool, 2019), but this observation also applies to emerging economies, which continue to recycle contaminated plastic waste (see also Chapter 23). Before China imposed an import ban on certain types of plastic waste in 2018, it amassed a variety of environmental problems,

with soil contamination with heavy metals being just one of them (Tang et al., 2015).

In the context of the implementation of a circular economy, it should perhaps be pointed out that attempts in mostly industrialised countries to recycle waste products have contributed to this development, which is certainly not part of a circular economy strategy. Of course, countries that import plastic waste and other types of waste want to benefit from the recovery of raw materials, for example by using plastic waste as a substitute for natural oil, or they will be paid to import and treat that waste. Exporting countries may not have enough recycling facilities, or shipping this waste to other countries is simply the better alternative given the regulatory environment.

This is indeed not an unusual economic model: comparative advantages, which provide an important economic reason for international trade, determine these activities (see e.g. Pindyck & Rubinfeld, 2018). So there is an interesting situation: environmental regulations on recycling lead to exports of certain types of waste to countries that have an interest in recovering resources, but possibly with outdated technologies for recycling.

The consequence of these observations is that recycling needs to be regulated more holistically. It is not enough to control the situation in the industrialised countries, export activities must also be taken into account. This situation could, of course, also be a consequence of recycling targets which are not economically reasonable (Section 11.1).

There is yet another issue associated with the economics of recycling: the commodities resulting from recycling are often not directly comparable to the original resources used for production. There is usually a "down-cycling", meaning that the original raw materials, metals, for example, end up by recycling as part of alloys or go directly to landfills. Ortego, Valero, Valero, and Iglesias (2018) analyse this situation for metal recycling techniques for end-of-life vehicles

(ELV). The recycling of plastic from household waste also leads to recycled plastics with reduced quality, due to contaminations with additives, for example. Eriksen, Christiansen, Daugaard, and Astrup (2019) are investigating the potential for closed-loop recycling of plastic waste.

What does the possibility of down-cycling mean from an economic point of view? An immediate consequence is that the recycled materials are not in a one-to-one relationship to the original materials, and they do not or do not completely replace the original material. This is the case, for example, for recycled plastics: although there are efforts in the U.S., with guidance by the Food & Drug Administration (FDA), and in the EU, regulated by the Framework on Food Contact Materials (see also Lara-Lledó, Yanini, Araque-Ferrer, Monedero, & Vidal, 2018), to increase the use of recycled materials in food-contact materials, the packaging industries still seem to prefer virgin plastics for their products (cf. BUND, 2019, p. 37).

This incomplete replacement refers also to recycled glass. In order to obtain qualities equivalent to newly produced glass, it is in general necessary to separate waste glass according to colour. Alternatively, recycled glass sand could be used as a partial replacement for sand in concrete to save high-quality sand needed for construction (Tamanna, Tuladhar, & Sivakugan, 2020).

Of course, it must be admitted that there are many research activities to expand and improve recycling. Therefore, these latest results and assessments could soon change. However, this could also affect other parts of the waste hierarchy – there is, thus, a need to thoroughly examine and analyse societal path dependencies.

18.4.2 Recycling and path dependencies

As already indicated, technical and societal path dependencies play an important role in the implementation of the waste hierarchy. This also applies to recycling, which is the most visible task of the waste hierarchy.

The problem is that recycling of waste is often seen as equivalent to prevention or reduction of waste: it is waste that does not end up in landfills, at least not directly and not completely. And this seems to correspond to the widely accepted societal need to deal with waste. As a result, more recycling plants with more advanced technologies will enable even better recycling, forgetting the priority goal of waste prevention. Various companies, especially drinks producers, take advantage of this societal path dependency and increase their share of drinks in one-way containers, mostly plastic bottles.

Instead of changing this mindset, rather than breaking this habit, these path dependencies are likely to be reinforced and will continue to govern waste management in many countries. The "systems change" necessary for the transition to a circular economy will take a long time in this way.

In addition to these societal path dependencies, there are also technical path dependencies affecting recycling activities. The presence of waste incineration plants may, for example, lead to the incineration of higher quantities of certain types of waste, as these plants are likely to require a certain minimum load for efficiency reasons. Conversely, the lack of suitable material or thermal recycling facilities can lead, for example, to the export of waste to developing countries, as was the case with plastic waste from industrialised countries. It is clear that these technical aspects can have a huge impact on waste management practices due to the time span required to plan and build new facilities, which may then lead to new path dependencies.

In summary, therefore, the waste hierarchy can be seen as a guideline, which must be adapted to the local conditions in a particular country or region. In order to implement a circular economy, waste management needs to be guided by environmental policies, which respect the technological framework, partially replace and augment the market mechanism and serve as an allocation mechanism for the environmental commodities relevant to waste management. The chapters of Part V take into account all the results achieved in the previous chapters and design IEPs, appropriate holistic policies to implement a circular economy.

References

BIO Intelligence Service (2014). *Development of guidance on extended producer responsibility (EPR). Report for the European Commission DG ENV. Retrieved from (2014).* https://ec.europa.eu/environment/archives/waste/eu_guidance/pdf/Guidance%20on%20EPR%20-%20Final%20Report.pdf.

BMU (2018). *Waste Management in Germany 2018.* Federal Ministry for the Environment, Nature Conservation and Nuclear Safety (BMU). Retrieved from (2018). https://www.bmu.de/fileadmin/Daten_BMU/Pools/Broschueren/abfallwirtschaft_2018_en_bf.pdf.

Boldoczki, S., Thorenz, A., & Tuma, A. (2020). The environmental impacts of preparation for reuse: A case study of WEEE reuse in Germany. *Journal of Cleaner Production, 252,* 119736. https://doi.org/10.1016/j.jclepro.2019.119736.

BUND (2019). *Plastics atlas 2019: Facts and figures about a world full of plastics.* Berlin: BUND, Heinrich-Böll-Stiftung. Retrieved from (2019). https://www.boell.de/de/2019/05/14/plastikatlas?dimension1=ds_plastikatlas.

Corvellec, H. (2016). A performative definition of waste prevention. *Waste Management, 52,* 3–13. https://doi.org/10.1016/j.wasman.2016.03.051.

Dias, P., Bernardes, A. M., & Huda, N. (2019). Ensuring best E-waste recycling practices in developed countries: An Australian example. *Journal of Cleaner Production, 209,* 846–854. https://doi.org/10.1016/j.jclepro.2018.10.306.

EEA (2019). *The European environment—State and outlook 2020 Knowledge for transition to a sustainable Europe.* https://doi.org/10.2800/96749.

Eriksen, M. K., Christiansen, J. D., Daugaard, A. E., & Astrup, T. F. (2019). Closing the loop for PET, PE and PP waste from households: Influence of material properties and product design for plastic recycling. *Waste Management, 96,* 75–85. https://doi.org/10.1016/j.wasman.2019.07.005.

Frondel, M., Horbach, J., & Rennings, K. (2007). End-of-pipe or cleaner production? An empirical comparison of environmental innovation decisions across OECD countries. *Business Strategy and the Environment, 16*(8), 571–584. https://doi.org/10.1002/bse.496.

Gharfalkar, M., Court, R., Campbell, C., Ali, Z., & Hillier, G. (2015). Analysis of waste hierarchy in the European waste directive 2008/98/EC. *Waste Management, 39,* 305–313. https://doi.org/10.1016/j.wasman.2015.02.007.

Hultman, J., & Corvellec, H. (2012). The European waste hierarchy: From the sociomateriality of waste to a politics of consumption. *Environment & Planning A, 44*(10), 2413–2427. https://doi.org/10.1068/a44668.

Lara-Lledó, M., Yanini, M., Araque-Ferrer, E., Monedero, F. M., & Vidal, C. R. (2018). *EU legislation on food contact materials.* Elsevier. https://doi.org/10.1016/B978-0-08-100596-5.21464-9.

OECD (2016). *Extended producer responsibility: Updated guidance for efficient waste management.* Paris: OECD Publishing. https://doi.org/10.1787/9789264256385-en.

Orlean, S. (2019, October 14). *The RealReal's online luxury consignment shop.* The New Yorker. Retrieved from (2019, October 14). https://www.newyorker.com/magazine/2019/10/21/therealreals-online-luxury-consignment-shop.

Ortego, A., Valero, A., Valero, A., & Iglesias, M. (2018). Downcycling in automobile recycling process: A thermodynamic assessment. *Resources, Conservation and Recycling, 136,* 24–32. https://doi.org/10.1016/j.resconrec.2018.04.006.

Pindyck, R., & Rubinfeld, D. (2018). *Microeconomics* (9th ed.). Pearson.

Pires, A., & Martinho, G. (2019). Waste hierarchy index for circular economy in waste management. *Waste Management, 95,* 298–305. https://doi.org/10.1016/j.wasman.2019.06.014.

Shahul Hamid, F., Bhatti, M. S., Anuar, N., Anuar, N., Mohan, P., & Periathamby, A. (2018). Worldwide distribution and abundance of microplastic: How dire is the situation? *Waste Management & Research, 36*(10), 873–897. https://doi.org/10.1177/0734242X18785730.

Sovacool, B. K. (2019). Toxic transitions in the lifecycle externalities of a digital society: The complex afterlives of electronic waste in Ghana. *Resources Policy, 64,* 101459. https://doi.org/10.1016/j.resourpol.2019.101459.

Tamanna, N., Tuladhar, R., & Sivakugan, N. (2020). Performance of recycled waste glass sand as partial replacement of sand in concrete. *Construction and Building Materials, 239,* 117804. https://doi.org/10.1016/j.conbuildmat.2019.117804.

Tang, Z., Zhang, L., Huang, Q., Yang, Y., Nie, Z., Cheng, J., … Chai, M. (2015). Contamination and risk of heavy metals in soils and sediments from a typical plastic waste recycling area in North China. *Ecotoxicology and Environmental Safety, 122,* 343–351. https://doi.org/10.1016/j.ecoenv.2015.08.006.

Tencati, A., Pogutz, S., Moda, B., Brambilla, M., & Cacia, C. (2016). Prevention policies addressing packaging and packaging waste: Some emerging trends. *Waste Management, 56,* 35–45. https://doi.org/10.1016/j.wasman.2016.06.025.

Van Ewijk, S., & Stegemann, J. A. (2016). Limitations of the waste hierarchy for achieving absolute reductions in material throughput. *Journal of Cleaner Production, 132*, 122–128. https://doi.org/10.1016/j.jclepro.2014.11.051.

Wilts, H. (2012). National waste prevention programs: Indicators on progress and barriers. *Waste Management & Research, 30*(9_suppl), 29–35. https://doi.org/10.1177/0734242X12453612.

Zink, T., & Geyer, R. (2017). Circular economy rebound. *Journal of Industrial Ecology, 21*(3), 593–602. https://doi.org/10.1111/jiec.12545.

PART V

Implementing a circular economy

The design of environmental policies for the implementation of a circular economy is the primary objective of the book, which we will achieve in this part, having prepared the ground in the previous parts. These special holistic policies, "Integrated Environmental Policies" (IEP), are characterised by constitutive elements, which reflect properties of the market mechanism and allow for partial decentralisation of decision-making. The constitutive elements refer to the locality principle, the adequate involvement of stakeholders and appropriate links between the policy tools (Chapter 17). The balance between economy-guidance and technology-guidance must also be observed (Chapter 14), which means that environmental policy has to respect the technological framework on the one hand, but also encourage a design for environment (DfE) on the other.

The preparation of the drafting of the policies requires agreement on environmental standards and the policy goals. Aspects of economic reasonableness (Section 11.1) and possible international issues must be taken into account. The structure of the holistic policies might depend on these details.

The first chapter takes stock, so to speak, and contains a number of key findings and conclusions from the previous chapters, only to prepare the ground for the definition and design of the policies. First, there is a brief overview of the roots of the circular economy, followed by an equally brief discussion of relevant information issues and structural properties of holistic environmental policies. A case study on Germany's progress towards a circular economy examines the details of these efforts and relates them to some of these key findings.

Thereafter, Integrated Environmental Policies (IEP) will be drafted for various "standard" categories of waste: packaging waste, waste electrical and electronic equipment (WEEE), and end-of-life vehicles (ELV). Since, unlike the market mechanism, there is no uniform approach, each area requires a slightly different policy. Nevertheless, the common structural background of these policies is the EPR principle, which, if properly implemented, creates incentives for designs for environment (DfE). There are, of course, various ways of adapting the policies or parts of them to local conditions. These chapters show the intricacies associated with the development of such policies and the requirements to make them incentive-compatible.

Another chapter relates to climate change mitigation. Similar to increasing waste volumes, global greenhouse gas emissions exceed the assimilation capacity of the environment. Appropriate holistic policies are needed to reduce these emissions and limit temperature rise. The reuse and recycling of greenhouse gases is not yet relevant in this context. Due to the global nature of greenhouse gas emissions, it is in principle possible to design a comparatively simple policy: to create a market for tradable emission certificates with an appropriate cap. This policy is very close to the market mechanism, therefore holistic. The only problem is the locality principle, which creates serious difficulties in implementing such a policy at global level. The individual perceived feedback may be too low to encourage the efforts from sufficiently many countries.

The following chapters cover other areas of relevance for the implementation of a circular economy: plastics and textiles, which have become more important in recent decades. They are of particular interest for a various reasons. First of all, they partly overlap with other policy areas. This refers to the mitigation of climate change, as both production of plastics and textiles and the incineration of plastic and textile waste lead to significant emissions of greenhouse gases. Plastics are, of course, also linked to packaging, and textiles are also made with synthetic fibres.

Another aspect is the international dimension of these areas. There is plastic waste in the rivers and the seas, and micro-plastics appears already in the food chain. This latter aspect points clearly to a global common, which needs to be protected. Perceived feedback on individual health may prove helpful in this regard. The international dimension of textiles arises from production in developing countries, which means that textile wastes from textile manufacturing pollute the environment especially in these countries. Waste textiles from consumption, which is driven by societal trends such as "fast fashion", shape the situation everywhere.

The design of these policies and their detailed presentation allow adaptation to other areas of waste management. The main challenges are to use appropriate policy tools to properly integrate the different stakeholder groups and to link these tools in such a way as to achieve as far as possible incentive-compatibility.

Where are we on the road to a circular economy?

Before discussing goals and principles, before introducing and linking instruments and policy tools to a holistic policy to implement a circular economy, it might be a good idea to reflect some of the issues raised in previous chapters. This offers more than a characterisation of the framework conditions, both from economics and from the technological environment, which must be taken into account. The concrete implementation of circular economy strategies depends on this initial or starting situation, and there are (societal) path dependencies. The following sections remind us of important topics, and a case study on Germany's road to a circular economy illustrates the challenges.

19.1 Roots of the circular economy

The concept of a circular economy or rather "the" circular economy, is deeply rooted in industrial ecology on the one hand, and in economics on the other. The economic background,

as characterised by Pearce and Turner (1989), points to the need to fully respect and maintain the fundamental functions of the environment: supplying natural resources, receiving waste and providing direct utility (Chapter 2). In practice, this mainly translates into compliance with the waste hierarchy, with particular emphasis on waste prevention, including saving natural resources and thus promoting sustainable development.

In addition to the positive effects of respecting the waste hierarchy, a market system signals scarcity of natural resources, thus motivating appropriate measures and adaptations. Nevertheless, the conditions that characterise the extraction of natural resources and/or the recovery of resources through recycling in a wide range of countries, developing countries and emerging economies in particular, need to be monitored – and not just in terms of environmental degradation. This is an essential part of "sustainability" for a country importing natural resources or exporting waste for recycling. And, although adherence to the waste hierarchy leads to less waste in the environment, the function of the environment as direct provider of utility should be protected – for example, through the involvement of citizens.

From a practical point of view, circular economy strategies are characterised by their technological background in industrial ecology or related schools of thought. This aspect of the circular economy is in particular promoted by the Ellen MacArthur Foundation, pointing to the important role that the technological environment plays in the implementation of a circular economy.

It needs to be stressed that there are many concepts, which are more or less closely linked to a circular economy. Sustainable, green or smart cities refer to the "hubs", or focus points, of circular economy strategies – with cities consuming a large part of natural resources and generating a lot of waste ending up in the environment. Sodiq et al. (2019) deal in particular with the development of sustainable cities, while Mora, Bolici, and Deakin (2017) investigate smart city research, and Yigitcanlar et al. (2019) and Ahvenniemi, Huovila, Pinto-Seppä, and Airaksinen (2017) look into the relationship between sustainable and smart cities. The concept of a green city is relevant in appropriate action plans to transform a city, for example in the corresponding program of the European Bank of Reconstruction and Development.

This diversity of related concepts, most of which aim in one way or the other to address sustainable development goals (SDGs), tends to divert attention from the core principles of a circular economy: adherence to the waste hierarchy and, consequently, the preservation of natural resources for sustainable development. These core principles form the reference point in this book for the implementation of a circular economy. Nevertheless, there are other slightly divergent approaches to the concept of a circular economy, as discussed in Chapter 3.

19.2 How to deal with information issues?

When characterising the initial or starting situation for the implementation of a circular economy, various facts and observations must be taken into account.

19.2.1 Facts and observations

The "enthusiasm" that characterises the transition to a circular economy varies widely across countries. External factors, such as abundance of natural resources or availability of land for landfilling waste, but also the level of environmental awareness, however this may be defined, seem to play a role. A strategic positioning as early-mover, also with regard to the export of environmental technologies, or as late-mover, in order to benefit from the efforts and experiences of others, comes in addition. This leads to a hierarchical structure with circular economy leaders and followers (Chapter 4), which is of particular importance for international efforts such as climate change mitigation.

Most of the current practical efforts to implement a circular economy are not fully satisfactory and/or successful, although, no doubt, progress has been made in recent decades. Nevertheless, a review of circular economy strategies points to shortcomings in the tools and policies applied (Chapter 5). The main concerns relate to the waste hierarchy, in particular to waste prevention. Although it is widely recognised as priority goal, the practical focus is clearly more on the recycling of waste with new technologies and rising recovery and recycling rates documenting continuous progress. The fact that waste prevention does not play much of a role in practical perceptions of a circular economy (Kirchherr, Reike, & Hekkert, 2017), emphasises this observation (see also Chapter 18).

In order to better understand and evaluate this observation for the successful implementation of a circular economy, it is advisable to take a closer look at the economic background.

19.2.2 Economy-guidance to a circular economy

When economics in general is driven by scarcity of resources and thus by scarcity of commodities produced with these resources, environmental economics is driven by perceived scarcity of environmental commodities. The extent to which scarcity is perceived depends on the specific situation of a country or region, its geographic, climatic and demographic condition, its economic wealth and perhaps additional geopolitical constellations.

One of the problematic issues is that the acceptance of measures to protect the environment seems to depend on this perceived scarcity, on environmental awareness in general. Moreover, as studies in behavioural economics show (Chapter 8), a missing or insufficient feedback from environmentally friendly actions can further reduce such efforts. This is in addition to the public good characteristics and external effects associated with environmental commodities, which trigger mechanisms such as the Tragedy of the Commons and the Prisoners' Dilemma, and significantly interfere with the fundamental allocation problems (see Chapter 6).

A lack of information and information asymmetries make it in practice difficult, if not impossible, to compensate for the shortcomings of the market mechanism in the adequate allocation of environmental commodities. Further complications arise from a potentially low level of perceived scarcity or an insufficient perceived feedback, or from inadequate and unexpected avoidance strategies of consumers and producers in response to certain measures. Societal path dependencies with the "social needs" of generally profitable business activities, such as in recycling of waste, can, on the one hand,

prevent "economically reasonable" environmental technologies, or, on the other hand, affect waste prevention (see Section 11.1).

As there is no easy way to "extend" the market mechanism to adequate coverage of environmental commodities and thus to guide an economy to a circular economy, it seems necessary to direct attention to the further development of the technological framework for such a guidance as in principle proposed by industrial ecology.

19.2.3 Technology-guidance to a circular economy

Implementing a circular economy is linked in many ways to the availability of appropriate environmental technologies and their further development, including a continuous stream of innovations in the context of a design for environment (DfE). This corresponds to industrial ecology, introducing and presenting a systems perspective, a holistic approach (Choudhary, 2012). The question is, whether this technological environment is sufficient to transform the economy into a circular economy? If waste is seen merely as a "design flaw", meaning that with appropriate technologies and designs waste could be prevented, then economics and policies would not really be necessary in this context, at least according to an opinion prevailing in some industrial ecology circles (Lifset & Graedel, 2002).

There is, for sure, an issue with the effectiveness of various environmental policies, as shown by the examples discussed in Chapter 5. There are, however, also various aspects related to environmental technologies, which have so far been of little relevance in the literature, but which nevertheless influence the possibilities to implement a circular economy. Insufficient acceptance of specific environmental technologies or processes can reduce the effectiveness of waste management (Gallardo, Bovea, Colomer, Prades, &

Carlos, 2010; Gallardo, Carlos, Colomer, & Edo-Alcón, 2017). The question of setting and, in particular, of adapting environmental standards, is also raised (Section 15.3).

These aspects, in turn, relate to incomplete information and information asymmetries that may affect the desideratum of a DfE (Section 11.2). Companies will then make use of their knowledge regarding a DfE if this in line with their legitimate business interests. Consequently, a technology-guided transformation of the economy (Section 14.1) alone need not provide relevant DfEs.

Of particular interest are rebound effects due to technological changes and changes in the markets for environmental commodities (Zink & Geyer, 2017), which can significantly reduce positive environmental effects of the new technologies. There is also the question of the economic reasonableness of some technologies, the question of the extent to which a technology is really needed, which is an adequate level of environmental standards. There are no straightforward answers to these questions. The fact that these responses depend not only on the technologies, but also on the local conditions in the economy (Section 11.1) is not helpful in this respect.

This last remark points to another observation: the provision of new environmental technologies, often depending on public initiatives. These technologies can be related to societal path dependencies, such as a focus on recycling, with further consequences for waste prevention and reuse (see Section 12.2 and Wilts, 2012), and they can themselves generate technological path dependencies. Candidates are technologies promoted through national innovation initiatives, with governments playing a leading role in selecting the technologies to be supported, and preparing them for competition in global markets. The promotion of e-vehicles in Germany is an example of this case (Buchal, Karl, & Sinn, 2019).

Finally, there is the digital transformation, which is crucial to the realisation of a circular economy on the one hand, and, on the other hand, involves a variety of features that do not necessarily promote the objectives of a circular economy (Chapter 13). But at the moment, we have probably only seen and experienced the beginning of a rapidly changing development with ambiguous effects on a circular economy (see Duan, Song, Qu, Dong, & Xu, 2019; Sarc et al., 2019).

In summary, the interaction between humans and technologies is not always without complications. The economic agents, consumers and producers, react or respond in many ways to environmental technologies with the possible consequence that the impact of these technologies is less than intended or different than expected.

These considerations show that also technology-guidance is not sufficiently goal-oriented to implement a circular economy (Section 14.1). Technologies and their further development are necessary for implementing a circular economy, but also guidance through the economic system is required. In order to overcome the information issues characterising the allocation of environmental commodities and the interaction between humans and technologies, and in order to deal with these societal path dependencies, it is necessary to use aspects of the market mechanism through appropriate environmental policies. At the same time, it is necessary to influence the provision of adequate environmental technologies. As these reflections and experiences show, this is not easy.

19.3 Holistic environmental policies

Thus, neither market-guidance nor technology-guidance alone can be expected to sufficiently support the transition to a circular economy. Environmental policies are necessary to augment, supplement or replace the market mechanism for allocating environmental commodities, for focusing the incentives and decisions of consumers and producers also on the objectives of a circular economy. Market-oriented policy tools such as the pollution tax or markets for certificates take advantage of decentralised decisions and thus exploit the individual knowledge of the economic agents (Chapter 16). But "tools" such as laissez-faire can also play a role: after all, insights from behavioural environmental economics point to certain behavioural aspects which can help mitigate the effects of the Tragedy of the Commons and the Prisoners' Dilemma. And simple command-and-control policies can, for example, be convenient to monitor and control the behaviour of small groups of producers.

More complex, holistic policies are needed to adequately address the multiple issues associated with implementing a circular economy, to handle all the information challenges, in particular information asymmetries. First of all, it is necessary to agree on the structural requirements of these policies. These requirements make it possible to "imitate" certain characteristics of the market mechanism, even to adopt them (Section 14.2). Decentralised decision-making is thereby of particular importance, and extended producer responsibility (EPR) policies provide an excellent starting point for developing appropriate policies. However, as much experience shows (see OECD, 2016; BIO Intelligence Service, 2014), they need a stronger focus on these structural requirements, the constitutive elements of an "Integrated Environmental Policy" (IEP), which is based on the EPR principle (Wiesmeth & Häckl, 2016).

Fig. 19.1 identifies the task of holistic environmental policies for implementing a circular economy, while markets alone and technologies alone have to contend with all kinds of information requirements. An appropriate combination of economy-guidance and technology-guidance can better address these issues.

The IEPs, these holistic approaches, integrate the environmental commodities into the fundamental economic allocation problems (Section 17.3) and can be designed to significantly support the transformation of the economy into a circular economy. However, different policy objectives, different environmental targets require different approaches to the different circular economy strategies. The study of the basics of holistic policies in waste management – collection systems and producer responsibility organisations (PRO) – in relation to the constitutive elements of an IEP, highlights the challenges (Section 17.4).

The case study in the following section draws attention to some of the peculiarities of the German efforts to introduce a circular economy.

19.4 Case study: Germany on the road to a circular economy

This case study puts together some aspects of Germany's efforts to achieve a circular economy. According to Ridder (2017), the research design could be described as "Social Construction of Reality": the analysis shows how results of various circular economy strategies emerge from social interactions – and thus lead to a constructed reality.

The core elements of Germany's approach to a circular economy – as set out in the Circular Economy Act of 2012 – include the "polluter pays principle, the five-tier waste hierarchy, and the principle of public and private responsibility for waste management" (see BMU, 2018, p. 8). The purpose of this Act is to conserve natural resources, and protect human health and the environment from the impacts associated with waste generation and management.

As far as the legal regulations are concerned, Germany seems to be well prepared. The question is what has happened, in particular, what has changed in the last 10 or 20 years? What needs to be done in near future?

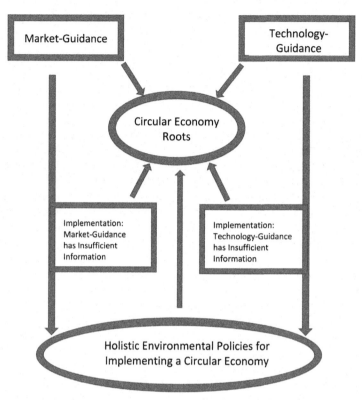

FIG. 19.1   Holistic policies to implement a circular economy. Neither market-guidance nor technology-guidance alone is sufficient for the implementation of a circular economy. © *Source: Own Drawing.*

19.4.1 Waste management in Germany

When it comes to waste prevention, an interesting reference is made to the reduced volumes of waste generated, which according to BMU (2018) is also one of the aims of waste prevention in Germany (p. 10). Fig. 19.2 shows that this volume has decreased significantly over the last 20 years. This "net" volume of waste excludes waste from waste treatment facilities.

However, a closer consideration shows that this decrease is largely due to a decrease in the volume of construction and demolition waste, which is responsible for around 60% of Germany's waste. Fig. 19.3 shows that in particular household waste has remained on the same level since 2000 and has therefore not increased with economic growth.

BMU (2018) points to many ways to prevent waste: "focusing on durable, lean, repairable products; avoiding unnecessary and short-lived items; purchasing services rather than goods; and using rather than owning" (p. 10). However, beyond the recommendation to "raising awareness and sensitising the general public to effective waste prevention", there are no further hints of how this can be achieved. Thus, there seems to be the aim to gradually establish waste prevention as a social norm – without, however, providing concrete guidelines.

On the other hand, Fig. 19.3 shows that, according to the German definition, waste prevention took place mainly in an area that depends on a variety of external factors which can hardly be influenced by individuals. To call this decreasing amount of waste as prevented

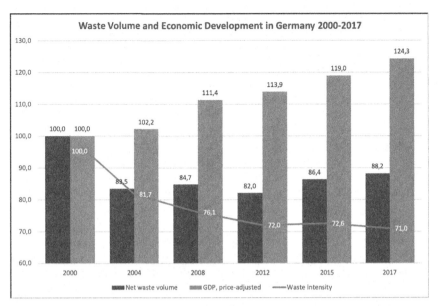

FIG. 19.2 Waste volume and economic growth in Germany 2000–2017. Waste intensity declined in the period 2000–2017. © *Own Drawing with Data from the German Federal Statistical Office 2019 (https://www.destatis.de/DE/Themen/Gesellschaft-Umwelt/ Umwelt/Abfallwirtschaft/Publikationen/Downloads-Abfallwirtschaft/abfallbilanz-xlsx-5321001.xlsx?__blob=publicationFile).*

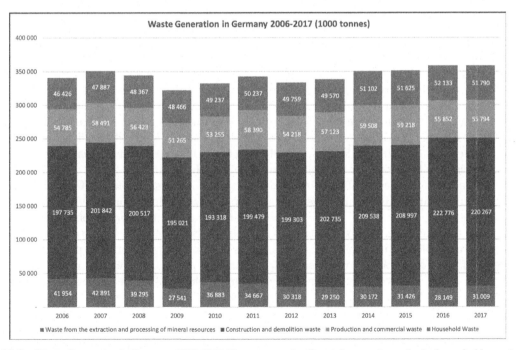

FIG. 19.3 Waste Generation in Germany 2006–2017. There are no significant changes, in particular household remained at the same level. © *Own Drawing with Data from the German Federal Statistical Office 2019 (https://www.destatis.de/DE/Themen/Gesell schaft-Umwelt/Umwelt/Abfallwirtschaft/Publikationen/DownloadsAbfallwirtschaft/abfallbilanz-xlsx-5321001.xlsx?__ blob=publicationFile).*

V. Implementing a circular economy

waste sounds therefore more like a "social construction of reality".

Packaging waste, as an essential part of household waste, is certainly more accessible for awareness-raising campaigns. But it is precisely in this area that most prevention efforts failed. As the German Environment Agency (UBA) points out, "the level of packaging consumption in Germany remains very high", packaging waste increased by 19% between 2000 and 2016, plastics packaging by 74%. This shows that Germany has not really succeeded in motivating consumers and producers to reduce packaging.

There is one exemption, however: the deposit system for one-way drinks packaging motivates consumers to return approximately 98% of empty one-way plastic bottles (see also the case study in Section 7.3). This is a surprisingly high number compared to countries with a similar environmental awareness, but without a mandatory deposit system such as Austria. On the other hand, the deposit system leads to a significant further increase in the quantity of one-way drinks packaging (see Table 7.2). The question that arises in this context is, whether the existence of the deposit system led to a change in the mindset, helped to establish a social norm, or whether compliance with the deposit system is used as "cheap excuse" (Engel & Szech, 2017) for the purchase of even more drinks in one-way containers? Since deposit fees can in principle be levied for the collection of other waste items such as waste electric and electronic equipment (WEEE), this question is of further interest.

Moreover, the existence of this deposit system, with its high collection and recycling rates, motivates various companies to increase their share of drinks in one-way containers, especially plastic bottles. Vested interests (see Section 17.4) may play a role in this context, too. As a result, the question of how consumers and/or producers react to the availability of certain

technologies is becoming increasingly important (see also Section 12.2 on path dependencies).

19.4.2 Other circular economy strategies in Germany

This subsection briefly deals with WEEE, end-of-life vehicles (ELV) and Germany's efforts to curb climate change. The various parts refer also to the chapters with additional information on these topics.

Waste Electrical and Electronic Equipment: With respect to WEEE, there is little to add as regards the situation in the European Union (EU) described in Section 5.1.2. As Fig. 13 in BMU (2018) shows, Germany exceeded the collection rate of 4 kg per inhabitant by far. In 2014, the 7.6 kg collected per inhabitant corresponded to 42.9% of electric and electronic equipment placed on the German market. In view of higher targets, this points to necessary improvements in near future. The wording that "Germany has far exceeded the EU's prescribed recovery and recycling quotas for waste electrical and electronic equipment every year to date" (see BMU, 2018, p. 28) is formally correct, but points also to a "social construction of reality", as this mainly refers to large and heavy appliances and less to small electronic devices.

End-of-Life Vehicles: There is a similar situation in Germany for ELV. The discussion in Section 5.1.3 shows that Germany performs quite well with regard to those ELVs which are returned and dismounted in Germany. Fig. 19.4 provides some details on the whereabouts of permanently decommissioned vehicles in Germany. The number of old cars sold to customers outside the EU is not insignificant and points to an understanding of "reuse", which is not completely compatible with "reuse" according to the waste hierarchy (see also Section 12.1 on rebound effects).

Mitigating Climate Change: As the discussion in Section 5.1.4 and the case study in Section 11.3

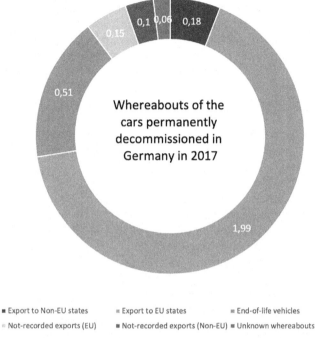

Permanently Decommissioned Cars in Germany in 2017

- ■ Export to Non-EU states
- ▪ Export to EU states
- ■ End-of-life vehicles
- ▫ Not-recorded exports (EU)
- ■ Not-recorded exports (Non-EU)
- ■ Unknown whereabouts

FIG. 19.4 Permanently decommissioned vehicles in Germany in 2017. Most old vehicles are exported to EU member states, but 180,000 leave the EU. © *Own Drawing with Data from "Jahresbericht über die Altfahrzeug-Verwertungsquoten in Deutschland im Jahr 2017" (BMU and UBA, Germany): https://www.bmu.de/fileadmin/Daten_BMU/Download_PDF/Abfallwirtschaft/jahresbericht_ altfahrzeug_2017_bf.pdf).*

show, Germany has also acted quite effectively in the transition to renewable energy sources. A significant amount of greenhouse gas emissions have been avoided in the last 20 years. Moreover, due to the Corona crisis, there is a chance that the goals for 2020 regarding the reduction of greenhouse gas emissions can be achieved.

One critical issue concerns the incomplete integration of German efforts into other regulations, such as the emission trading system of the EU. In this way, the greenhouse gas emissions avoided in Germany are in some way left to other emitters in the EU. This indicates insufficient coordination with regard to the energy policy of the EU, which is necessary to successfully mitigate climate change.

19.4.3 Germany's progress to a circular economy

The research design of this case study is "Social Construction of Reality". According to Ridder (2017), this assumes that "there is no unique 'real world' that preexists independently of human mental activity and symbolic language". Rather, "the world is a product of socially and historically related interchanges amongst people (social construction)" (p. 288).

This characterises the path of Germany, but probably not only of Germany, to a circular economy quite well. The situation that has emerged in recent years has in fact resulted from a variety of societal path dependencies. This concerns, for example, the management of waste in general, which focuses mainly on recycling with its many job opportunities. This also applies to the high return rates of empty drinks containers, with a possible moral self-licensing for buying more drinks in one-way packaging, and this relates to the sale of used cars to customers outside the EU and their interpretation as "reuse" within the meaning of the waste hierarchy.

There is, in principle, nothing really wrong with this development or with this attitude, which is probably not much different in other industrialised countries. However, these observations show that it is necessary to "align" the social constructions, the interchanges amongst people, with the goals of a circular economy: the reality arising from social construction should correspond to the objectives of a circular economy.

Wilts (2016) points in particular to the following issues, which characterise Germany's approach or rather its current starting point for a circular economy (p. 19f):

- The circular economy debate in Germany still focuses too much on waste management: the optimised separation of recyclable materials and recycling continue to play an important role.
- In Germany, more efforts should be made for DfEs: this requires the integration and the cooperation of resource producers, product designers, merchants, consumers and actors in waste management.
- A circular economy needs a clear regulatory framework: in Germany, only talking about possible business opportunities, about possible economic savings, is not helpful, if less waste will threaten existing businesses.
- The implementation of a circular economy needs an appropriate policy mix: current policies in Germany seem inconsistent.

So much for Wilts (2016), which shows once again that Germany is still too dependent on societal path dependencies. It will probably take some time to steer the economy more towards a circular economy.

Of course, the situation in other countries is likely to be characterised by similar or related challenges, and sometimes additional issues, such as the Corona crisis, change the situation. In view of the locality principle (Section 17.3.1), it is necessary to pay attention to reality. This will be relevant for the IEPs for the implementation of a circular economy to be designed in the following chapters.

References

Ahvenniemi, H., Huovila, A., Pinto-Seppä, I., & Airaksinen, M. (2017). What are the differences between sustainable and smart cities? *Cities, 60*, 234–245. https://doi.org/10.1016/j.cities.2016.09.009.

BIO Intelligence Service (2014). *Development of Guidance on Extended Producer Responsibility (EPR). Report for the European Commission DG ENV.* European Commission.(2014). http://ec.europa.eu/environment/waste/pdf/target_review/Guidance on EPR - Final Report.pdf.

BMU (2018). *Waste management in Germany 2018.* Berlin: Federal Ministry for the Environment, Nature Conservation and Nuclear Safety (BMU). Retrieved from https://www.bmu.de/fileadmin/Daten_BMU/Pools/Broschueren/abfallwirtschaft_2018_en_bf.pdf.

Buchal, C., Karl, H.-D., & Sinn, H.-W. (2019). Kohlemotoren, Windmotoren und Dieselmotoren: Was zeigt die CO2-Bilanz? *Ifo Schnelldienst, 72*(8). Retrieved from https://www.ifo.de/DocDL/sd-2019-08-sinn-karl-buchal-motoren-2019-04-25.pdf.

Choudhary, C. (2012). Industrial ecology: Concepts, system view and approaches. https://www.ijert.org/research/industrial-ecology-concepts-system-view-and-approaches-IJERTV1IS9396.pdf.

Duan, H., Song, G., Qu, S., Dong, X., & Xu, M. (2019). Post-consumer packaging waste from express delivery in China. *Resources, Conservation and Recycling, 144*, 137–143. https://doi.org/10.1016/j.resconrec.2019.01.037.

Engel, J., & Szech, N. (2017). *Little good is good enough: Ethical consumption, cheap excuses, and moral self-licensing. In Discussion Paper No. 17-28 GEABA.* Retrieved from https://www.ijert.org/research/industrial-ecology-concepts-system-view-and-approaches-IJERTV1IS9396.pdf.

Gallardo, A., Bovea, M. D., Colomer, F. J., Prades, M., & Carlos, M. (2010). Comparison of different collection systems for sorted household waste in Spain. *Waste*

Management, *30*(12), 2430–2439. https://doi.org/10.1016/j.wasman.2010.05.026.

Gallardo, A., Carlos, M., Colomer, F., & Edo-Alcón, N. (2017). Analysis of the waste selective collection at drop-off systems: Case study including the income level and the seasonal variation. *Waste Management & Research*, *36*(1), 30–38. https://doi.org/10.1177/0734242X17733539.

Kirchherr, J., Reike, D., & Hekkert, M. (2017). Conceptualizing the circular economy: An analysis of 114 definitions. *Resources, Conservation and Recycling*, *127*, 221–232. https://doi.org/10.1016/j.resconrec.2017.09.005.

Lifset, R., & Graedel, T. E. (2002). Industrial ecology: Goals and definitions. In R. U. Ayres, & L. W. Ayres (Eds.), *A handbook of industrial ecology*. Cheltenham: Edward Elgar Publishing. https://doi.org/10.4337/9781843765479.00009.

Mora, L., Bolici, R., & Deakin, M. (2017). The first two decades of smart-city research: A bibliometric analysis. *Journal of Urban Technology*, *24*(1), 3–27. https://doi.org/10.1080/10630732.2017.1285123.

OECD (2016). *Extended producer responsibility: Updated guidance for efficient waste management*. Paris: OECD Publishing. https://doi.org/10.1787/9789264256385-en.

Pearce, D. W., & Turner, R. K. (1989). *Economics of natural resources and the environment*. The Johns Hopkins University Press.

Ridder, H.-G. (2017). The theory contribution of case study research designs. *Business Research*, *10*(2), 281–305. https://doi.org/10.1007/s40685-017-0045-z.

Sarc, R., Curtis, A., Kandlbauer, L., Khodier, K., Lorber, K. E., & Pomberger, R. (2019). Digitalisation and intelligent robotics in value chain of circular economy oriented waste management—A review. *Waste Management*, *95*, 476–492. (2019). https://doi.org/10.1016/j.wasman.2019.06.035.

Sodiq, A., Baloch, A. A. B., Khan, S. A., Sezer, N., Mahmoud, S., Jama, M., & Abdelaal, A. (2019). Towards modern sustainable cities: Review of sustainability principles and trends. *Journal of Cleaner Production*, *227*, 972–1001. https://doi.org/10.1016/j.jclepro.2019.04.106.

Wiesmeth, H., & Häckl, D. (2016). Integrated environmental policy: A review of economic analysis. *Waste Management & Research*, *35*(4), 332–345. https://doi.org/10.1177/0734242X16672319.

Wilts, H. (2016). *Germany on the road to a circular economy?* Bonn: Friedrich-Ebert-Stiftung. Retrieved from https://library.fes.de/pdf-files/wiso/12622.pdf.

Wilts, H. (2012). National waste prevention programs: Indicators on progress and barriers. *Waste Management & Research*, *30*(9), 29–35. https://doi.org/10.1177/0734242X12453612.

Yigitcanlar, T., Kamruzzaman, M. D., Foth, M., Sabatini-Marques, J., da Costa, E., & Ioppolo, G. (2019). Can cities become smart without being sustainable? A systematic review of the literature. *Sustainable Cities and Society*, *45*, 348–365. https://doi.org/10.1016/j.scs.2018.11.033.

Zink, T., & Geyer, R. (2017). Circular economy rebound. *Journal of Industrial Ecology*, *21*(3), 593–602. https://doi.org/10.1111/jiec.12545.

Packaging waste in a circular economy

Packaging waste accounts for a large and increasing share of household waste: between 30% to 35% of total municipal waste in industrialised countries, and 15% to 20% in general, with developing countries and transition countries catching up rapidly (Tencati, Pogutz, Moda, Brambilla, & Cacia, 2016). The packaging waste statistics of the European Union (EU) also shows a still increasing consumption of packaging – from 156.3 kg per inhabitant in 2012 to 169.7 kg per inhabitant in 2016. In fact, for 2016, packaging waste represents almost 35% of municipal solid waste generated in the EU. Fig. 20.1 shows the evolution of the volumes of the various packaging materials in the EU in kg per inhabitant.

These numbers result from a variety of factors, influencing the generation of packaging waste: the increasing demand for packaging due to a general increase in consumption as a consequence of economic growth, or the digital transformation with its manifold of influences on shopping behaviour. In addition to these factors, which rather tend to increase packaging consumption, there are the other factors that tend to reduce packaging: the various environmental policies of the EU and its member states

among them, and, perhaps, also increasing environmental awareness.

Nevertheless, as detailed in Section 5.1, there seems to be room for improvement, especially regarding prevention of packaging waste and, closely related to it, designs for environment (DfE). According to Tencati et al. (2016), for example, recent attempts to prevent waste, and packaging waste in particular, have not been too successful (p. 36). The "Design for Recycling" (DfR), which is promoted by practitioners in waste management (see, for example, the "Der Grüne Punkt" in Germany), obviously focuses on recyclability and not so much on prevention of waste through an appropriate design. The DfR thus perfectly corresponds to the societal path dependencies affecting waste management in most countries. Again, this is less about criticising a DfR, it is more about emphasising the need for a DfE in the context of implementing a circular economy.

In this chapter we are drafting an "Integrated Environmental Policy" (IEP) for packaging waste. This holistic policy, based on the extended producer responsibility (EPR) principle, supports the development towards a circular economy (see Section 17.4). As far as the

(kg per inhabitant)

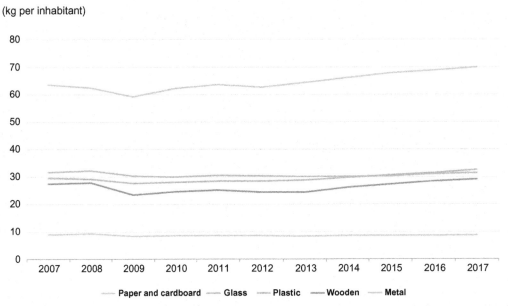

FIG. 20.1   Development of packaging generated, per inhabitant in the EU: 2007–2017. Volume of packaging material in kg per EU inhabitant. © *From EuroStat. Packaging waste statistics. Retrieved April 12, 2020 from https://ec.europa.eu/eurostat/statistics-explained/index.php?title=File:Development_of_packaging_generated_per_inhabitant,_EU,_2006%E2%80%932017_(kg_per_inhabitant).png (Original work published 2020).*

achievement of the environmental objectives is concerned, such a policy should be, to the greatest extent, incentive-compatible. Of course, it is neither economically nor politically feasible to control the behaviour of each consumer, of each producer. Then there are these societal path dependencies, which can, for example, interfere with certain desirable regulations on the export and import of waste. Therefore, implementing the technology-oriented EPR principle by an IEP is an attempt to bridge the gap between command-and-control policies and a free market system. We know that neither free markets nor command-and-control policies are fully operational as allocation tools in this context (see Section 14.1).

The following details of an IEP for packaging waste reveal the crucial aspects, which need closer examination and perhaps more careful monitoring. Here, too, the initial situation, the starting point may vary from country to country.

20.1 Setting the targets of the policy

Environmental policies in general and IEPs in particular, need to have quantitative and qualitative goals, such as targets for collecting and recycling of waste, the application of "best available techniques", the development of suitable product designs. As discussed in the context of the economics of the waste hierarchy (see Section 18.1), these goals have to respect the relevant framework conditions in the country concerned – in accordance with Constitutive Element I of an IEP (Section 17.3.1).

Therefore, for designing an IEP economically reasonable quantitative targets are assumed to be given (see Section 11.1). The situation is somewhat more complicated in terms of qualitative objectives, such as a suitable product design, which are difficult to measure and equally difficult to control for reasons related to insufficient or asymmetric information (see Section 11.2). Anyway, the more important question is then, how these "soft" objectives can be achieved?

20.2 Auxiliary environmental regulations

The effectiveness of environmental policies depends also on the avoidance possibilities, which stakeholders, in particular consumers and producers can take in order to avoid higher costs or pay pollution taxes, etc. These avoidance possibilities are welcome if they are more environmentally friendly. They should be excluded, if they do not meet the objectives of the policies and perhaps lead to an even worse environmental situation.

For this reason, it might be a good idea to close some avoidance possibilities, which are obviously not in the interest of the environmental policy. This idea has some relationships with behavioural economics: one way to break old habits is to reduce or even take away the possibilities of following these old habits (see Chapter 8) – the landfilling of biologically or chemically active waste, for example for the situation considered here.

20.2.1 Reducing the opportunities for landfilling

The waste hierarchy, in whichever formulation, ranks landfilling last, it is the least preferred option for waste management and should therefore be kept to a minimum. Under the EU Waste Framework Directive of 1999 and its Amendment of 2014, waste must be treated before landfilling. Moreover, "by 1 January 2025, recyclable waste including plastics, metals, glass, paper and cardboard, and other biodegradable waste" will not be accepted in landfills for non-hazardous waste.

The consequences of these regulations are obvious: not only landfilling waste is becoming more expensive, but also the production of waste, which needs to be recycled at increasing costs, with potentially insufficient returns from the sale of recycled materials. Therefore, in order to avoid higher costs, preventing waste by reducing packaging and packaging material could be an adequate strategy.

There are, of course, other avoidance possibilities, which need to be kept in mind. Exporting waste, plastic waste, for example to developing countries or emerging economies for recycling, was one of these avoidance strategies, which were thought to be economically beneficial in view of the benefits of international trade and which were expected not to cause environmental harm. However, lower environmental awareness, outdated recycling technologies and perhaps unskilled workers in the importing countries have had the opposite effect (Tang et al., 2015).

To control the shipment of waste is, apart from some obvious regulations on mainly hazardous waste, not a simple issue: societal path dependencies resulting from an economic context, in particular international trade, seem to be in the way of stricter regulations. On the other hand, more and more developing countries and emerging economies, such as China, Malaysia, Indonesia and others are recognising the real and potential environmental degradation associated with these import activities. It is therefore increasingly understood that the economic benefits are more than outweighed by the environmental pollution.

20.2.2 General take back requirements

Take back requirements have meanwhile turned into a preferred policy tool, both as part of more complex environmental policies, such as policies for waste electrical and electronic equipment (WEEE) and end-of-life vehicles (ELV) policies, but also as separate regulation of certain areas of waste management. This application of this tool will now briefly be addressed.

If consumers are allowed to leave any excess packaging in the shop, or can return it to the shop, they are likely to act accordingly, if they otherwise have to return this packaging to separate collection points or if they have to pay for the collection of this waste. Then the shops take care of this waste packaging. They may reuse it, as is the case with some transport packaging, or they may have to consign it to recycling at their own expenses. Exactly this last alternative will motivate shops to negotiate with the producers of the various commodities to reduce packaging or cover the cost of recycling. Then the aim of the producers is to reduce the cost of packaging and costly recycling. But, of course, such regulations include also some psychological aspects: perhaps they make consumers and producers aware of the need to reduce packaging – beyond the possibility to reduce costs.

It is important to understand the links between the various stakeholders in this case: the options available to consumers put some pressure on sellers and producers to reduce package or packaging material. As part of a holistic policy, such links correspond exactly to Constitutive Element III of an IEP (see Section 17.3.3). This shows again that the adequate integration of all stakeholders into complex policies, including links between them, is required in order to achieve certain objectives (see also the discussion of the case study in Section 7.3).

Auxiliary regulations could also be useful from an environmental point of view in the context of trading waste, especially when waste or packaging waste is shipped to developing countries or emerging economies. This sustainability aspect depends on the specific situation in the exporting and importing countries according to Constitutive Element I, the locality principle.

20.2.3 Limiting trade with packaging waste

This subsection refers to international trade in waste in general and packaging waste in particular. The motives for these export and import activities are manifold: Germany, for example, imports a considerable quantity of "combustible waste" for its incineration plants. On the other hand, there are also sizeable amounts of exports of "mixed municipal waste" from Germany, and the export of packaging waste amounted to 10.9% in 2016, including 10.6% of plastic packaging, according to the German Environment Agency (UBA).

Environmental concerns are associated with exports of waste to countries with only limited technical possibilities for making adequate use of this waste, for recycling it without further degrading the environment. There is certainly some economic relevance of these trade activities, otherwise they would not happen. However, since environmental awareness or perceived scarcity of environmental commodities can vary significantly between exporting and importing countries, especially if developing countries are to be involved, it might be a good idea to restrict these trade activities for sustainability reasons. The experiences of and with China regarding plastic waste is proof of this (Tang et al., 2015).

Fortunately, again from an environmental point of view, various emerging economies are now restricting the import of (packaging) waste, at least, if it is highly contaminated. As contaminated waste, plastic or paper, is more costly to recycle in the industrialised countries, this development could help to support more efforts regarding DfEs.

In general, it might be difficult to restrict import and export of waste beyond the existing regulations regarding hazardous waste. This results from the trade regulations, which, as part of societal and economic path dependencies, are not always optimal with respect to environmental issues, as experience shows.

In the following section, the collection and separation of packaging waste is investigated – important for the implementation of the EPR principle.

20.3 Collection and separation of packaging waste

Packaging waste consists of many different materials: glass, paper, cardboard, wood and metal. There are large pieces, small pieces, which need all be collected, and, if possible, already separated at the source. Pre-sorting waste at home can help to reduce waste management costs (Bartolacci, Paolini, Quaranta, & Soverchia, 2018) and can lead to higher recycling levels, as separately collected fractions of waste are usually sent to recycling and recovery.

20.3.1 Relevance of socio-economic, cultural and technical factors

But the more active participation of the households in waste management is perhaps even more important: for implementing a circular economy it is necessary to influence consumer and household behaviour in the larger area of waste management. One way to achieve this is, in view of behavioural environmental economics, to raise awareness for the challenges. Returning empty packaging due to deposit fees or segregate waste can provide good opportunities. In this context, Knickmeyer (2020) points out the importance of understanding the social factors influencing household behaviour in waste management. Also Agovino, Cerciello, and Musella (2019) examine "the effects of neighbour influence and cultural consumption on separate waste collection", and Valenzuela-Levi (2019) is interested in the potential impact of income inequality on waste separation.

Apart from these cultural factors and questions of behavioural economics, a variety of technical issues seem to play a role in the separation of waste. Thus, Leeabai, Suzuki, Jiang, Dilixiati, and Takahashi (2019) investigate the impact of the various settings of waste bins on waste separation. In addition, there is the vast literature on technical aspects of waste collection and the collection rate, which has already been addressed

in Section 17.4.1. The topics in the literature include bring systems, kerbside collection and others (see, for example, OECD, 2016; Seyring, Dollhofer, Weißenbacher, Bakas, & McKinnon, 2016; Hahladakis, Purnell, Iacovidou, Velis, & Atseyinku, 2018). Consequently, it is necessary to offer the optimal system of separate waste collection, according to the latest studies. Again, the system must depend on local conditions, and systems preferred in one area do not have to be optimal for others.

Among those issues, which are not yet handled in an optimal way, is the labelling of the packaging. The "Green Dot", dating back to the beginnings of labelling and collecting packaging waste for recycling, has lost its original importance due to modified regulations. Nevertheless, many households in Germany still interpret the Green Dot on a package as evidence that the empty packaging should be consigned for recycling, which is true. However, it does not mean that all these empty packages, no matter what material, should go to the same waste bin for recyclable waste. For example, paper, plastic, glass and metal packaging can be found in the same waste bin, although there are separate containers for glass and paper. This is an issue, at least in Germany, asking for an innovative solution to increase the volume of carefully separated recyclable packaging waste.

As far as packaging waste is concerned, all these aspects are of relevance, but there remains the possibility of introducing monetary incentives for households for either reducing packaging waste or increasing collection rates.

20.3.2 Collection systems with monetary incentives

There are various structurally different possibilities for introducing financial incentives into waste collection systems (see also Section 17.4.1). A "deposit system", to be considered first, introduces a mandatory fee to be paid upon the purchase of the product, such as a drink, and refunded upon return of the empty packaging. Such systems are usually applied for one-way drinks containers, which otherwise can end up in the environment. Examples are the California Bottle Bill (see Section 5.1.6), or the German Deposit System (cf. Groth, 2008 for a review), two examples among many others.

Such deposit systems establish some individual product responsibility and motivate the return of empty packaging. They require a more or less sophisticated technical and organisational infrastructure to label drinks packaging as protection against fraud, to collect the deposit fee, to collect the returned packaging, and to clear collecting and refunding the deposit fee (see the German DPG Deposit Scheme for more details). In the context of packaging waste in general, such administrative and organisational efforts are unlikely to be justified as economically reasonable, and societal acceptance is not self-evident.

The second option in terms of monetary incentives focuses on waste reduction. The aim is to charge a fee for waste in general and packaging waste in particular, depending on the amount of waste returned for collection. These "pay-as-you-throw" (PAYT) systems were introduced in the 1990s (see e.g. Bilitewski, 2008; Elia, Gnoni, & Tornese, 2015) in order to "provide waste generators with individual incentives for waste diversion efforts" (Reichenbach, 2008).

However, PAYT systems can only be implemented, if there is enough support from the local population. Otherwise, there is the risk that household will take advantage of some obvious avoidance strategies and dispose of their waste elsewhere, in the forests, for example. PAYT system thus can provide incentives for reducing residual waste, for carefully separating recyclable waste.

Finally, there are "advanced disposal fees", already discussed in Section 17.4.1, which could be levied on packaging in order to cover the expected costs of collecting and recycling these items. Although they may be useful as surcharge on the prices of items, which are used

in some events, such as fireworks for celebrating New Year, they do not provide any incentives for the collection itself. And even for the example just mentioned, it might be better not to announce the existence of this fee to the general public, because this could change the behaviour of the partying people, and lead them to return even less waste to the waste bins. Advanced disposal fees will therefore not further be considered in this context.

20.4 Applications of the EPR principle

The analysis now returns to the question of how to organise the various responsibilities in managing packaging waste in order to arrive at an incentive-compatible policy, which motivates packaging producers in particular for a DfE?

The discussion of this topic in Section 17.4 focused mainly on various systems of PROs, of "producer responsibility organisations" for the implementation of the EPR principle. There were individual systems, controlled by one producer, there were collective systems, organised as associations and controlled by a group, an association of producers, and there were collective systems with independent PROs operating in a competitive context and providing services for many producers. Moreover, there was the issue of for-profit or not-for-profit PROs and incentive-compatibility with respect to the objectives of the EPR policy.

20.4.1 How to address the various stakeholders?

In accordance with Constitutive Element II of an IEP, the relevant groups of stakeholders need to be appropriately addressed. Several topics are important:

- Larger groups of people, the households, for example, should not only be addressed by means of a command-and-control policy, as their behaviour cannot be adequately monitored and controlled. In addition to that, the attempt at permanent surveillance is in many countries politically not feasible, and can lead to behavioural responses that are detrimental to the goals of the policy (see Liao, 2018; Taylor, 2000; Thaler & Sunstein, 2008), among others, for more details on the role of command-and-control policies).

- Framework conditions with some degrees of freedom to adapt behaviour to environmental goals are preferable in this case. These framework conditions, also market-based policies, provide incentives and do not impose a ban on some activity (Thaler & Sunstein, 2008).
- Command-and-control polices should therefore only be used for small groups in very general contexts, which are easier to control. The mandatory participation of packaging producers in a PRO, for example, is part of such regulations.

So much for these considerations, how different stakeholders can be addressed. Nevertheless, the policy should provide incentives for consumers and producers to focus on the objectives of the policy and not be distracted by other issues, such as vested interests.

20.4.2 Vested interests

Vested interests are a common phenomenon that accompanies, perhaps even guides, all forms of societal developments. Boehmer-Christiansen (1990), for example, examines the consequences of the possibility that "public opinion is not a determining factor of energy policy (…) but a dependent variable". In this case, "important conclusions would follow about the role of government in raising environmental awareness, the credibility of expert advice and the significance of decision-making structures and processes". These conclusions

seem to be somewhat related to the German government's efforts to promote e-vehicles, which may arise from vested interests (see also Section 10.2.3).

Vested interests are usually negatively connoted. They can be an obstacle to innovation in general (Bridgman, Livshits, & MacGee, 2007) and to the introduction of specific technologies in particular (Haukkala, 2015). However, vested interests may also have positive effects, also in the environmental context. Boute and Zhikharev (2019), for example, point to vested interests as drivers of a clean energy transition.

Under the EPR principle, producers are expected to be interested in a DfE, and vested interests could imply weakening their focus on this goal. Combined with societal and technical path dependencies, the original focus could then more or less disappear completely.

As indicated in Section 17.4.2, vested interests are mainly associated with particular structures of the EPR policies, with certain implementations of the EPR principle. Vested interests become relevant in the field of packaging waste, whenever producers who are involved in packaging, in whatever capacity, also have business interests in the collection and recycling of their packaging waste.

The typical examples include individual PROs where a producer organises the collection and recycling of the packaging waste associated with that company's products. Of course, this producer can also entrust a company, a PRO, with the fulfilment of these tasks: it will not change the possibilities of the producer to intervene in these obligations. A similar situation exists when a group of producers, typically from the same industry, drinks producers, car manufacturers, or producers and importers of electronic equipment, organises these environmental tasks in an association of which they are members. Here, too, they can outsource these tasks to a PRO.

What are these vested interests? Why can they be detrimental to the environmental goals of an EPR policy? There are some straightforward examples that are already given in Section 17.4.2. Consider a producer with an individual system offering products in a certain packaging, drinks in one-way PET bottles, for example. Assume that, for whatever reason, recycling of PET bottles provides a return, which covers at least part of the firm-specific collection costs – after all, consumers are meant to support collection and segregation activities and may thereby reduce collection costs for the company. In such a situation, it might be of advantage for a producer to increase the amount of drinks sold in these PET bottles, if this step is supported by customers.

Fig. 20.2 provides some information on market prices for recycled plastics in the EU, which indicate possible positive returns from plastics recycling.

Even if recycling of packaging waste is not profitable, then the producer could still deal with recycling and recycled materials in one way or the other. In this case, one has to accept that there are probably some options for producers, some avoidance strategies to reduce the financial burden, thereby reducing the incentives for a DfE. Of course, a similar argument applies to producers organised in an association to organise the collection and recycling of packaging waste.

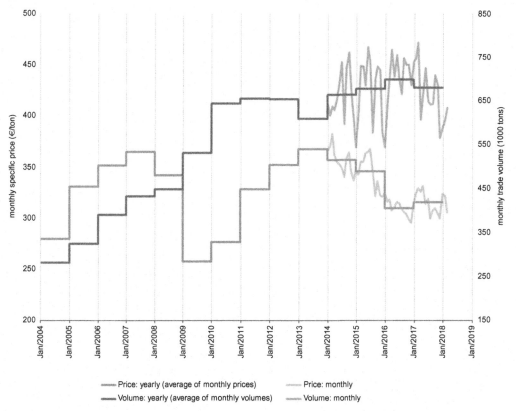

FIG. 20.2   Price Indicator and Trade Volume for Plastic Waste in EU-20. Development of prices and volumes from in the period 2004–2018. © *From Eurostat. Recycling - secondary material price indicator. Retrieved April 16, 2020 from https://ec.europa.eu/ eurostat/statistics-explained/index.php?title=File:Figure_3_Price_indicator_and_trade_volume_for_waste_plastic_in_EU-28_until_ March_2018.png (Original work published 2020).*

In such EPR systems the incentives of the companies to prevent or reduce waste through a DfE are likely to be limited. On the contrary, there could be incentives to extend the use of certain packaging materials. For this reason, individual PROs and PROs based on associations should be excluded, if possible in view of the local conditions.

This leaves PROs where individual companies cannot influence decisions on the collection and recycling of packaging waste – in order to exclude vested interests, environmental decisions must be separated from the regular managerial decisions of the companies. A prominent way to do this is to introduce PROs that are financed by the companies in one way or the other, but are also independent of these companies in terms of their decisions.

However, as the following considerations will show, vested interests at the level of the compliance schemes cannot be completely ruled out. This is the result, as has already been indicated, of bridging the gap between command-and-control policies on the one hand, and market approaches with self-interested consumers and producers on the other. The extent of these vested interests and their interference in environmental policy depend on the local situation and must be monitored.

20.4.3 For-profit compliance schemes

Assume that there are several PROs, and that the companies producing packaging must join one of these PROs and pay a "licence fee" depending on packaging material and quantity. Suppose that these PROs, compliance schemes, are for-profit and compete for companies to join them.

Such a constellation has the following advantages in the context of the implementation of a circular economy:

- Companies need to join one of these compliance schemes, but cannot influence collection and recycling decisions.

- Compliance schemes are in competition with licence fees as "equilibrium prices" – it is therefore to be expected that companies are offered services at reasonable prices.
- Conversely, compliance schemes must offer an attractive service at reasonable prices to attract companies.
- Companies can reduce the licence fees by changing their packaging, modifying or reducing packaging material through DfEs.

Sometimes these compliance schemes are not-for-profit. It is, however, clear from the above arguments that there may be problems with the quality of the services. Moreover, not-for-profit does not provide much incentive to reduce costs. Thus, there is a variety of issues, which tend to arise without competition as a coordinating instrument.

There is yet another issue, which needs to be addressed here: the compliance schemes have their legitimate business interests, which also focus on making profits. This should be done in connection with the collection and recycling waste from their licensees. But EPR policies should also motivate producers for a DfE, which, however, does not necessarily have to be in the interest of the compliance schemes, as they may lose business with less recycling waste (Van Ewijk & Stegemann, 2016). This is likely to be difficult to control, as this development is in line with the focus on recycling and not predominantly on waste prevention, which, as a societal path dependency, can be observed in many countries.

Fig. 20.2, for example, shows the prices of recycled plastics. Depending on the collection costs it could be interesting for compliance schemes to extend these activities, to motivate packaging distributors to a DfR instead to a DfE. However, prices for recycled materials fluctuate. This is especially true for recycled plastics, as the world market price for crude oil has a significant impact on this price. Therefore, only a detailed analysis can provide more insight into this issue.

The results of the discussion in the preceding sections now allow the proposal for a policy for packaging waste, which contributes to waste prevention and motivates companies to a DfE.

20.5 An IEP for packaging waste

The discussion in the last sections has at least shown that the design of an incentive-compatible policy for packaging waste is far from simple. This section brings together the various parts of the policy and links these parts appropriately with the objectives of the policy, according to Constitutive Element III of an IEP.

20.5.1 Households

With the obligation to separate waste, households play an important role in the IEP policy. Since their behaviour cannot really be monitored and controlled, they should be adequately integrated into the policy according to Constitutive Element II of an IEP. As mentioned earlier, the active participation of the households in the context of waste separation can already have a positive and stimulating effect (Agovino et al., 2019). Households could also be addressed through public campaigns, but children's education also seems to be relevant, as adults tend to be more concerned about the environment in the presence of children (Moqbel, El-tah, & Hassad, 2019).

The separate collection of the packaging waste could, of course, be supported by some auxiliary regulations such as a ban on landfilling non-inert waste. Moreover, support from an appropriate technical organisation of the collection system could also prove helpful (Hahladakis et al., 2018). A clear and unambiguous labelling of the package waste seems to be important – at least for some countries, including Germany. The reason for this is that as much packaging waste as possible should be collected and consigned for recycling in order to increase the overall costs for the management of packaging waste and motivate DfEs.

20.5.2 Producers of packaging

The comparatively small number of producers and importers of packaging can be addressed by means of a command-and-control policy. They have to join one of the compliance scheme to licence their packaging material and quantity, and are free to join that compliance scheme, which offers them the best conditions. Of course, experience shows that there are incentives for free-riding, and therefore the compliance with these regulations has to be monitored and controlled.

Technical details on how this monitoring could be carried out can be found, for example, in the German Packaging Act of 2019. Moreover, there are various digital guides (see e.g. 1&1 IONOS).

20.5.3 Compliance schemes

The existence and the operation of these compliance schemes as private for-profit companies must be guaranteed by the IEP: they have to offer services related to the collection and recycling of packaging waste, and these services should be good enough in terms of both price and quality to attract, in a competitive framework, sufficiently many companies as licensees for a profitable business. For an example, Table 20.1 contains licence fees for various packaging materials from two of the compliance schemes operating in Germany.

Some of these services could also support their clients' efforts for a DfE regarding packaging. So far, however, compliance schemes seem to prefer to develop DfRs, designs for recycling. This, of course, fits perfectly with how the waste hierarchy is interpreted in many industrialised countries, the member states of the EU, for example (OECD, 2016). This was examined and discussed in more detail in Section 18.1.1. Moreover, one must not forget that recycling is probably more in the business interests of the compliance schemes than waste prevention (Van Ewijk & Stegemann, 2016).

TABLE 20.1   Licence fees for packaging material (Compliance Schemes in Germany, 2020).

	Interseroh GmbH	Veolia Umweltservice Dual GmbH
10 tons of glass	605 Euro	540 Euro
1 ton of plastic	1020 Euro	790 Euro
10 tons of paper/cardboard	1915 Euro	1600 Euro

The prices (excl. VAT) refer to contracts with business companies to be concluded in 2020, and are subject to change anytime.
© *Source: Own Drawing with data taken from: https://www.lizenzero.de/verpackungsmengen-kalkulator/ Interseroh GmbH and https:// portal.veolia.de/b2bdualregistration/product for Veolia Umweltservice Dual GmbH.*

Not to be misunderstood: a DfR likely points in the direction of a DfE, however, there certainly is less emphasis on waste prevention. Thus, there are also some vested interests of the compliance schemes, which may lead to results not corresponding to the waste hierarchy.

These potential vested interests of the compliance schemes can probably be reduced, if the quantities of packaging waste collected and consigned for recycling increase. In this case licence fees will also increase if there are no appropriate design changes for packaging. Thus, the interest of the licensees for a DfE increases with the fees they have to pay (see also Table 20.1).

For this reason, it is necessary to set sufficiently high, but economically reasonable targets for the collection of packaging waste in order to strengthen this relationship. In addition, auxiliary regulations restricting the export of packaging waste, for example, continue to put pressure on companies for a DfE. But it seems necessary to closely monitor the development of the business activities of the PROs.

If we bring together all these policy tools, we will come to an IEP for packaging waste, which will support the implementation of a circular economy.

20.6 Putting everything together

This final section examines the links between the various policy tools, also in view of Constitutive Element III of an IEP.

As has already been indicated, there need to be high collection rates for packaging waste in order to restrict these vested interests of the compliance schemes, which are in favour of more recycling and less prevention of waste. High and rising collection targets, without many opportunities to export packaging waste tend to increase licence fees. This puts pressure on companies and thus compliance schemes, which are in a competitive framework, to consider not only DfRs but also DfEs. The companies, the licensees, can, of course, reduce their licence fees and thus their costs if they use less packaging material, either through reusable packaging, refillable bottles or through different designs. They may also switch to different materials, of course. In this sense, the integration of the households into the EPR policy is important for achieving the objectives of the policy.

Current policies on packaging waste sometimes do not sufficiently integrate households. In Germany, the somewhat misleading labelling of the packaging leads to a rather incomplete separation of packaging waste, so that a large part of it is incinerated. However, waste incineration is also necessary, otherwise Germany would have to import even more waste for its incineration plants from neighbouring European countries – also a consequence of technical path dependencies.

Another problem with current policies is the existence of individual EPR systems and collective EPR systems based on associations, opening up opportunities for the pursuit of vested interests. Deposit systems for drinks packaging, for example, are sometimes only incompletely integrated into the EPR principle. In Germany, producers of drinks packaging in the deposit system do not have to licence the packaging with the compliance schemes, implying a focus on the collection and recycling of empty packaging, but not so much on DfEs. This, too, fits seamlessly into the common "interpretation" of the waste hierarchy, with its apparent neglect of waste prevention. Wiesmeth, Shavgulidze, and Tevzadze (2018), however, design an IEP policy with a mandatory deposit fee for drinks packaging.

The proposed IEP for packaging waste consists of various policy tools: there is some degree of laissez-faire (Section 14.1.1) in relation to separation of waste, supported by technology-guidance and appropriate campaigns to raise environmental awareness, thus to establish waste separation as a "social norm"

in the sense of behavioural economics (Section 8.2.3). There are, moreover, various instances of a command-and-control policy (Section 14.1.3). For example, the take back requirement for packaging, the regulations, which allow the establishment of PROs, or the obligation to join, as a manufacturer or distributor of packaging, a compliance scheme to licence the packaging. Then there are these policy tools that decentralise relevant decisions: companies are free to join anyone of the compliance schemes, the compliance schemes are in competition and base their decisions on this competitive environment. This applies in particular to contracts with collection and recycling companies and includes again technology-guidance (Section 14.1.2).

Fig. 20.3 shows the details of the IEP, the various policy tools used in this holistic approach, and indicates the required links. Vested interests

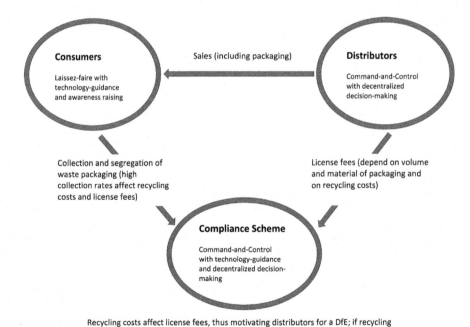

FIG. 20.3   Structure of the IEP for packaging waste. The various policy tools have to be linked appropriately to motivate distributors of packaging for a DfE. © *Source: Own Drawing.*

of the compliance schemes can arise with profitable recycling.

It remains necessary to reduce or rule out vested interests by separating decisions on the handling of packaging waste from the business interests of the distributors putting packaging into circulation. This requires a careful structure of the policy tools – in this case by establishing independent compliance schemes. Moreover, another crucial part of designing the policy consists in linking these policy tools in an appropriate way, thus in carefully observing Constitutive Element III of an IEP (Section 17.3.3). Only this last step leads to higher costs for packaging distributors and motivates them for a DfE in order to reduce packaging costs.

This IEP thus guides the economy through "interchanges amongst people" to a "reality" that corresponds to the objectives of a circular economy. It reduces the environmental impact of waste by respecting the waste hierarchy and the technological environment. This is the combination of economy-guidance and technology-guidance required for the implementation of a circular economy.

The following chapter continues this discussion and applies these considerations to IEPs for WEEE and ELVs.

References

Agovino, M., Cerciello, M., & Musella, G. (2019). The effects of neighbour influence and cultural consumption on separate waste collection. Theoretical framework and empirical investigation. *Ecological Economics*, 166, 106440. https://doi.org/10.1016/j.ecolecon.2019.106440.

Bartolacci, F., Paolini, A., Quaranta, A. G., & Soverchia, M. (2018). Assessing factors that influence waste management financial sustainability. *Waste Management*, 79, 571–579. https://doi.org/10.1016/j.wasman.2018.07.050.

Bilitewski, B. (2008). Pay-as-you-throw—A tool for urban waste management. *Waste Management*, 28(12), 2759. https://doi.org/10.1016/j.wasman.2008.08.001.

Boehmer-Christiansen, S. A. (1990). Energy policy and public opinion manipulation of environmental threats by vested interests in the UK and West Germany. *Energy Policy*, 18 (9), 828–837. https://doi.org/10.1016/0301-4215(90)90062-9.

Boute, A., & Zhikharev, A. (2019). Vested interests as driver of the clean energy transition: Evidence from Russia's solar energy policy. *Energy Policy*, 133, 110910. https://doi.org/10.1016/j.enpol.2019.110910.

Bridgman, B. R., Livshits, I. D., & MacGee, J. C. (2007). Vested interests and technology adoption. *Journal of Monetary Economics*, 54(3), 649–666. https://doi.org/10.1016/j.jmoneco.2006.01.007.

Elia, V., Gnoni, M. G., & Tornese, F. (2015). Designing pay-as-you-throw schemes in municipal waste management services: A holistic approach. *Waste Management*, 44, 188–195. https://doi.org/10.1016/j.wasman.2015.07.040.

Groth, M. (2008). *A review of the German mandatory deposit for one-way drinks packaging and drinks packaging taxes in Europe. In University of Lüneburg Working Paper Series in Economics, No. 87*, Retrieved from (2008). https://core.ac.uk/download/pdf/6781128.pdf.

Hahladakis, J. N., Purnell, P., Iacovidou, E., Velis, C. A., & Atseyinku, M. (2018). Post-consumer plastic packaging waste in England: Assessing the yield of multiple collection-recycling schemes. *Waste Management*, 75, 149–159. https://doi.org/10.1016/j.wasman.2018.02.009.

Haukkala, T. (2015). Does the sun shine in the high north? Vested interests as a barrier to solar energy deployment in Finland. *Energy Research & Social Science*, 6, 50–58. https://doi.org/10.1016/j.erss.2014.11.005.

Knickmeyer, D. (2020). Social factors influencing household waste separation: A literature review on good practices to improve the recycling performance of urban areas. *Journal of Cleaner Production*, 245, 118605. https://doi.org/10.1016/j.jclepro.2019.118605.

Leeabai, N., Suzuki, S., Jiang, Q., Dilixiati, D., & Takahashi, F. (2019). The effects of setting conditions of trash bins on waste collection performance and waste separation behaviors; distance from walking path, separated setting, and arrangements. *Waste Management*, 94, 58–67. https://doi.org/10.1016/j.wasman.2019.05.039.

Liao, Z. (2018). Environmental policy instruments, environmental innovation and the reputation of enterprises. *Journal of Cleaner Production*, 171, 1111–1117. https://doi.org/10.1016/j.jclepro.2017.10.126.

Moqbel, S., El-tah, Z., & Hassad, A. (2019). Littering in developing countries: The case of Jordan. *Polish Journal of Environmental Studies*, 28(5), 3819–3827. https://doi.org/10.15244/pjoes/94811.

OECD (2016). *Extended producer responsibility: Updated guidance for efficient waste management.* Paris: OECD Publishing. https://doi.org/10.1787/9789264256385-en.

Reichenbach, J. (2008). Status and prospects of pay-as-you-throw in Europe—A review of pilot research and implementation studies. *Waste Management*, *28*(12), 2809–2814. https://doi.org/10.1016/j.wasman.2008.07.008.

Seyring, N., Dollhofer, M., Weißenbacher, J., Bakas, I., & McKinnon, D. (2016). Assessment of collection schemes for packaging and other recyclable waste in European Union-28 member states and capital cities. *Waste Management & Research*, *34*(9), 947–956. https://doi.org/10.1177/0734242X16650516.

Tang, Z., Zhang, L., Huang, Q., Yang, Y., Nie, Z., Cheng, J., ... Chai, M. (2015). Contamination and risk of heavy metals in soils and sediments from a typical plastic waste recycling area in North China. *Ecotoxicology and Environmental Safety*, *122*, 343–351. https://doi.org/10.1016/j.ecoenv.2015.08.006.

Taylor, D. C. (2000). Policy incentives to minimize generation of municipal solid waste. *Waste Management & Research*, *18*(5), 406–419. https://doi.org/10.1177/0734242X0001800502.

Tencati, A., Pogutz, S., Moda, B., Brambilla, M., & Cacia, C. (2016). Prevention policies addressing packaging and packaging waste: Some emerging trends. *Waste Management*, *56*, 35–45. https://doi.org/10.1016/j.wasman.2016.06.025.

Thaler, R. H., & Sunstein, C. (2008). *Nudge: Improving decisions about health, wealth, and happiness.* New Haven, CT: Yale University Press. https://doi.org/10.1002/bdm.652.

Valenzuela-Levi, N. (2019). Do the rich recycle more? Understanding the link between income inequality and separate waste collection within metropolitan areas. *Journal of Cleaner Production*, *213*, 440–450. https://doi.org/10.1016/j.jclepro.2018.12.195.

Van Ewijk, S., & Stegemann, J. A. (2016). Limitations of the waste hierarchy for achieving absolute reductions in material throughput. *Journal of Cleaner Production*, *132*, 122–128. https://doi.org/10.1016/j.jclepro.2014.11.051.

Wiesmeth, H., Shavgulidze, N., & Tevzadze, N. (2018). Environmental policies for drinks packaging in Georgia: A mini-review of EPR policies with a focus on incentive compatibility. *Waste Management & Research*, *36*(11), 1004–1015. https://doi.org/10.1177/0734242X18792606.

WEEE and ELV in a circular economy

The main regulations of the Directive on WEEE (waste electrical and electronic equipment) of the European Union (EU) refer to product design (Art. 4), separate collection of WEEE including a mandatory, free of charge take-back system (Art. 5) and proper treatment of waste equipment (Art. 8). These four policy areas are, in fact, characteristic for most WEEE regulations in industrialised countries. Differences concern some details of the collection system and the structure of the take-back system. It is interesting to note that the resources for manufacturing EEE, in particular electronic equipment, is usually not explicitly addressed. Nevertheless, the social and environmental circumstances of mining special natural resources in various developing countries should be taken into account with a view to sustainable development.

In their final report for the European Commission on "WEEE compliance promotion exercise", BiPRO (2018) point to the challenges of implementing the EPR principle in general, but also to a close cooperation between public and private stakeholders in particular. They further consider collection of WEEE and preparation for reuse as challenges for all EU countries (see p. 10ff). Apart from the need to strengthen control procedures for illegal activities related to the export of WEEE, these critical remarks refer to issues associated with the waste hierarchy and thus with an "Integrated Environmental Policy" (IEP).

Similarly, the ELV legislation (end-of-life vehicles) of the EU provides that "vehicle and equipment manufacturers must factor in the dismantling, reuse and recovery of the vehicles when designing and producing their products". They also must provide systems to collect ELVs at no cost to the vehicle owner. Of course, there are quantitative targets for reuse and recovery of ELVs.

On the other hand, Eunomia (2019) point to a "significant number of vehicles of unknown whereabouts and those being exported as used rather than waste" (p. 42), and Mehlhart, Kosińska, Baron, and Hermann (2018) refer to the more than one million of used vehicles exported 2014 to countries outside the EU, and to the many more with "unknown whereabouts" (p. 14). Therefore, there seems to be an issue with the reuse of cars outside the EU, in countries that are likely to be without maintenance and recycling requirements to protect human health and the environment. This is, of course, of relevance for the waste hierarchy.

Implementing the Circular Economy for Sustainable Development
https://doi.org/10.1016/B978-0-12-821798-6.00021-1

WEEE and ELV are therefore two areas where an IEP is needed, with a particular focus on the reuse of old electrical and electronic equipment, and vehicles.

21.1 An integrated environmental policy for WEEE

The structural properties of an IEP for WEEE are consistent with the IEP for packaging waste: there is the general requirement of observing the constitutive elements as mentioned earlier (see also Wiesmeth & Häckl, 2016). Therefore, the following comments refer mainly to some specific issues related to the design of IEPs in these areas.

21.1.1 Collection of WEEE

The return and collection of large waste electrical equipment such as refrigerators or washing machines, is certainly less of a problem than the return of small WEEE such as batteries or mobile phones. Fig. 21.1 shows that 18 EU member states as well as Liechtenstein and Norway surpassed the 45% target for 2016 referring to electrical and electronic equipment (EEE) put on the market. Whether they will also reach the 65% target for 2019 remains to be seen. However, attaining this target is also dependent on the previous endowment with large appliances. And it is very likely that this endowment was below average in most of the newer member states of the EU. In addition, these weight targets obviously focus on the return of large appliances and not so much on small WEEE (see also BiPRO, 2018, Table 2.7 and Fig. 2.10).

In 2014, according to BiPRO (2018), half of the member states of the EU did not separately report WEEE amounts prepared for reuse, seven member states reported shares between 4.6% and 1.4%, and the remaining seven member states between 1.3% and 0.03% (p. 43). Thus, there is a large share of WEEE neither collected nor officially prepared for reuse. In Germany, households tend to hoard small WEEE: according to a survey by Schmiedel, Löhle, and Bartnik (2018), a significant share of owners of WEEE dispose of their old equipment with a considerable time delay (p. 12). However, the review reports of BIO Intelligence Service (2014) and OECD (2016) point to comparable results in other parts of the world. Nowakowski (2019) investigates the reasons "for storage of WEEE by residents" and notes that "the most frequent reason for stockpiling is intended possible use of the equipment in the future".

As high collection and recycling rates of WEEE are of relevance for motivating manufacturers for a DfE, collection of WEEE should be encouraged. In addition, official markets for the reuse of EEE should be promoted. These considerations lead to the following proposals for the introduction of a refund system for WEEE (see also Wiesmeth & Häckl, 2016 for some further aspects).

21.1.2 A refund or deposit system for WEEE?

Suppose owners of WEEE get a "refund" for returning WEEE to an official collection point. The refund should depend on the category and other characteristics of the old equipment. Of particular importance is the dependence of the refund on DfE: the refund should increase with the difficulty to disassemble the product. The market value of the recoverable materials could as well play a role. The refunds have to be paid by the manufacturers or importers of EEE and could be managed by the compliance schemes. The refund returned to the consumers need not coincide with the fee paid by the manufacturer on top of the licence fees. In fact this fee could turn into a subsidy, if the product specifications are particularly environmentally friendly. Of course, this system of fees and returns needs to be balanced over time.

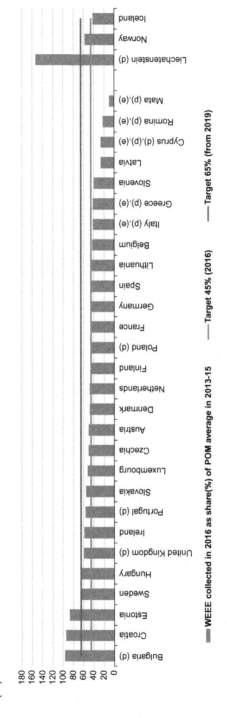

Rate of total collection of waste electrical and electronic equipment in 2016 in relation to the average weight of EEE put on the market in the three preceding years (2013-2015)

(%)

Note: Ranked on 'Share of WEEE collected...' data.
(d) definition differs, see meatdata
(e) estimated
(p) provisional

Source: Eurostat (online data code: env_waselee)

▬ WEEE collected in 2016 as share(%) of POM average in 2013-15   ▬ Target 45% (2016)   ▬ Target 65% (from 2019)

eurostat ▣

FIG. 21.1 Collection of WEEE in European Countries. Rate of collection in comparison to the targets referring to the weight of equipment put on the market. © *From Eurostat. Waste statistics – Electrical and electronic equipment. Retrieved April 17, 2020 from https://ec.europa.eu/eurostat/statistics-explained/images/1/18/Figure_5_Rate_of_total_collection_for_WEEE_in_2016_in_relation_to_the_average_weight_of_EEE_POM_2013-2015_percentage.png (Original work published 2019).*

There is some practical experience with refunds for WEEE. Zhu et al. (2012) report on the "trade-in policy for home appliances and electronics" implemented in China in 2009. Under this policy, (registered) consumers in the pilot provinces and cities can receive a 10% subsidy on a new product if they provide an old one (p. 1217). Besides stimulating demand for new products, this trade-in policy effectively promotes the collection of WEEE (see Table 4, p. 1218). However, for an IEP this special trade-in policy alone is not yet sufficient (p. 1220). In particular, there is no stimulus for DfE, and some consumers started to exploit the system by buying cheap second-hand products just to get the subsidy on a new product (p. 1219). So not all parts of this policy offer the right incentives, there are obviously vested interests. In particular, a refund system could attract WEEE from other regions if precautions are not taken.

One could, of course, also introduce a deposit fee on EEE to foster the separate collection of WEEE – similar to the mandatory deposit fee on one-way drinks packaging in various countries (see Wiesmeth & Häckl, 2016, Example 4.3b). According to the German Advisory Council on the Environment (SRU, 2012), deposit schemes are an effective tool, in particular for small EEE such as mobile phones and computers. There is also some experience with deposit schemes for EEE in various countries, including Austria, Italy and the U.S. (see Wilts & von Gries, 2016, Chapter 3). Moreover, Zuo, Wang, and Sun (2020) analyse new business models for the collection of WEEE based on the Internet, and Corsini, Rizzi, and Frey (2017) examine and compare different organisational systems for WEEE collection – and point to some technology-guidance.

Needless to say that refund and deposit systems require a sophisticated technical and organisational infrastructure to prevent fraud. Experience with deposit systems for drinks packaging could certainly provide guidance

for setting up such a system, and the EU regulation for EEE registration establishes a database, which could be useful in this context.

Whether collection of WEEE is based on laissez-faire with technology-guidance through appropriate collection points, or on a more or less open refund system, or on a separate deposit system has to be made dependent on the local situation, in particular on the willingness to invest in the necessary technical and organisational infrastructure, but also on the expected acceptance by the population. As indicated in Section 17.4.1, the possibility of using such a system as a "cheap excuse" for less environmentally friendly behaviour in other areas should also be taken into account.

Nevertheless, there are differences regarding the environmental effectiveness. A laissez-faire system can benefit from a high level of environmental awareness, although the Tragedy of the Commons continues to induce households to dispose of small WEEE in household waste. A deposit system establishes some "individual product responsibility" and motivates consumers to return waste products to collection points, a refund system requires manufacturers to include the refund fee in the prices of the products. As this fee can become a subsidy for equipment with a DfE, potentially lowering prices and increasing demand and profits, there is an additional incentive for a DfE.

21.1.3 An IEP for implementing the EPR principle

Similar to the implementation of the EPR principle in the IEP proposed for packaging waste, manufacturers and importers of EEE have to join a compliance scheme to licence their products. These schemes, which are assumed to operate in a competitive environment, take back WEEE, consign it to treatment and recycling (see Walls, 2006, p. 35ff, for the WEEE policy in South Korea).

Independent compliance schemes for packaging waste can also serve as PROs for WEEE. In this sense, the "Green Dot", one of the compliance schemes for packaging waste in Germany, offers these services. There is, thus, no need to set up additional PROs for different types of waste, and economies of scope can be used.

The licence fees to be paid by the manufacturers and importers of EEE, depend on characteristics of the products and should include ecological aspects such as DfE. This is already stipulated in Art. 21 of the German Packaging Act. Sousa, Agante, Cerejeira, and Portela (2018) examine the practice of charging EEE fees, which include licence fees to be paid to PROs, in the member states of the EU.

Such environmental product characteristics could be energy efficiency specifications that have already been introduced for various household appliances, such as refrigerators, washing machines, etc. Some of these energy efficiency regulations extend also to electronic products such as televisions and computers, although technical specifications with a focus on environmental issues seem still to be missing.

However, after the introduction of certain basic specifications, the regulations of the IEP will likely induce manufacturers to provide such parameters of their products, as this may imply a lower licence fee. Moreover, if there is a refund system, then the fee to be paid by the manufacturers on top of the licence fee, can turn into a subsidy if their product is above these basic specifications.

The licence fees depend on certain characteristics of the EEE, affecting the prices of the new products: there will be higher prices for products without DfE, and manufacturers will experience higher licence fees due to higher recycling costs. A further increase in licence fees results from high collection and recycling rates, but also from a possible refund system. In addition, higher fees for products with a low level of DfE are expected to lead to higher prices and declining demand for these products.

Consequently, if a high level of DfE reduces lifetime costs and stimulates demand, producers have a stronger incentive to change the design of their products – without intensive monitoring, which is not very effective anyway.

This implementation of the EPR principle leads to various links between the decisions of the producers, the consumers and the compliance schemes: the consumers have a stronger incentive to return WEEE to official collection points, at least if there is a refund or deposit system, and "leakage" to export markets decreases because treatment decisions are made by the independent compliance schemes and not by the manufacturers with their vested interest. However, the compliance schemes can also have vested interests – depending on the market situation for the recovered materials and other framework conditions. It is therefore necessary to monitor closely the relevant developments (see also the comments in Section 20.6).

The extensive literature includes analyses of WEEE regulations for all kinds of WEEE and for a variety of countries. This is justified by Constitutive Element I of an IEP, stipulating the consideration of the local conditions. These case studies allow a structural comparison with the IEP for WEEE proposed here. Important are always the incentives provided by the various policies.

Kalimo, Lifset, Atasu, Van Rossem, and Van Wassenhove (2015) address these issues to some extent when they conclude that "in order to divide the responsibilities to create incentives, EPR requires an effective regulatory framework", and "...public authorities have a responsibility to intervene with guidance and regulation in case of market failures" (p. 53). The above approach to an IEP for WEEE establishes such a regulatory framework, thus following Constitutive Element III.

Collective EPR systems, which share the costs of collection, treatment and recycling, seem to characterise the practice of handling WEEE (see BIO Intelligence Service, 2014; OECD,

2016). However, as there are no independent compliance schemes, there is the risk of vested interests, which may, among other things, lead to excessive exports of used equipment to developing countries with all the associated environmental impacts (see Section 20.4.2, but also Atasu & Subramanian, 2012; Sovacool, 2019). Cost allocation mechanisms, which allow recycling costs to be apportioned according to the returned WEEE of the participating manufacturers, are investigated by Gui, Atasu, Ergun, and Toktay (2015). Overall, practical EPR systems for WEEE do not fully meet the requirements of an IEP. The consequences are the shortcomings of the policies reviewed in the literature (see also Chapter 5).

The following subsection deals with the issue of reusing EEE, of particular importance for an IEP in the context of the waste hierarchy.

21.1.4 Reusing and remanufacturing electrical and electronic equipment

Empirical investigations show that reusing EEE is not yet a preferred activity regarding the waste hierarchy (see, for example, BiPRO, 2018, p. 57, for the case of the EU). A surprisingly large share of appliances, such as refrigerators and washing machines, but also TV sets and personal computers, is neither returned for recycling, nor offered for reuse. According to the German Environment Agency (UBA), only 41% of all large household appliances were collected in accordance with the regulations in 2016. "Nearly 465,000 tonnes of refrigerators, washing machines and dishwashers disappear through illegal disposal channels or into non-certified processing plants. This is a pure waste of resources, not to mention the grave damage it does to the environment when hazardous substances are disposed of improperly or not at all" (M. Krautzberger, former President of the UBA). Of course, it is also likely that part of this older equipment is just retained and still used together with new, more efficient appliances,

thus proving the relevance of the rebound effect (Section 12.1).

Small equipment, such as mobile phones, is stored at home, for whatever reason (Nowakowski, 2019), and some batteries probably end up in household waste (Janz & Bilitewski, 2009). In addition, sizeable quantities of WEEE are likely to be exported to developing countries, declared to be reusable, but then "recycled" in a way that threatens both human health and the environment (Schnoor, 2012). The question therefore arises as to how to create functioning markets for reusing EEE? In principle, such markets could be set up and operated by the compliance schemes.

"Remanufacturing" used EEE seems to be developing into an interesting business activity (Govindan, Jiménez-Parra, Rubio, & Vicente-Molina, 2019). There is already talk about a "Design for Remanufacturing" (Singhal, Tripathy, & Jena, 2020), and Pazoki & Samarghandi (2020) ask the question, whether take-back regulations favour remanufacturing or eco-design? Is therefore "remanufacturing" on the way to replace "reusing"? According to the Ellen MacArthur Foundation, "remanufacturing conserves the value of machine products and components, returning them to a high condition so they can be used again and again".

Of course, remanufacturing helps to save material and energy, probably even in comparison to recycling, and provides affordable components for customers. It is an important aspect of the implementation of a circular economy, although "affordable" components point to the rebound effect (Section 12.1). There is just one caveat: remanufacturing opens up interesting market potentials, but exactly these potentials may threaten traditional reuse activities, that have hitherto been characterised by limited market potential. Similar to the case of recycling, societal path dependencies divert business interests away from waste prevention. Therefore, the development of remanufacturing – as an additional part of the waste hierarchy – needs to be monitored in the context of implementing a circular economy.

Monitoring compliance schemes could help to mitigate all kinds of vested interests that manufacturers, importers and the compliance schemes themselves may have. This is not only about exporting perhaps too much old equipment to developing countries, it is also about the fact that reusing old equipment is not really beneficial to the legitimate business interests of manufacturers and importers of EEE.

In the context of returning old equipment to the compliance schemes, appliances and small EEE could be purchased for preparation for reuse or for remanufacturing and offered to potential customers. Boldoczki, Thorenz, and Tuma (2020) analyse the reusability of large household appliances, and in the Czech Republic, some PROs encourage consumers to offer a used mobile phone, computer or household appliances. Similar activities exist in other member states of the EU (BiPRO, 2018, p. 77f). The existence of a refund or a deposit system for WEEE is likely to facilitate such efforts, as owners of old equipment are now more inclined to return these used commodities. It is then possible to refund the households and even pay a provision for the equipment, in case there is some chance to resell or remanufacture it. Refunds and deposit fees have, of course, to be passed on to the buyers.

It remains to be clarified to what extent reusing or remanufacturing of EEE helps to save natural resources – the sustainability context of the waste hierarchy. If EEE is reused by individuals, who otherwise would have bought new equipment, then there is certainly a "displacement" of newly produced commodities through used ones. If, however, this used or remanufactured equipment is sold to people, who otherwise would not have bought new equipment for lack of financial resources, then there is obviously no such displacement, and reuse or remanufacturing is functioning more like a rebound effect.

But then there is the ethical question: can we deny access to these technologies?

Therefore, current practices of reusing EEE, i.e. exports to developing countries, sales to non-authorised scrap metal dealers, but also continued domestic use alongside new equipment, are not without problems in implementing the waste hierarchy and a circular economy. Whether refunds or deposits and official markets for used EEE established and operated by the compliance schemes can be helpful in this regard, remains to be seen. As already mentioned, some activities in the EU seem to be moving in this direction.

Fig. 21.2 indicates the various policy tools of an IEP for WEEE, the main obligations of the stakeholders and the links between them.

21.2 An integrated environmental policy for old cars and ELV

The structural properties of an IEP for ELV coincide also with the IEP for packaging waste: there is the general requirement to observe the constitutive elements, as already mentioned (Wiesmeth & Häckl, 2016). Therefore, the following subsections will mainly refer to some specific issues with respect to designing an IEP in this area. Similar to the WEEE policy, the reuse of cars is one of the important aspects, with the additional dimension of the air pollution created by poorly maintained second-hand vehicles in developing countries and emerging economies. For this reason, the focus of an IEP should also be on the reuse of old cars, and not just recycling. Moreover, according to the Ellen MacArthur Foundation, remanufacturing ELV is becoming increasingly important.

21.2.1 Collection of ELV

Since the recycling of ELV at the expense of the manufacturers is relevant for a DfE, regulations should provide sufficient motivation for a return of a large part of ELV to the manufacturers.

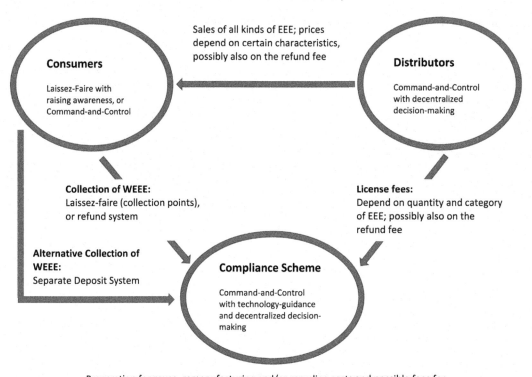

Preparation for reuse, remanufacturing and/or recycling costs and possible fees for the refund system affect license fees, thus motivating distributors for a DfE; societal path dependencies and profitable activities can play a role

FIG. 21.2 Structure of an IEP for WEEE. The links between the various policy tools are of importance for the IEP. © *Source: Own Drawing.*

As old cars cannot be placed in waste bins and not just left abandoned in the streets, a convenient policy allows the costless return of ELV to the manufacturer or a corresponding collection point – with official documentation for the last owner of the ELV. A take-back requirement obliges car manufacturers to consign their ELV for recycling at their own expense. A high share of the ELV returned will then motivate manufacturers for a DfE to reduce recycling costs.

Regarding this simple policy approach, there are a few issues, which need a more careful consideration. First of all, it is not always clear, whether an old vehicle is a scrap car or whether it is still reusable: in a low-income country with lower social and perhaps also lower environmental standards, for example. Another problem is closely related: an

old car, exported to a country outside the EU need not be costly recycled in the EU and helps to reduce the overall recycling costs for the car manufacturers. But if there is no serious maintenance of these cars, they then contribute to air pollution in the major cities of the import countries.

These aspects are not yet fully integrated into current policies. The challenge is to take these developments into account in an IEP for ELV – with reuse and remanufacturing of old cars, of course.

21.2.2 An IEP for implementing the EPR principle

A complete implementation of the EPR principle would obviously require that the design of cars should also focus on prevention of waste, not just recycling. The currently existing policies seem to concentrate on a DfR, a design for recycling, and, more recently, also on a design for remanufacturing (Singhal et al., 2020). At the moment, there are not many explicit incentives to prevent waste, although the focus on a DfR certainly helps to minimise environmental pollution associated with recycling cars (Gerrard & Kandlikar, 2007), similar to the focus on a design for remanufacturing. There is a vast literature on all kinds of technical aspects related to the recycling of ELV (see, for example, D'Adamo, Gastaldi, & Rosa, 2020), and remanufacturing is also gaining more and more attention.

What could a policy to promote a DfE for vehicles look like? Similar to the other IEPs introduced so far, car manufacturers and importers have to join an independent compliance scheme in order to restrict their vested interests. Individual PROs or PROs controlled by associations seem to grant too many degrees of freedom for vested interests of the car manufacturers – regarding exports of used cars, for example (see the report of Mehlhart et al., 2018 for the EU).

Car manufacturers and importers then have to licence their cars with the licence fees covering collection and recycling of the particular vehicle types. These fees therefore include environmental parameters of the cars, aspects of a DfE, for example. A large share of returned ELV then raises overall collection and recycling costs with a focus not only on the cost average for all vehicle types of a manufacturer, but also on the cost contributions of each type.

Cars with a poor environmental record would then become more expensive, perhaps leading both to efforts for a DfE and a decreasing demand, which could close the links between collection and a DfE, as intended by the EPR principle. The actual development, however, depends on additional parameters and behavioural aspects: the price elasticity of demand and the preferences of the customers, in general. Buyers of larger vehicles may be insensitive to price increases and price increases due to higher

licence fees need not lead to significant changes in the sale of these vehicles. It could then be necessary to intervene in car manufacturing, similar and in addition to the increasing standards regarding the fuel economy (see Section 15.4.1).

A further restriction of such policies arises from second-hand vehicles, which are an important part of the business of car dealerships.

21.2.3 Reusing and remanufacturing vehicles

Unlike EEE, reusing vehicles is widespread and an important part of the business of car dealerships. The question arises as to the extent to which the current practise of reusing cars meets the objectives of the waste hierarchy – and thus of an IEP?

As far as the waste hierarchy is concerned, the question of replacement or rather the displacement of new cars by second-hand cars is relevant. In addition, maintenance of used cars might play a role in air pollution, and the recycling of ELV should not, of course, harm human health and the environment.

If used cars are sold to individuals, who, for whatever reason, would not have bought a new car, then obviously there is no one-on-one displacement. This situation becomes more critical, when used cars are sold outside the domain of the IEP. In this case, neither collection nor recycling of the later ELV might be regulated – with possible environmental consequences, as already mentioned. The fact that the fate of many vehicles is simply not known and not sufficiently documented, does not make the situation any easier (see Mehlhart et al., 2018, p. 2, for some numbers regarding the EU).

How to regulate behaviour of consumers and producers to align them better with the goals of the waste hierarchy and to arrive at an IEP? First of all, it is certainly neither socially nor politically correct, to deny someone legal access to a car, be it a new or a used one. Perhaps one possibility is to restrict selling used cars, which are already close to a scrap car – which implies a consent on an adequate definition. This decision should be made by independent institutions – before the actual sale of the car. This could help to reduce some environmental pollution.

However, inadequate maintenance, contaminated gasoline, problematic recycling, etc. continue to pollute the environment. The cooperation of the export countries is therefore necessary: they can help to establish a mandatory and periodical check-up of the cars, and they can supervise recycling activities. Support from the international car manufacturers is expected, as functional markets for second-hand vehicles are also in their interest. This applies in particular to sales of used cars outside the domain of the regulations.

Again, remanufacturing requires some additional thoughts. Regarding cars, it is an additional step in the waste hierarchy before recycling ELVs and probably helps to save material, energy and resources. Perhaps, as investigated by Govindan et al. (2019), the requirement of a "design for remanufacturing" should complement the "design for environment" for the car manufacturing?

21.3 Some final remarks regarding these policies

These policy proposals are again based on the constitutive elements of an IEP. However, there are some challenges regarding their implementation.

There are, first of all, certain behavioural aspects of the customers, which interfere with the policies in one way or the other. Regarding WEEE, there is the "hoarding" of WEEE and old equipment, either for personal (re)use, thus increasing the rebound effect, or for other purposes. Not surprisingly, markets for used EEE are not really established. This hoarding

behaviour of the households may both be a consequence of non-existing markets for used EEE, or it may constitute a reason for their non-existence.

Regarding vehicles, the situation is different in view of fully functional markets for used cars. This is probably also a consequence of the different levels of saturation of the markets with rapid technological changes regarding EEE and somewhat less frequent changes regarding cars. But, there is, both for EEE and for vehicles, an obvious societal preference for the latest equipment, implying the necessity to replace used commodities with new ones.

There is this special context of reusing EEE or vehicles. In both situations, current reuse activities are not really in accordance with the philosophy of the waste hierarchy. A desirable one-on-one displacement of new commodities by used ones is hardly realisable. Societal path dependencies seem to give the impression that current reuse activities regarding remanufactured parts are fully compatible with circular economy strategies. The Ellen MacArthur Foundation is one of the advocates of this position. Of course, this helps to save energy and resources, but it affects the waste hierarchy: remanufacturing might assume the role of reuse for WEEE, for ELV it is already developing into an additional option in the waste hierarchy – before recycling. Sitcharangsie, Ijomah, and Wong (2019) review some remanufacturing practices, thereby allowing a comparison with traditional reuse. It is certainly an interesting development which deserves further attention.

The proposed policies try to "correct" some problematic observations and developments. In particular, they motivate owners of WEEE and old equipment to return it, either for remanufacturing and recycling, or preparation for reuse. Moreover, they put some financial pressure on manufacturers and importers of EEE and vehicles to motivate them for a DfE, and they prevent, or at least reduce, vested interests. But, these policies show also that regarding international trade the cooperation of the participating countries is required. At this point the locality principle, Constitutive Element I of an IEP, comes into view, possibly creating further difficulties in view of different levels of environmental awareness.

These IEPs are also composed of various tools: there are various aspects of command-and-control-policies such as the take-back requirements, the compliance schemes, the licence fees, etc. But there is also decentralised decision-making in order to make use of the expertise of the manufacturers for a DfE. Vested interests are largely neutralised, although the compliance schemes may not have much of an interest in expanding reuse activities. Moreover, it remains to be seen to what extent remanufacturing can partially assume the role of reuse – interpreted in the sense of the waste hierarchy.

Finally, these policies allow a comparison with current policies, thereby pointing to some of their shortcomings: insufficient collection, focus on DfR instead of DfE, export of sizeable quantities of old equipment and cars to domains outside the regulations, probably a consequence of vested interests. All these observations are not in accordance with a circular economy. In this sense, the proposed IEPs are trying to remedy some of these deficiencies.

The following chapter on efforts to mitigate climate change requires intensive cooperation between the countries, which are major emitters of greenhouse gases.

References

Atasu, A., & Subramanian, R. (2012). Extended producer responsibility for E-waste: Individual or collective producer responsibility? *Production and Operations Management*, *21*(6), 1042–1059. https://doi.org/10.1111/j.1937-5956.2012.01327.x.

BIO Intelligence Service (2014). *Development of guidance on extended producer responsibility (EPR). In Report for the European commission DG ENV*(2014). https://ec.europa.eu/environment/archives/waste/eu_guidance/pdf/report.pdf.

BiPRO (2018). WEEE compliance promotion exercise. In *Final report for the European commission*. Luxembourg: Publications Office of the EU. https://doi.org/10.2779/918821.

Boldoczki, S., Thorenz, A., & Tuma, A. (2020). The environmental impacts of preparation for reuse: A case study of WEEE reuse in Germany. *Journal of Cleaner Production*, *252*, 119736. https://doi.org/10.1016/j.jclepro.2019.119736.

Corsini, F., Rizzi, F., & Frey, M. (2017). Extended producer responsibility: The impact of organizational dimensions on WEEE collection from households. *Waste Management*, *59*, 23–29. https://doi.org/10.1016/j.wasman.2016.10.046.

D'Adamo, I., Gastaldi, M., & Rosa, P. (2020). Recycling of end-of-life vehicles: Assessing trends and performances in Europe. *Technological Forecasting and Social Change*, *152*, 119887. https://doi.org/10.1016/j.techfore.2019.119887.

Eunomia (2019). *Final report on the implementation of directive 2000/53/EC on end-of-life vehicles. For the period 2014–2017. Retrieved from (2019)*. https://ec.europa.eu/environment/waste/elv/pdf/ELV%20Directive%20Implementation%20Report%202014-2017_Final%2027.06.19.pdf.

Gerrard, J., & Kandlikar, M. (2007). Is European end-of-life vehicle legislation living up to expectations? Assessing the impact of the ELV directive on 'green' innovation and vehicle recovery. *Journal of Cleaner Production*, *15*(1), 17–27. https://doi.org/10.1016/j.jclepro.2005.06.004.

Govindan, K., Jiménez-Parra, B., Rubio, S., & Vicente-Molina, M.-A. (2019). Marketing issues for remanufactured products. *Journal of Cleaner Production*, *227*, 890–899. https://doi.org/10.1016/j.jclepro.2019.03.305.

Gui, L., Atasu, A., Ergun, Ö., & Toktay, L. B. (2015). Efficient implementation of collective extended producer responsibility legislation. *Management Science*, *62*(4), 1098–1123. https://doi.org/10.1287/mnsc.2015.2163.

Janz, A., & Bilitewski, B. (2009). WEEE in and outside Europe – Hazards, challenges and limits. In P. Lechner (Ed.), *Prosperity waste and waste resources* (pp. 113–122). Vienna: BOKU-University of Natural Resources and Applied Life Sciences.

Kalimo, H., Lifset, R., Atasu, A., Van Rossem, C., & Van Wassenhove, L. (2015). What roles for which stakeholders under extended producer responsibility? *Review of European, Comparative & International Environmental Law*, *24*(1), 40–57. https://doi.org/10.1111/reel.12087.

Mehlhart, G., Kosińska, I., Baron, Y., & Hermann, A. (2018). Assessment of the implementation of the ELV directive with emphasis on ELVs of unknown whereabouts. In *Report for DG environment of the EU. Prepared by institute for applied ecology*. https://doi.org/10.2779/446025.

Nowakowski, P. (2019). Investigating the reasons for storage of WEEE by residents – A potential for removal from households. *Waste Management*, 87, 192–203. https://doi.org/10.1016/j.wasman.2019.02.008.

OECD (2016). *Extended producer responsibility: Updated guidance for efficient waste management.* Paris: OECD Publishing. https://doi.org/10.1787/9789264256385-en.

Pazoki, M., & Samarghandi, H. (2020). Take-back regulation: Remanufacturing or eco-design? *International Journal of Production Economics*, 227, 107674. https://doi.org/10.1016/j.ijpe.2020.107674.

Schmiedel, U., Löhle, S., & Bartnik, S. (2018). *Verbraucherumfrage zum Entsorgungsverhalten von Elektro(nik)altgeräten (Vol. 92/2018).* Umweltbundesamt. Retrieved from (2018). https://www.umweltbundesamt.de/sites/default/files/medien/1410/publikationen/20180705-bericht_ap3-fkz_3717_34_345_0_verbraucherumfrage_barrierefrei_pb2.pdf.

Schnoor, J. L. (2012). Extended producer responsibility for E-waste. *Environmental Science & Technology*, 46(15), 7927. https://doi.org/10.1021/es302070w.

Singhal, D., Tripathy, S., & Jena, S. K. (2020). Remanufacturing for the circular economy: Study and evaluation of critical factors. *Resources, Conservation and Recycling*, 156, 104681. https://doi.org/10.1016/j.resconrec.2020.104681.

Sitcharangsie, S., Ijomah, W., & Wong, T. C. (2019). Decision makings in key remanufacturing activities to optimise remanufacturing outcomes: A review. *Journal of Cleaner Production*, 232, 1465–1481. https://doi.org/10.1016/j.jclepro.2019.05.204.

Sousa, R., Agante, E., Cerejeira, J., & Portela, M. (2018). EEE fees and the WEEE system—A model of efficiency and income in European countries. *Waste Management*, 79, 770–780. https://doi.org/10.1016/j.wasman.2018.08.008.

Sovacool, B. K. (2019). Toxic transitions in the lifecycle externalities of a digital society: The complex afterlives of electronic waste in Ghana. *Resources Policy*, 64, 101459. https://doi.org/10.1016/j.resourpol.2019.101459.

SRU (2012). *Environmental report 2012: Responsibility in a finite world.* Erich Schmidt Verlag. Retrieved from (2012). https://www.umweltrat.de/SharedDocs/Downloads/EN/01_Environmental_Reports/2012_05_Environmental_Report_summary.pdf?__blob=publicationFile&v=5.

Walls, M. (2006). *Extended producer responsibility and product design: Economic theory and selected case studies. In Discussion Papers, Resources for the Future,* (dp-06-08) Retrieved from (2006). https://ideas.repec.org/p/rff/dpaper/dp-06-08.html.

Wiesmeth, H., & Häckl, D. (2016). Integrated environmental policy: A review of economic analysis. *Waste Management & Research*, 35(4), 332–345. https://doi.org/10.1177/0734242X16672319.

Wilts, H., & von Gries, N. (2016). Increasing the use of secondary plastics in electrical and electronic equipment and extending products lifetime—Instruments and concepts. In F.-C. Mihai (Ed.), *E-waste in transition—From pollution to resource.* InTechOpen. https://doi.org/10.5772/62778.

Zhu, S., He, W., Li, G., Zhuang, X., Huang, J., Liang, H., & Han, Y. (2012). Estimating the impact of the home appliances trade-in policy on WEEE management in China. *Waste Management & Research*, 30(11), 1213–1221. https://doi.org/10.1177/0734242X12437568.

Zuo, L., Wang, C., & Sun, Q. (2020). Sustaining WEEE collection business in China: The case of online to offline (O2O) development strategies. *Waste Management*, 101, 222–230. https://doi.org/10.1016/j.wasman.2019.10.008.

22

Climate change mitigation in a circular economy

Mitigating climate and implementing a circular economy are linked in a variety of ways. First of all, the traditional "take-make-waste" economy is resource-intensive, also with respect to fossil fuels. Any attempt to save resources through a DfE can therefore also help to reduce greenhouse gas emissions. The Ellen MacArthur Foundation, referring to estimates of the IPCC for 2010, points to the "45% of global greenhouse gas emissions that can be attributed to the production of materials, products, and food, as well as the management of land" (p. 16). This corresponds to 22.1 Gt CO_2, with energy for transportation contributing another 6.7 Gt. However, one must not forget that global CO_2 emissions from fuel combustion keep on rising (IEA, 2019).

It is possible, to look at the emission of greenhouse gases yet in a slightly different way. Anthropogenic greenhouse gas emissions into the atmosphere can be considered as waste, and it is necessary to reduce these emissions so as not to exceed the assimilative capacity of the atmosphere in order to limit the rise in temperature.

This is the result of ongoing research, despite the existing political and economic polarisation regarding this issue (Ballew, Pearson, Goldberg, Rosenthal, & Leiserowitz, 2020).

This chapter therefore considers greenhouse gas emissions to be waste and examines, from an economic point of view, some of the global and national attempts to reduce and thus prevent these emissions according to the waste hierarchy. It is not surprising that the lack of international cooperation so far prevents a major breakthrough. This points to mechanisms such as the Tragedy of the Commons and the Prisoners' Dilemma, which are firmly rooted in environmental economics (Section 7.1). In addition, aspects of historical fairness and distributional justice play a role, making necessary negotiations even more difficult.

Other relevant steps of the waste hierarchy in this context are "recycling" and "landfilling" of greenhouse gases. Using forests as "carbon sinks" for recycling, and "carbon capture and storage" (CCS) technologies for depositing greenhouse gases point in this direction.

22.1 Global attempts to mitigate climate change

This section provides a short overview of the various initiatives of the United Nations (UN) in recent decades. As indicated, the focus will be on economic aspects and mechanisms.

22.1.1 The Kyoto protocol

In addition to the "Rio Declaration" with its reference to sustainable development, the outcome of the United Nations Conference on Environment and Development (UNCED) in Rio de Janeiro in 1992, the "Earth Summit", included also the United Nations Framework Convention on Climate Change (UNFCCC). The main objective of the UNFCCC is the "stabilisation of greenhouse gas concentrations in the atmosphere at a level that would prevent dangerous anthropogenic interference with the climate system" (Art. 2).

There are some guiding principles of the UNFCCC. The "precautionary principle" says that the lack of full scientific certainty should not be used as an excuse to postpone action when serious or irreversible damage is imminent. The "principle of the common but differentiated responsibilities of states" gives the lead in combatting climate change to developed countries. Other principles address the specific needs of developing countries and the importance of promoting sustainable development.

When the UNFCCC with its general commitments encouraged industrialised countries to reduce greenhouse gas emissions, the Kyoto Protocol committed them to do so, with a heavier burden placed on the Annex I Parties, the industrialised countries. The Kyoto Protocol was adopted on December 11, 1997, and entered into force on February 16, 2005, after being ratified by at least 55 Parties to the UNFCCC, which accounted in total for at least 55% of the CO_2 emissions of the Annex I Parties in 1990.

In order to allow cost-effective reductions of greenhouse gas emissions, "flexible mechanisms" were introduced, among them "international emissions trading". Participating countries, which account for around 18% of global emissions, have agreed to reduce their greenhouse gas emissions by an average of 5% below 1990 levels by the 2008–2012 corridor, and the at that time 15 member states of the European Union (EU) committed themselves to a reduction of 8%.

The results of the first commitment period of the Kyoto Protocol are mixed and can be interpreted in different ways. All Parties with targets succeeded, at least according to some analyses, to a reduction of 12.5%. However, the collapse of the Soviet Union in 1991 led to a dramatic economic recession in Russia and Ukraine, both Parties to the Kyoto Protocol, with the consequence of a significant reduction of greenhouse gases in these countries after 1991. In fact, Russia and Ukraine showed reductions of 32.4% in 2012 compared to their 1990 emission levels. Of course, also Germany "benefited" in a similar way from the breakdown of the industry in the former German Democratic Republic (GDR) after reunification in 1990. Not surprisingly, without Russia and Ukraine, the remaining Parties reduced emissions only by 2.7%. And taking into account the reductions of more than 11% in the 15 member states of the EU, then many of the other Parties obviously increased their emissions – indeed, some countries, Australia, Greece and Spain among them, even had positive emission targets.

Therefore, the "success" of the Kyoto Protocol in the first commitment period till 2012 is mainly due to a particular effect, namely the collapse of the Soviet Union in 1991. Nevertheless, it was one of the first major attempts to encourage participating countries to cooperate in this field– of importance for a sustainable development and thus also for implementing a circular economy. For sure, it has created some awareness for the necessity to make all reasonable efforts to mitigate climate change. Przychodzen and Przychodzen (2020), for example, found that the implementation of the Kyoto Protocol increased the share of renewable energy sources.

The Prisoners' Dilemma can be held responsible for these not too positive developments, at least to some extent. A particular Annex I country might expect a disadvantage with respect to the competitive edge of its industry, if it moves too quickly to reach its target level for the reduction of the CO_2 emissions. If it slows down the process, it can benefit from two things: its industry will retain or gain an international competitive advantage, and it will also benefit from other countries' emission reductions. Therefore, slowing down the process of complying with the Kyoto Protocol seems to be a dominant strategy that at least produces characteristics predicted by the Prisoners' Dilemma (see also Wiesmeth, 2011, Chapters 3.2 and 5.3). There is no doubt that additional effects, such as the general economic situation of a country, also play a role in this context.

22.1.2 Copenhagen, Paris, Madrid and then?

Towards the end of the first commitment period of the Kyoto Protocol, international climate negotiations focused on aspects of a possible second commitment period. There was the Copenhagen Accord in 2009, a legally nonbinding political agreement with quantified emission targets submitted by the Parties, which were, however, not generally deemed sufficient to support the 2 °C target, and various commitments were contingent on other developments.

A couple of further Conferences of the Parties (COP) followed, with no substantial progress. Only COP 21 in Paris in 2015 led to a consensus among the Parties. The Paris Agreement was ratified by 187 of the 197 Parties to the UNFCCC and entered into force in late 2016. The Paris Agreement obliges all Parties to declare their "nationally determined contribution" (NDC) regarding their efforts to limit temperature increase to 1.5 °C. These individual targets should be adjusted every 5 years to set more ambitious goals. The current NDCs are considered too low for the 1.5° target, and adjusted NDCs are required for 2020 when the second commitment period of the Kyoto Protocol ends. Fig. 22.1 shows the current situation and what needs to be done in order to limit global warming to 1.5°.

However, according to general assessment, COP 25 in Madrid in 2019 did not turn out as expected and brought only a minimal agreement that the Parties' efforts to mitigate climate change should be stepped up the following year. There was no agreement on international emissions trading, and there were no significant

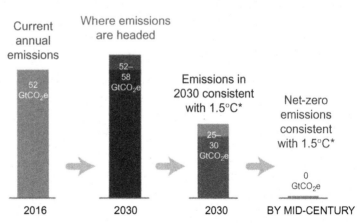

FIG. 22.1   The world is not on track to limit temperature rise to 1.5°.It is still technically feasible to avoid a 1.5 °C rise in temperature, but... © *From Levin, K. 8 Things you Need to Know About the IPCC 1.5 °C Report. Retrieved May 6, 2020 from https://www.wri.org/blog/2018/10/8-things-you-need-know-about-ipcc-15-c-report (Original work published 2018).*

commitments for further funds to support poorer countries to adapt to climate change.

The Prisoners' Dilemma continues to show its power: once a treaty is signed and ratified, there is a chance that efforts to protect the climate will be reduced and the time horizon for effective action will be extended. In addition, developing countries and emerging economies defend their levels of greenhouse gas emissions and point to historical fairness.

Of course, other considerations are likely to play a role in such a context. An optimal "positioning" of a country for further negotiations, or stimulating innovations for environmental technologies, or even increased efforts to motivate other countries to follow suit to step up their efforts to mitigate climate change (Shahnazi & Dehghan Shabani, 2020): these and other motives and aspects of behavioural economics – together with the Prisoners' Dilemma – explain developments in relation to the Conferences of the Parties.

As long as none of the major "players" in this game is clearly affected by global warming, not much could change – but then it might be too late. International negotiations must therefore continue, accompanied by efforts of various countries to limit global temperature rises.

22.2 Renewable energy sources and emissions trading

As a great deal of greenhouse gas emissions come from fossil fuels – 80% of all EU greenhouse gas emissions, for example (see EEA, 2019, p. 164), many countries, member states of the EU among them, make increasingly use of energy from renewable sources. In 2017, more than 17% of the EU's energy came from renewable sources. Of course, these changes in generating electricity and heat also helped to comply with the various obligations of the Kyoto Protocol.

Moreover, additional efforts to mitigate climate change refer to systems of tradable emission certificates. There is, of course, the EU Emission Trading System (EU ETS), a cap and trade system set up in 2005, as the world's first international emissions trading system and still accounting for three-quarters of international emissions trading. UBA (2019) reviews and analyses various climate policy instruments in different countries for their suitability to serve as a basis for establishing emission trading systems (see Section 16.2).

The following subsections highlight some aspects of these efforts. As developments in both renewable energy and emissions trading systems are still ongoing, these considerations cannot constitute a final appraisal of all related activities.

22.2.1 Renewable energy sources

Renewable energy sources to generate electricity and heat is an important means to reduce greenhouse gas emissions. Although this is an international issue due to the public good characteristics of mitigating climate change, it needs to be addressed at national, if not regional level: every consumer, every producer contributes to the emission of greenhouse gases. For this reason, the efforts of many countries to switch to renewable energy sources can be explained in this context. The case study on the promotion of renewable energy sources in Section 11.3 presents details on Germany's renewable energy policy.

Fig. 22.2 illustrates the situation with energy from renewable sources in the EU. The Corona crisis could help most member states meet the ambitious 2020 targets.

The continuing efforts of the EU to ambitiously reduce greenhouse gas emissions – according to the EU Climate and Energy Package by at least 40% in 2030 compared to 1990 levels – are dependent on a further increase in the share of electricity from alternative

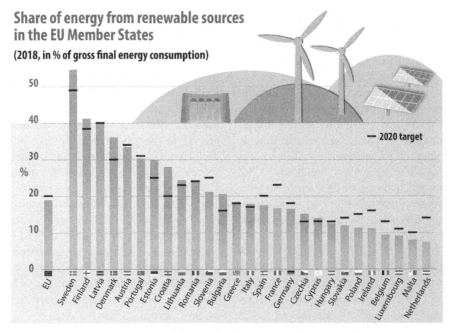

FIG. 22.2 Share of energy from renewable sources in the EU member states. The numbers, % of gross final energy consumption, refer to 2018. © *From Eurostat. Retrieved May 6, 2020 from https://ec.europa.eu/eurostat/documents/2995521/10335458/8-23012020-AP-EN.pdf/292cf2e5-8870-4525-7ad7-188864ba0c29 (Original work published 2020).*

sources. However, due to the resistance of the local residents, it is becoming increasingly difficult to build more wind turbines. On the other hand, the installation of wind turbines on the open sea does not yet provide a sustainable solution: long transmission lines need to be built to supply the industrial centres with electricity.

In addition, there is the still largely open question of how to store excess electricity from renewable sources to use it, when renewable energy is not available. Although addressed by many scientists (see, for example, Miller, 2020), backup-systems with conventional power plants are currently required to ensure continuous power supply in industrialised countries.

It remains to be seen whether all these challenges can be met in the near future. But the EU seems ready to move on to the next step to give emissions trading a greater role in curbing climate change.

22.2.2 Emissions trading systems

Economists generally agree that emissions trading systems (ETS) can be effective policy tools to limit greenhouse gas emissions. Moreover, if there is a general cap on emissions, then total emissions can be controlled and necessary reductions can be made at lowest costs. This "cap" is an environmental standard, which should be economically reasonable (see Section 11.1.1). Many countries are currently using various policy tools to limit greenhouse gas emissions, and some of them are aiming at introducing an ETS (UBA, 2019). As ETS are discussed in Section 16.2, in particular the ETS of the EU, this subsection briefly considers some international approaches. The international context is, of course, of special importance for an "Integrated Environmental Policy" (IEP) to mitigate climate change.

ETS are used in different countries: there is the California Cap and Trade, which was launched in 2013 and is, according to its own statement, the fourth largest in the world, after the ETS of the EU, of the Republic of Korea, and of the Chinese province of Guangdong. California's program covers some 85% of the state's greenhouse gas emissions and is linked to similar programs in Canadian provinces. The state's goal is to reduce greenhouse gas emissions to 40% below 1990 levels by 2030. Auction prices for a tonne of CO_2 were around 16 USD (United States Dollar) in November 2019, according the information of the California Air Resources Board.

The EU ETS, established in 2005, covers approximately close to half of the EU's emissions of CO_2 (see also Section 16.2.2). The EU's goal is to reduce greenhouse gas emissions to 40% below 1990 levels, with the EU ETS sectors contributing a reduction of 43% relative to 2005. Thus, the EU ETS is one of the corner stones to mitigate climate change. Auction prices for a tonne were around 25 Euro in January 2020.

The Korean Emissions Trading System (KETS) was launched in 2015 and covers about 68% of the national greenhouse gas emissions. The overall reduction target for greenhouse gas emissions in South Korea for 2030 refers to 37% below "business as usual". As for the period 2018–2020 participants will receive almost all allowances (97%) for free, there is not yet a reliable price to indicate scarcity.

China and the EU recognise their crucial roles in combating climate change and to collaborate in this context for a sustainable economic and social development. In the EU-China Joint Statement on Climate Change of 2015 they agree to, among other things, of course, to "work together in the years ahead on the issues related to carbon emissions trading". The project "Platform for Policy Dialogue and Cooperation between EU and China on Emissions Trading" aims to provide capacity building and training to support Chinese authorities in their efforts to implement and further develop the Chinese national ETS.

Finally, there is Germany, which is integrated into the EU ETS, but there are still additional efforts to extend carbon pricing: the goal is to include the transport and building heating sectors, which accounted for about 32% of Germany's greenhouse gas emissions in 2018. There will be fixed, but yearly increasing prices per tonne of CO_2 equivalents, starting form 25 Euro in 2021 and rising to 55 Euro in 2025. In 2026 there will be auctions with a price corridor of 55 to 65 Euro; from 2027 the system will be gradually included in the EU ETS, with an option for price corridors, however. According to the German Climate Action Plan 2050, the medium-term target is to cut greenhouse gas emissions by at least 55% by 2030 compared to 1990 levels. The "long-term goal is to become extensively greenhouse gas-neutral by 2050", also in view of "the country's particular responsibility as a leading industrialised nation and the EU's strongest economy".

The implementation of the EPR principle to outline the characteristics of an IEP to mitigate climate change remains to be specified.

22.3 Implementing the EPR principle

Regarding climate change, the EPR principle has to be considered in a broader context: all people, consumers and producers, contribute to global warming – with their production and consumption activities, with their transport activities, with their buildings, etc. Therefore, all people must be integrated into an IEP in order to combat climate change. The first comments refer to a local, national approach.

22.3.1 A national ETS for implementing the EPR principle

On a national level, implementing the EPR principle can be achieved in a more or less straightforward way. An ETS which includes all relevant sectors can effectively reduce greenhouse gas emissions, in particular, if the cap is adjusted regularly (see Section 15.3), taking into account Constitutive Element I, the locality principle of an IEP. From an economic point of view, it would for a first step suffice to include the consumption of fossil fuels into the ETS, as this is one of the main sources of anthropogenic greenhouse gas emissions. Such an ETS affects all kinds of transport and construction activities, it influences generation of electricity and heat from fossil fuels and also production activities.

Interestingly, it is possible to "recycle" greenhouse gases by means of carbon sinks such as reforestation. After all, greenhouse gases are waste materials that are thus "reprocessed into products, materials or substances whether for the original or other purposes", referring to the term "recycling" in the EU Waste Directive. They can be used to increase the cap of the ETS and thus reduce the price of the allowances. However, Kallio, Solberg, Käär, and Päivinen (2018) investigate the economic impact of EU plans regarding new goals for forest carbon sinks and point to the possibility of carbon leakage, which would reduce overall benefits of the forest carbon sinks from the EU policy.

Carbon capture and storage (CCS) as a technology for "depositing" greenhouse gases from the use of fossil fuels has attracted much attention. However, as Tcvetkov, Cherepovitsyn, and Fedoseev (2019) point out in their review, "public awareness of the technology is extremely low". Therefore, a broad application of this technology seems unlikely in the near future.

The integration of the agriculture remains a challenge, at least beyond activities such as consumption of fossil fuels for agricultural equipment and production of chemical fertilisers. Rice farming and cattle raising are among the agricultural processes, which produce substantial amounts of greenhouse gases. Needless to say, this field is full of cultural traditions and historical associations, societal path dependencies, so to say, and it will be challenging to reduce the emission of greenhouse gases from livestock, for example. However, there are mitigation potentials, as pointed out by Kipling, Taft, Chadwick, Styles, and Moorby (2019). An IEP for reducing food waste should therefore also address emissions of greenhouse gases.

There are some additional issues regarding a national ETS. A significantly higher price of fossil fuels will for sure spark debate about social fairness: will there be a compensation for households living in the colder parts of the country? Will there be compensations for rural residents for commuting to the centre of the city? Economically, these compensations are possible. They result in a difference between a substitution and an income effect, which will point in the right direction, if the compensation is not exorbitant. However, to find the politically correct level of compensation could already be a challenge for policy-makers.

Another issue refers to the "readiness" of a country for the economic pressures associated with an ETS. The example is the EU, and in particular Germany: the country has established renewable energies for generating electricity, the country is phasing in transport and buildings into the EU ETS, and the car manufacturers are obliged to reduce fleet emissions of greenhouse gases, which probably can only be achieved through an increasing share of e-mobility. That is how the country's decades-long efforts could pay off. The question remains whether this "benefit" is accepted or tolerated by other countries in the international context.

The reason for this is that what is still manageable at national level leads to additional difficulties at global level. And, because of the public good characteristics of mitigating climate change, this international level is necessary – most countries are simply too small to have a tangible impact on global warming with their own actions.

22.3.2 An international ETS for implementing the EPR principle

Moving to an ETS with international participants leads immediately to the question of the appropriate consideration of the local conditions in the various countries. Thus, there could be national targets or caps for greenhouse gas emissions, which lead to different prices for the allowances, or there could be one global cap. But then it is necessary to somehow adjust prices for allowances: 50 Euro for one tonne of CO_2 in the EU is different from 50 Euro for one tonne of CO_2 in India, for example. The political debate on this issue will be a challenge.

All these modifications are possible in principle, but of course they lead to further complications in international trade: trade activities or decisions on foreign direct investments are dependent on these national differences. Therefore, adequately linking the existing cap-and-trade systems is necessary for establishing a functioning global ETS system. The EU, referring to the International Carbon Market, is working on this issue and considers "levelling the international playing field by harmonising carbon prices across jurisdictions" as one of the potential benefits of linking ETS. So far, there are only few agreements linking ETS: for example, the California ETS is linked to ETS in some Canadian provinces, and the EU ETS is linked to the ETS in Switzerland. Mutual recognition of

emission allowances from the EU and Switzerland is probably an easy task compared to establishing such links between ETS in developed and developing countries.

Moreover, the debate on forests and other carbon sinks as a means of "recycling" greenhouse gases will continue: to what extent and how should the conservation of such carbon sinks in a country be taken into account in the cap of the national or international ETS? Is it economically and environmentally reasonable to support reforestation programmes? How to deal with countries, which burn their forests and then start reforestation? These are just a few questions, which need to be answered.

In addition to the different local circumstances, it should not be forgotten that mechanisms such as the Prisoner's Dilemma could play a role in this context: delaying one's own actions, letting others go ahead, and perhaps benefiting from their activities. Of course, strategic behaviours such as taking first-mover or late-mover advantages, could also have an impact on these decisions.

22.4 An integrated environmental policy for combatting climate change?

Due to the public good characteristics of mitigating climate change, each IEP in this context must achieve some cooperation between sufficiently many countries, which jointly emit a significant share of the global greenhouse gases. An international ETS certainly has the advantage that it helps to adequately address all stakeholders – in accordance with Constitutive Element II of an IEP.

This is not too surprising, because an ETS is structurally very close to the market mechanism, only the target or the cap is or has to be determined exogenously. Then, of course, also the links between the stakeholders and with the goals of the policy are provided by the ETS, again similar to the market system – confirming Constitutive Element III of an IEP.

However, Constitutive Element I, the consideration of the local situations in the participating countries, requires more attention and seems to be the trouble maker. As indicated above, for a variety of reasons, it will remain difficult to take adequately care of this issue. Section 6.1.1 briefly described the environmental situation in Saxony some 300 years ago. There was a substantial reaction on deforestation only, when relevant stakeholders felt the consequences personally. And it has taken a long time for the environment to recover, for the forests to regrow. Interestingly, regarding forests we are in a similar situation today, but for different reason.

Nordhaus (2015) also points to the difficulties to forge international agreements because of free-riding, as indicated above. He then examines the club as a model for international climate policy, and he finds that "a regime with small trade penalties on non-participants, a Climate Club, can induce a large stable coalition with high levels of abatement".

Perhaps, approaches from behavioural environmental economics can help to create adequate social norms. Diekert, Eymess, Luomba, & Waichman (2020) want to examine the "creation of social norms under weak institutions" in order to overcome the Tragedy of the Commons as "a key challenge for the sustainable management of natural resources". Their focus is on settings where voluntary cooperation is required because of "weak state capacities". Barrett and Dannenberg (2012) are interested in the question, how "uncertainty about 'dangerous' climate change affects the prospects for international cooperation", thereby referring to the Prisoners' Dilemma. They can show "that the fear of crossing a dangerous threshold can turn climate negotiations into a coordination game, making collective action to avoid a dangerous threshold

virtually assured". This corresponds to the experience in Saxony some 300 years ago: an immediate threat motivates action.

Of interest are E. Ostrom's "8 Principles for Managing a Commons". Ostrom offers these principles "for how commons can be governed sustainably and equitably in a community" (Walljasper, 2011):

1. Define clear group boundaries.
2. Match rules governing use of common goods to local needs and conditions.
3. Ensure that those affected by the rules can participate in modifying the rules.
4. Make sure the rule-making rights of community members are respected by outside authorities.
5. Develop a system, carried out by community members, for monitoring members' behaviour.
6. Use graduated sanctions for rule violators.
7. Provide accessible, low-cost means for dispute resolution.
8. Build responsibility for governing the common resource in nested tiers from the lowest level up to the entire interconnected system.

Applying these rules to efforts mitigating climate change remains challenging, however. Especially, but not only, the issues of sanctions and means for dispute resolution are difficult to handle in this global context with completely different and independent countries.

Therefore, an IEP to combat climate change seems to be dependent on making an appropriate proposal on how to deal with Constitutive Element I. Without a "world government" that overcomes the mechanisms of the Tragedy of the Commons and the Prisoners' Dilemma, appropriate social norms must be created, which may take some time. Or "the fear of crossing a dangerous threshold" (Diekert et al., 2020) turns climate negotiations into cooperation.

References

Ballew, M. T., Pearson, A. R., Goldberg, M. H., Rosenthal, S. A., & Leiserowitz, A. (2020). Does socioeconomic status moderate the political divide on climate change? The roles of education, income, and individualism. *Global Environmental Change, 60,* 102024. https://doi.org/10.1016/j.gloenvcha.2019.102024.

Barrett, S., & Dannenberg, A. (2012). Climate negotiations under scientific uncertainty. *Proceedings of the National Academy of Sciences of the United States of America, 109,* 17372–17376. https://doi.org/10.1073/pnas.1208417109.

Diekert, F., Eymess, T., Luomba, J., & Waichman, I. (2020). *The creation of social norms under weak institutions.* University of Heidelberg, Department of Economics, Discussion Paper Series No. 684. Retrieved from(2020). https://www.uni-heidelberg.de/md/awi/forschung/dp684.pdf.

EEA (2019). *The European environment—State and outlook 2020: Knowledge for transition to a sustainable Europe.* Luxembourg: Publications Office of the European Union https://doi.org/10.2800/96749.

IEA (2019). *Global energy & CO$_2$ status report 2019.* Paris: IEA. Retrieved from https://www.iea.org/reports/global-energy-and-co2-status-report-2019.

Kallio, A. M. I., Solberg, B., Käär, L., & Päivinen, R. (2018). Economic impacts of setting reference levels for the forest carbon sinks in the EU on the European forest sector. *Forest Policy and Economics, 92,* 193–201. https://doi.org/10.1016/j.forpol.2018.04.010.

Kipling, R. P., Taft, H. E., Chadwick, D. R., Styles, D., & Moorby, J. (2019). Implementation solutions for greenhouse gas mitigation measures in livestock agriculture: A framework for coherent strategy. *Environmental Science & Policy, 101,* 232–244. https://doi.org/10.1016/j.envsci.2019.08.015.

Miller, G. (2020). Beyond 100% renewable: Policy and practical pathways to 24/7 renewable energy procurement. *The Electricity Journal, 33*(2), 106695. https://doi.org/10.1016/j.tej.2019.106695.

Nordhaus, W. (2015). Climate clubs: Overcoming free-riding in international climate policy. *American Economic Review, 105*(4), 1339–1370. https://doi.org/10.1257/aer.15000001.

Przychodzen, W., & Przychodzen, J. (2020). Determinants of renewable energy production in transition economies: A panel data approach. *Energy, 191,* 116583. https://doi.org/10.1016/j.energy.2019.116583.

Shahnazi, R., & Dehghan Shabani, Z. (2020). Do renewable energy production spillovers matter in the EU? *Renewable Energy, 150,* 786–796. https://doi.org/10.1016/j.renene.2019.12.123.

Tcvetkov, P., Cherepovitsyn, A., & Fedoseev, S. (2019). Public perception of carbon capture and storage: A state-of-the-art overview. *Heliyon, 5*(12), e02845. https://doi.org/10.1016/j.heliyon.2019.e02845.

UBA (2019). *How can existing national climate policy instruments contribute to ETS development?* German Environment Agency. Retrieved from https://www.umweltbundesamt.de/sites/default/files/medien/1410/publikationen/2019-05-15_cc-11-2019_development-ets_v2.pdf.

Walljasper, J. (2011). *Elinor Ostrom's 8 principles for managing A commmons.* Retrieved April 19, 2020, (Original work published 2011)(2011). https://www.onthecommons.org/magazine/elinor-ostroms-8-principles-managing-commmons.

Wiesmeth, H. (2011). *Environmental economics: Theory and policy in equilibrium. Springer texts in business and economics.* Springer Nature https://doi.org/10.1007/978-3-642-24514-5.

23

Plastics in a circular economy

Under the title "A European Strategy for Plastics in a Circular Economy" the European Union (EU) lists key issues, which need to be addressed in the EU, in the industrialised countries, but also with a view to emerging economies and developing countries.

Above all, there is the important reference to the variety of plastic products that are ubiquitous in our daily lives and help to save resources as insulation materials, as light and innovative materials in cars and airplanes. In packaging, plastics helps to ensure food safety and reduce food waste, and it is of relevance in all kinds of innovations. It is therefore not surprising that global demand for plastics continues to rise: it has increased twentyfold since the 1960s, reaching 322 million tonnes in 2015, and is, according to the EU, expected to double again in the next 20 years. The plastics sector currently employs 1.5 million people in the EU and generated sales of 340 billion Euro in 2015, but growth rates in other parts of the world are already higher.

The EU report points yet to another aspect: the reuse and recycling of end-of-life plastics remains very low. Of the around 30 million tonnes of plastic waste generated in the EU each year, 17% is collected for reuse or recycling, and a significant share leaves the EU to be treated in third countries with potentially lower environmental standards. 57% of plastic waste is lost through landfilling and energy recovery, implying in particular that 83% lost to the economy (see Fig. 23.1). Due to down-cycling, demand for recycled plastics accounts for only around 6% of plastics demand in the EU.

These are the facts from the EU strategy for plastics. As far as the circular economy is concerned, plastics is on both sides: it is of great importance for our daily lives, but it is also responsible for serious waste problems. Moreover, due to exports and imports of plastic products and plastic waste, due to plastic waste ending up in rivers and seas, and due to micro-plastics showing up in the food chain, plastics and plastic waste have acquired an international dimension, which is of relevance because of the economic and social differences between the countries and the differences in environmental awareness.

Of course, it should not be said that "plastics are responsible" for the environmental degradations associated with its use. The problems arise from handling plastics in an irresponsible way, from largely lacking coherent environmental policies, especially in the international context. Societal path dependencies play a role, as do

Implementing the Circular Economy for Sustainable Development
https://doi.org/10.1016/B978-0-12-821798-6.00023-5

277

FIG. 23.1   Resource efficiency and plastic waste.In the EU, only a small amount of plastic waste generated is recycled. ©
From EEA. Resource Efficiency and Waste. Retrieved April 21, 2020 from https://www.eea.europa.eu/media/infographics/plastic-waste
(Original work published 2019).

the well-known mechanisms of the Tragedy of the Commons and the Prisoners' Dilemma.

The aim of this chapter is to provide an insight into the cornerstones of an "Integrated Environmental Policy" (IEP) for plastics and plastic waste, thereby supporting the implementation of a circular economy. It is necessary to deal with plastics and plastic waste at national level, but not to forget the international dimension.

Compared to local measures to combat climate change, there is a higher level of perceived feedback from efforts to reduce plastic waste, as micro-plastics is already appearing in the food chain, threatening human health

(see Section 6.1.3). This supports the implementation of local policies and thus also contributes to reducing the pollution of global commons.

The following section briefly explains some prominent types of plastics, including their environmental properties.

23.1 One type of plastic is not like another

Talking about environmental risks associated with plastics and plastic waste, one has to take into account that there is a vast variety of different types of plastics for different uses, characterised by their special properties – and the environmental hazards, associated with some of them.

23.1.1 Common types of plastics

In the "Plastics Atlas 2019" of the German Heinrich Böll Stiftung, "The Green Political Foundation", the authors refer to the differences regarding the lifespan of certain plastic products (BUND, 2019). In particular, plastic products for the construction sector account for about 16% of the more than 400 million tonnes of plastics produced worldwide each year (p. 15). These products such as tubes and pipes, but also doors and windows can last as long as 35 years and more. Polyvinyl chloride (PVC), a plastic polymer, is typically used for these products. Various additives, among them phthalate plasticisers, ultraviolet (UV) and heat stabilisers, and flame retardants, are required to achieve certain properties of the end products.

With more than 35%, however, plastic packaging represents the largest share of plastic products (p. 15) – with a generally very short period of usage, often labelled as single-use products. Polyethylene (PE) and related polymers such as polypropylene (PP), polystyrene (PS), polyethylene terephthalate (PET), and polyurethane (PU), are the most common types of plastic. They are primarily used for plastic bags, plastic films, textiles (PP), and bottles (PET), but also kitchenware, beverage crates (PE), elastomeric wheels, high-performance adhesives, insulation material (PU) as well as protective packaging and isolation gaskets (PS).

Another issue of relevance for recycling activities is the fact that plastic packaging often relies on different types of plastic: PET bottles can have PE caps, and labels of yet another type of plastic. Despite their chemical proximity, these materials have their specific properties, which need different recycling methods – an issue of relevance for dealing with and recycling plastic waste. In its report on "Plastic Waste in the Environment" the EU gives a survey on the polymers used in a variety of applications (p. 42), which have to be respected when recycling plastic waste. Ragossnig and Schneider (2017) provide additional insights into plastics recycling when they ask the question about the "right level of recycling of plastic waste?".

23.1.2 Some critical issues regarding the handling and recycling of plastics

Environmentally relevant issues include, for example, the phthalate plasticisers added to many plastic products (Pivnenko, Eriksen, Martín-Fernández, Eriksson, & Astrup, 2016). They are criticised mainly for their alleged interaction with the human hormonal system, potentially impairing functions of the body. As they are often found in PVC products, it is recommended to avoid these products, which are also characterised through a variety of other additives. Environmental organisations claim that PVC causes environmental problems at all stages: from production to recycling to landfilling (see BUND, 2019, p. 16f, or the EU Report on Plastic Waste: Ecological and Human Health Impacts, p. 18f).

PET, which is most commonly used for plastic bottles, is suspected of containing this kind of substances, which pollute water and harm people. Nevertheless, PET bottles dominate drinks packaging, also in transition and emerging economies, and in developing countries (Wiesmeth, Shavgulidze, & Tevzadze, 2018). Economic growth and a changing lifestyle can further contribute to the apparent preference of drinks in one-way plastic containers.

The next larger group of polymers are the polycarbonates (PC), which find many applications for strong, sometimes optically transparent plastic products such as drink bottles, microwave dishes and others. The issue is that PC releases Bisphenol A, which again is hormonally active and could therefore be problematic in connection with food contact (Lara-Lledó,

Yanini, Araque-Ferrer, Monedero, & Vidal, 2018).

These environmental issues refer mainly to the use of plastic products. But they are of course also of relevance for landfilling, incinerating or mechanical recycling (washing, grounding, melting). PS and PU seem to cause more problems in terms of recycling, and the combustion of these polymers can lead to the emission of dioxins if adequate precautions regarding emission control are not taken. There is only limited experience for each of these possible "end-of-life" activities. For this reason, it is advisable to handle plastics and especially plastic waste with all required thoughtfulness. Unfortunately, chemical recycling (breaking down the polymers into their chemical constituents) is not yet sufficiently developed to provide a large-scale alternative for mechanical recycling (Ragaert, Delva, & Van Geem, 2017; Tullo, 2019).

The fact that plastic waste pollutes the oceans, that plastic particles and micro-plastics appear in water, food, animals and human beings should sound the alarm. Analysis of a typical piece of agricultural area in Germany revealed that the soil pollution with micro-plastics is between four and 23 times as high as the corresponding water pollution (see BUND, 2019, p. 20).

The recycling of plastics is not without problems, both for the environment and for human health. This refers, in particular, to countries with less stringent regulations. As is well known from primitive recycling activities in some developing countries, the lack of suitable facilities can have serious negative consequences. This refers as well to waste incineration plants, also in industrialised countries: according to García-Pérez et al. (2013), cancer mortality rises in Spanish towns in the vicinity of incinerators, handling hazardous waste, among which they count the recycling of plastic industrial packaging (p. 32).

Tang et al. (2015) investigate the soils and sediments in a typical plastic recycling area in northern China. Their findings show that in more than half of the soils there is considerable, even high potential ecological risk and in almost all sediments there is a high potential risk through heavy metals from poorly controlled recycling of highly contaminated plastics.

As far as the recycling of plastics is concerned, it is important to separate and sort the different types of polymers, otherwise the residues left for landfilling can have a significant environmental impact. But in view of the experience of emerging economies (including China), and industrialised countries (including Spain), the recycling of plastics, in particular plastic waste, is not comparable to the recycling of paper or aluminium in terms of health and environmental risks.

23.2 A look at the global situation

According to BUND (2019), p. 36, around 9.2 billion tonnes of plastic goods have been produced worldwide since the 1950s, of which probably only 0.6 billion tonnes have been or will be recycled (see Fig. 23.2). 14% of plastic packaging is currently recycled (mostly a down-cycling), about 40% is landfilled and 14% is incinerated, and the remaining 32% end up in the environment, which may lead to serious environmental problems.

In addition to the emergence of increasing quantities of plastic waste, there are also imports and exports. In various industrialised countries, exports are divided into regular, domestically-produced exports and re-exports. The following subsection examines international trade in plastic waste.

23.2.1 Import and export of plastic waste

There are certainly many and quite different reasons for exporting or importing plastic waste. These trade activities are legal as long as they are in line with the Basel Convention on the Control

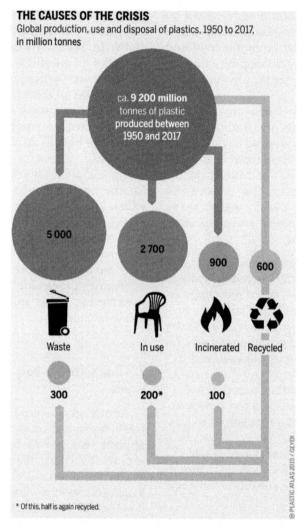

THE CAUSES OF THE CRISIS
Global production, use and disposal of plastics, 1950 to 2017, in million tonnes

ca. **9 200 million** tonnes of plastic produced between 1950 and 2017

5 000

2 700

900

600

Waste   In use   Incinerated   Recycled

300   200*   100

* Of this, half is again recycled.

© PLASTIC ATLAS 2019 / GEYER

FIG. 23.2 Global production, use and disposal of plastics. The development from 1950 to 2017, in million tonnes. © *From PLASTIC ATLAS | Appenzeller/Hecher/Sack CC-BY-4.0. (2019). Plastic Atlas 2019(2 nd). Retrieved from https://www.boell.de/sites/default/files/2020-01/Plastic%20Atlas%202019%202nd%20Edition.pdf?dimension1=ds_plastic_atlas, p. 37.*

of Transboundary Movements of Hazardous Wastes and their Disposal and related regulations of the OECD or the EU, for example.

It must therefore be assumed that the export and import of plastic waste is beneficial for the participating countries or rather the companies involved in these activities. However, the concept of "benefit" needs to be seen in the context of the different countries, and since environmental issues are important, perceived scarcity regarding environmental commodities and the level of environmental awareness can vary considerably. As a result, trade in waste in general and plastic waste in particular can benefit businesses but harm the environment of the countries.

Such a development should be ruled out by the regulations on trade in plastic waste, but it seems quite common that environmental and social standards in rapidly developing countries and emerging economies are not always easy to verify.

As already indicated, there are probably many operational reasons for trading plastic waste. In addition, they depend on changing technological, technical, economic and legal framework conditions and, of course, on the "quality" of plastic waste. The reasons for exporting or importing plastic waste, many countries are both exporters and importers, may include the following:

- Due to increasing collection and recycling rates, there are not enough recycling facilities.
- The export of plastic waste for recycling can help to achieve high recycling targets.
- The incineration of plastic waste in a country leads to the emission of greenhouse gases, for which additional costly certificates need to bought.
- The treatment (incineration, recycling) of contaminated plastic waste is too expensive under the environmental regulations in a country.
- A country wants to import plastic waste because it wants to use the recycled materials for its own industry as a substitute for virgin materials.

There may be additional reasons for the export or import of plastic waste. Some of them, such as the export of poorly sorted plastic waste, appear to be close to the legal limits of environmental legislation. Velis (2015) refers to the "least environmental standards and lowest wages global pathway" that can arise in these global recycling networks, which according to Crang, Hughes, Gregson, Norris, and Ahamed (2013) "operate not through adding value, but by connecting different regimes of value".

According to Brooks, Wang, and Jambeck (2018), the United States (U.S.) is the leading

exporter of PVC, Germany is the leading exporter of PE, and Japan is the leading exporter of PS. In 2018, Germany exported more than 740,000 t of plastic waste. Malaysia received 132,000 t, the Netherlands 109,000 t and Hong Kong 73,000 t. Of course, both the Netherlands and Hong Kong serve mainly as transshipment points, even as entrepôts for re-exporting plastic waste (see BUND, 2019, p. 39) to various Asian countries.

Fig. 23.3 shows the importance of Hong Kong for the import of plastic waste from the EU, which was then probably re-exported to China. Logistics is not the only reason for this observation. As Feenstra and Hanson (2004) point out, information costs can be lower, when intermediaries are present. Moreover, there may be some "outward processing" in Hong Kong for re-exporting these still incompletely recycled waste to China.

23.2.2 China's ban on imports of plastic waste

Annual global imports and exports of plastic waste began to increase rapidly in 1993. By 2016, imports had grown by more than 700% and exports by more than 800% (see Fig. 23.4). According to Brooks et al. (2018), about half of the plastic waste destined for recycling (14.1 million tonnes) was exported from 123 countries in 2016 alone, with China taking most of it (7.35 million tonnes). Since 1992, China has imported 106 million tonnes of plastic waste, making up 45.1% of all cumulative imports. The main exporters were the high-income countries, including the G7 (p. 2).

Velis (2014) provides additional insight into the shifts of plastics production to Asian countries, in particular China. The drivers were, according to his view, "increasing local demand and lower costs – mainly labour, but also lower environmental and health and safety costs, due to the initial absence of regulations and/or their

Export of plastic waste for recycling form the EU to receiving countries, 2015 to March 2018

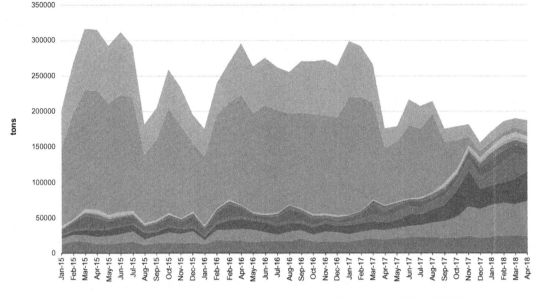

Source: Eurostat COMEXT

Extraction from the Foreign Trade Statistics:
10 July 2018, General Disclaimer of the EC

eurostat

FIG. 23.3 Export of plastic waste from the EU. Export for recycling from the EU to receiving countries, 2015 to 2013. © *From Eurostat. Export of Plastic Waste from the EU. Retrieved April 22, 2020 from https://ec.europa.eu/eurostat/statistics-explained/images/ 8/8d/Figure_10_Export_of_plastic_waste_for_recycling_from_the_EU_to_receiving_countries%2C_2015_to_March_2018.png (Original work published 2019).*

implementation in both manufacturing and reprocessing", and this shift in production is also responsible for all aspects of plastic waste.

After being the main importer of plastic waste from countries around the world, China banned the imports of various categories of solid waste from 2018, including highly contaminated plastic waste, unsorted scrap paper and waste textiles.

What were the reasons for this development in China? Until, 2008, China needed all kinds of raw materials, also secondary plastics to fuel

the economic growth (Velis, 2014), and resources obtainable from recycling waste products was certainly a cheap alternative to using natural oil. The enormous and cheap labor force in China made this possible, but created also an enormous environmental pollution, due to the inexperience of the workers and the non-availability of suitable technologies. In 2013, for example, the total consumption of plastics in China was up to 70 million tonnes, and the total amount of plastic waste was more than 35 million tonnes (Li, Jiang, & Zhang, 2015).

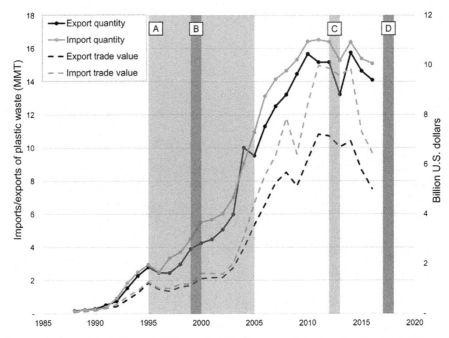

FIG. 23.4 Export and import of plastic waste. Volume of plastics in mass and trade value. *Reprinted with permission of AAAS from Brooks, A. L., Wang, S., & Jambeck, J. R. (2018). The Chinese import ban and its impact on global plastic waste trade. Science Advances, https://doi.org/10.1126/sciadv.aat0131. © Brooks et al., some rights reserved; exclusive licensee American Association for the Advancement of Science. Distributed under a Creative Commons Attribution NonCommercial License 4.0 (CC BY-NC) https:// creativecommons.org/licenses/by-nc/4.0/.*

Entire towns and some 10,000 companies in China were engaged in recycling, helping China to become "one of the most effective countries for plastic recycling in the world", according to the systematic assessment of the environmental impact of recycling plastic waste in China by Li et al. (2015). All residents were more or less exposed to dangerous working conditions and to toxic chemicals. Dirty wastes and even hazardous wastes were mixed with the solid waste consigned to recycling. This contributed further to the enormous pollution of the environment, because heavily contaminated waste could not be recycled and ended up in landfills.

But, in general, it is difficult to find details of this pollution and the associated health impacts. Scientific articles, when they address the issue, soon turn to "optimisations" to reduce the environmental pollution (Li et al., 2015), or they consider "only" the greenhouse gas emissions

as part of a lifecycle assessment (Cong et al., 2017).

There seem to be various reasons for China's ban. The documentary "Plastic China" is informative in this regard. The ban is, however, not a complete ban. It refers to low-quality plastic waste with a contamination degree above 0.5%, as opposed to the previous level of 1.5%. It seems difficult to achieve such a high quality of plastic waste, since the contamination levels are often above 15% according to (BUND, 2019) (p. 38).

The ban is also likely to be a consequence of China's enormous economic growth in recent decades, which has led to increased environmental awareness among the population and is forcing the government to take better care of environmental pollution. China's circular economy strategy points exactly in this direction: the Circular Economy Promotion Law dates from 2008 (China, 2008). Yuan, Bi, and

Moriguichi (2006), Yong (2007) and Su, Heshmati, Geng, and Yu (2013), among others, characterise this increasing emphasis on environmental issues, which corresponds to the gradual ban on imports of contaminated (plastic) waste (see again Brooks et al., 2018 for more details and additional background information).

Behind this decision is an additional economic argument: wages in China are rising, making labour-intensive recycling activities increasingly expensive. For this reason, China is currently investing in appropriate technology, relying less on labour, and focusing on recycling its own plastic waste. It's now about time for many workers in the recycling business to either get additional qualifications or to look for alternative jobs. Rising wages seem to signal the availability of adequate jobs, so the ban should not lead to sustained high unemployment in the recycling areas of China.

23.2.3 Global streams of plastic waste after China's ban

To sum up, the high growth rates in China in recent decades, which have also been fuelled by the imports of waste products for recycling, have increased wages, but likely also environmental awareness. These developments make the recycling of contaminated (plastic) waste now less attractive both economically and environmentally. Nevertheless, China still imports recycled plastics from neighbouring Southeast Asian countries such as Malaysia – from Chinese companies in those countries, which apparently hope to benefit from China's import ban. However, Malaysia, together with several other countries, also plans to restrict imports of contaminated plastic waste. So the question remains what will happen to the streams of plastic waste from industrialised countries following China's ban?

Of course, there are and will be diversions for plastic waste, mainly due to short-term unavailable recycling and incineration facilities. And the prices for contaminated plastic waste tend

to be negative (Stephenson, 2018). Exporting countries will have to look for adequate countries to accept their plastic waste (see also Fig. 23.3 for recent plastic waste diversions).

As already indicated, the final "equilibrium" regarding trade in plastic waste has not yet been reached. It will take some time, for example, for EU member states to adopt a stricter approach regarding plastic waste. Until then, new havens for plastic waste could be needed, threatening the environment in other developing countries and emerging economies. An "Integrated Environmental Policy" (IEP) for plastics and plastic waste is therefore urgently needed.

23.3 Cornerstones of an IEP for plastics and plastic waste

This section covers and examines some aspects of an IEP for plastics and plastic waste. Also because of Constitutive Element I, the locality principle, the elaboration of an IEP will be a difficult task.

23.3.1 Fundamental aspects of an IEP

The forgoing considerations have shown that plastic waste has become an international environmental issue. Not just because of its ubiquity, rather because of the pollution of rivers and seas with plastic waste, and because of micro-plastics showing up in the food chain. An IEP for plastics and plastic waste has to take this fact into account.

Another aspect refers to the income effect: due to economic growth a rising demand for most commodities will likely include plastic products and, in particular, an increasing volume of packaging material with plastic packaging leading the list. This is part of the rebound effect, which can get even more strength, if packaging gets cheaper through new technologies, thinner wrappings or other developments. As already indicated, this rising demand is being exacerbated by lifestyle changes, with more attention being paid, for example, to online shopping and food deliveries. Undoubtedly, lifestyle changes can also mean more protection of the environment, but experience teaches us that these positive changes may outweigh others which do more harm to the environment.

As far as the recycling of plastic waste is concerned, there is again a kind of societal path dependency: the fact that a country has a seemingly functioning recycling system for plastics seems to attract companies: some drinks producers tend to increase their volume of drinks in one-way packaging, arguing that there is such a good recycling system. This happens in Germany, where, for example, Coca Cola and various discounters want to increase the proportion of their drinks in non-refillable containers. Some of these companies have their own PRO system and, taking into account the demand for drinks in one-way plastic bottles, increase their share of drinks in one-way plastic packaging, as long as their own recycling of these PET containers is the preferred business option (see also Wilts, 2012).

In addition, the technologies are continuously being developed. There are already technologies that seem to make it possible to convert non-recyclable mixed plastic waste streams into environmentally friendly composite products. Whether these technologies can deliver on what they promise remains to be seen.

As already mentioned, the global recycling of plastics is low, probably only 15% in plastic packaging (BUND, 2019, p. 36), although recycling seems economically sensible at first glance: it helps to save the consumption of fossil fuels when recycled plastics can be used as a substitute for new plastics.

However, this is only partially possible: in order to be competitive with new plastics, used plastics must be sorted and carefully prepared for recycling (see Ragossnig & Schneider, 2017 for more details on the economics of recycling). If the price of natural oil is low, then a qualitative

and high-level recycling of plastics is expensive, more expensive than producing new plastic from natural oil, at least in industrialised countries with their high wages and costly technologies. As a result, the operation of recycling plants should not necessarily be regarded as a profitable business, at least not in high-wage countries and for high-quality "food-grade plastics". Currently only a very small share of recycled PET can be used in new PET bottles. As a rule, recycled plastics is usually not usable in contact with food (Lara-Lledó et al., 2018). Recycled plastics cannot therefore be seen as a substitute for new plastics, down-cycling seems to remain the rule rather the exception. As already mentioned, chemically recycling plastics is not yet available on a large scale (see Ragaert et al., 2017; Tullo, 2019).

Many countries understandably try to operate recycling plants with the aim of establishing a profitable business. However, this depends, among other things, on the very volatile world market price for natural oil. When expected profits suddenly turn into losses, then environmental concerns are usually ignored, just to protect business and jobs depending on it. Therefore, the operation of recycling facilities for plastic waste should be based on cooperation between public institutions and private companies with a carefully balanced distribution of obligations.

23.3.2 The constitutive elements of an IEP

23.3.2.1 Constitutive element I

Due to the global dimension of plastic waste, international cooperation is needed to address this issue. The considerations in Chapter 22 have, however, shown that the requirements of Constitutive Element I, namely adequate compliance with local conditions, may cause some difficulties, for example by free-riding or cheap-riding linked to the Prisoners' Dilemma.

Due to "the least environmental pathway", characterising parts of the global recycling network (Velis, 2015), the industrialised countries at the top of this pathway are particularly in charge of modifying this situation. The industrialised countries are, of course, also in charge of increasing efforts to combat climate change. In this case, however, the industrialised countries have fewer possibilities to influence the global situation, as various emerging economies are claiming their rights for further economic development.

Therefore, in order to comply with the first constitutive element, industrialised countries, but also more and more emerging economies, have to characterise their local situation in order to find the right approach to implement the waste hierarchy regarding plastic waste.

As usual, there has to be a focus on prevention of plastic waste, which has only recently gained more prominence, with Design for Recycling (DfR) still dominating the picture. As the discussion of previous IEPs has shown, local targets for the separate collection and recycling of plastic waste are of relevance in this regard. This implies that export of plastic waste to third countries, in particular emerging economies and developing countries, has to be reduced – a goal of the EU Strategy for Plastics and other political initiatives. More plastics will then have to be recycled locally, with financial consequences for producers. The costs of collection and recycling, including a potential ecological surcharge, have to be passed on to the producer, an issue of Constitutive Element III.

Given an appropriate infrastructure, collection activities of plastic waste can be intensified by a deposit system – on one-way drinks packaging, for example. Also, a fee for returning larger plastic products, tubes and pipes and others, seems to be possible under certain circumstances. In addition, banning certain plastic products, in particular single-use plastics such as plastic bags or disposable dishes, could help to prevent plastic waste. In this case, avoidance possibilities have to be investigated (see Section 16.1.2). What kind of shopping bags, for example, could replace plastic bags? Paper or jute bags, or rather bags made from bioplastics? According to a study by Wagner (2017), a ban on plastic bags without a fee on paper bags is likely to increase the consumption of single-use paper bags, which, as a single-use product, need not be much better from an environmental point of view. Although these bans typically apply to small quantities of plastics, they could still play a role in changing household environmental behaviour, similar to the collection of comparatively low deposit fees (see also Chapter 8).

Consequently, each country or region must, in accordance with Constitutive Element I, find the appropriate framework conditions for the establishment of an IEP for plastics and plastic waste, depending on the economic, geographic, climatic, etc. situation.

However, it seems necessary to coordinate national policies in order to cope with the international dimension. Developed countries must support emerging, developing, and transition countries. Societal path dependencies that focus on international trade can lead to even more trade in plastic waste, with potentially further pollution. Given the experience gained, there should be restrictions – also to protect global commons, and there should be efforts to raise environmental awareness in relation to plastic waste.

23.3.2.2 Constitutive element II

The second constitutive element relates to integrating all relevant stakeholders into the policy and addressing them appropriately. Above all, it is obvious that all individuals have their share in "plastics". Nevertheless, manufacturers of plastic products, such as packaging material, must be given a particular emphasis.

Manufacturers of plastic products could be addressed through a command-and-control policy that determines their participation in a competitive licensing scheme, as set out in the policy proposals in Chapters 20 and 21. Households and other manufacturers are obliged to separate plastic waste and return it to collection points, perhaps additionally motivated by a take-back requirement or a refund (see Section 21.1). Of course, various technical aspects of the collection system could improve the integration of these individuals into the policy (Seyring, Dollhofer, Weißenbacher, Bakas, & McKinnon, 2016).

As mentioned, trade in plastic waste should be regulated in order to prevent further pollution and to reduce health hazards, especially in developing countries. This requires appropriate command-and-control policies, optimally at a global level.

23.3.2.3 Constitutive element III

The various parts of the policy have to be linked in order to achieve the goals of the policy, implementing the waste hierarchy for plastic waste. As explained in previous examples of IEPs, the licence fees for plastic products must be paid to a compliance scheme, depending on the quantity and various characteristics of the plastic material. These fees increase as collection and recycling rates increase. Therefore, as these fees have to be paid by the manufacturers of the plastic products, they will be passed on to consumers at least to some extent, and this could provide incentives for a design for environment (DfE), incentives to reduce some quantities of plastics in order to reduce costs and thus prevent plastic waste.

There remain the parts "reuse" of plastic products and "recycling" of plastic waste of the waste hierarchy. This is still changing as it depends on critical technological developments, especially in terms of material recovery and recycling.

Reuse in the true sense of the concept is carried out with the known case of refillable plastic bottles for all types of drinks or to some extent reusable plastic containers for food (Accorsi, Cascini, Cholette, Manzini, & Mora, 2014). This

certainly helps to prevent further plastic waste. Beyond that the literature mainly talks about "reusing waste plastics" as building material (see Mansour & Ali, 2015; Hita, Pérez-Gálvez, Morales-Conde, & Pedreño-Rojas, 2018, for example). These activities are more similar to the recovery of materials from used plastic products and do not lead to a prevention of plastic waste if these items do not not displace new plastics. Certainly, it delays environmental issues until these buildings, including plastic parts, need to be "recycled".

Reuse in the sense of refillable drinks containers, for example, can be stimulated through a deposit-system for one-way containers and the licensing of packaging with a compliance scheme, making one-way systems more expensive. The "reuse" of recovered materials from plastic waste is not "reuse" in the definition of the waste hierarchy. It is, as already indicated, more like recycling, perhaps also remanufacturing.

There are many technologies for recycling plastics and each day additional ones are introduced. Among the extensive literature, Thomas et al. (2020) presents a "study of the current pathways for recycling of bulk plastics through mechanical, biological, and chemical recycling". However, regarding the implementation of a circular economy, there is a critical issue, which needs some further thoughts.

Of course, plastic products with their many advantages will remain important for our daily lives. Consequently, beyond some reuse activities, plastic waste needs to be recycled, and new, more efficient and less polluting recycling technologies have to be developed, and additional possibilities for "reusing" recovered materials have to be explored. A market-oriented approach with private business companies to look for profitable business ideas is certainly of advantage in this context, as it integrates all potential companies and start-ups to develop innovative technologies.

That is one side of the story. On the other hand, once there are profitable recycling and recovery activities, there will also be incentives to extend these operations and to forget about the prevention of plastic waste as priority of the waste hierarchy.

This is in principle a consequence of societal path dependencies. The obvious advantage of a market system, namely to motivate suitable companies to develop innovative technologies and products, is based on the ability to make profits. However, it is precisely this motivation to earn profits that may conflict with the intentions and requirements of the waste hierarchy.

In fact, the above considerations regarding "reuse" of plastics as well as the continuous development of new recycling technologies point in this direction. Moreover, Coca-Cola with its vision, even for Africa, of a World without Waste (Africa), refers to 100% recyclable packaging. In Germany, a country with a long-time tradition to raise the share of refillable drinks packaging, the company's World without Waste (Germany) strategy relies "on a mixture of single use deposit bottles and reusable packaging in a variety of sizes".

Of course, Coca-Cola is not alone in this context with activities, which support this more or less global trend to putting recycling ahead of preventing waste. It will be difficult to gradually change this development, these societal path dependencies. The implementation of Constitutive Element III becomes a little more complicated.

23.4 Where are we with the IEP for plastics and plastic waste?

Various challenges affect the implementation of the EPR principle in the design of an IEP for plastics and plastic waste. There are, above all, these societal path dependencies, that seem to focus on profitable recycling technologies. On the one hand, this is a good thing, as it attracts business. On the other hand, however, this could contribute to an excessive expansion of recycling activities and to the neglect of prevention of plastic waste. Therefore these issues need to be taken into account in an IEP.

Transferring the tasks of handling plastic waste to independent compliance schemes is the recommended proposal. Of course, the compliance schemes should operate in a competitive environment, and their activities should be monitored in view of vested interests.

Another issue refers to the international dimension: if industrialised countries adopt this kind of policies, and are motivated or even forced to limit their exports of plastic waste to third countries, then part of the problem with the "least environmental pathway" regarding plastic waste gets under control. However, there is still a need to convince emerging economies and developing countries to follow suit. Some Asian countries show some promising developments – unfortunately due to poor experience. However, much remains to be done to reduce the pollution of rivers and seas with plastic waste, although, as indicated, the increased perceived feedback due to microplastics in the food chain, could be helpful in this regard.

References

Accorsi, R., Cascini, A., Cholette, S., Manzini, R., & Mora, C. (2014). Economic and environmental assessment of reusable plastic containers: A food catering supply chain case study. *Sustainable Food Supply Chain Management*, *152*, 88–101. https://doi.org/10.1016/j.ijpe.2013.12.014.

Brooks, A. L., Wang, S., & Jambeck, J. R. (2018). The Chinese import ban and its impact on global plastic waste trade. *Science Advances*. *4*(6), eaat0131. https://doi.org/10.1126/sciadv.aat0131.

BUND (2019). *Plastics Atlas 2019: Facts and figures about a world full of plastics*. Berlin: BUND, Heinrich-Böll-Stiftung. (2019). *https://www.boell.de/sites/default/files/2020-01/Plastic%20Atlas%202019%202nd%20Edition.pdf?dimension1=ds_plastic_atlas*.

China (2008). *Circular economy promotion law of the people's Republic of China*. Retrieved from (2008). http://www.fdi.gov.cn/1800000121_39_597_0_7.html.

Cong, R., Matsumoto, T., Li, W., Xu, H., Hayashi, T., & Wang, C. (2017). Spatial simulation and LCA evaluation on the plastic waste recycling system in Tianjin. *Journal of Material Cycles and Waste Management*, *19*(4), 1423–1436. https://doi.org/10.1007/s10163-016-0538-4.

Crang, M., Hughes, A., Gregson, N., Norris, L., & Ahamed, F. (2013). Rethinking governance and value in commodity chains through global recycling networks. *Transactions of the Institute of British Geographers, 38*(1), 12–24. https://doi.org/10.1111/j.1475-5661.2012.00515.x.

Feenstra, R. C., & Hanson, G. H. (2004). Intermediaries in Entrepôt trade: Hong Kong re-exports of Chinese goods. *Journal of Economics & Management Strategy, 13*(1), 3–35. https://doi.org/10.1111/j.1430-9134.2004.00002.x.

García-Pérez, J., Fernández-Navarro, P., Castelló, A., López-Cima, M. F., Ramis, R., Boldo, E., & López-Abente, G. (2013). Cancer mortality in towns in the vicinity of incinerators and installations for the recovery or disposal of hazardous waste. *Environment International, 51*, 31–44. https://doi.org/10.1016/j.envint.2012.10.003.

Hita, P. R., Pérez-Gálvez, F., Morales-Conde, M. J., & Pedreño-Rojas, M. A. (2018). Reuse of plastic waste of mixed polypropylene as aggregate in mortars for the manufacture of pieces for restoring jack arch floors with timber beams. *Journal of Cleaner Production, 198*, 1515–1525. https://doi.org/10.1016/j.jclepro.2018.07.065.

Lara-Lledó, M., Yanini, M., Araque-Ferrer, E., Monedero, F. M., & Vidal, C. R. (2018). *EU legislation on food contact materials.* Elsevier. https://doi.org/10.1016/B978-0-08-100596-5.21464-9.

Li, S. Y., Jiang, J. J., & Zhang, B. (2015). Research on environmental impact evaluation and recycling systematic assessment of plastic waste in China. *Applied Mechanics and Materials, 768*, 240–248. https://doi.org/10.4028/www.scientific.net/AMM.768.240.

Mansour, A. M. H., & Ali, S. A. (2015). Reusing waste plastic bottles as an alternative sustainable building material. *Energy for Sustainable Development, 24*, 79–85. https://doi.org/10.1016/j.esd.2014.11.001.

Pivnenko, K., Eriksen, M. K., Martín-Fernández, J. A., Eriksson, E., & Astrup, T. F. (2016). Recycling of plastic waste: Presence of phthalates in plastics from households and industry. *Waste Management, 54*, 44–52. https://doi.org/10.1016/j.wasman.2016.05.014.

Ragaert, K., Delva, L., & Van Geem, K. (2017). Mechanical and chemical recycling of solid plastic waste. *Waste Management, 69*, 24–58. https://doi.org/10.1016/j.wasman.2017.07.044.

Ragossnig, A. M., & Schneider, D. R. (2017). What is the right level of recycling of plastic waste? *Waste Management & Research, 35*(2), 129–131. https://doi.org/10.1177/0734242X16687928.

Seyring, N., Dollhofer, M., Weißenbacher, J., Bakas, I., & McKinnon, D. (2016). Assessment of collection schemes for packaging and other recyclable waste in European Union-28 member states and capital cities. *Waste Management & Research, 34*(9), 947–956. https://doi.org/10.1177/0734242X16650516.

Su, B., Heshmati, A., Geng, Y., & Yu, X. (2013). A review of the circular economy in China: Moving from rhetoric to implementation. *Journal of Cleaner Production, 42*, 215–227. https://doi.org/10.1016/j.jclepro.2012.11.020.

Tang, Z., Zhang, L., Huang, Q., Yang, Y., Nie, Z., Cheng, J., … Chai, M. (2015). Contamination and risk of heavy metals in soils and sediments from a typical plastic waste recycling area in North China. *Ecotoxicology and Environmental Safety, 122*, 343–351. https://doi.org/10.1016/j.ecoenv.2015.08.006.

Thomas, P., Rumjit, N. P., Lai, C. W., Johan, M. R. B., Saravanakumar, M. P., Hashmi, S., & Choudhury, I. A. (2020). *Polymer-recycling of bulk plastics* (pp. 432–454). Oxford: Elsevier. https://doi.org/10.1016/B978-0-12-803581-8.10765-9.

Tullo, A. H. (2019). Plastic has a problem; is chemical recycling the solution? *Chemical & Engineering News, 97*(39). Retrieved from (2019). https://cen.acs.org/environment/recycling/Plastic-problem-chemical-recycling-solution/97/i39.

Velis, C. A. (2014). *Global recycling markets—Plastic waste: A story for one player—China. Report prepared by FUELogy and formatted by D-waste on behalf of international solid waste association—Globalisation and waste management task force.* Vienna: ISWA. Retrieved from (2014). *https://www.iswa.org/fileadmin/galleries/Task_Forces/TFGWM_Report_GRM_Plastic_China_LR.pdf.*

Velis, C. A. (2015). Circular economy and global secondary material supply chains. *Waste Management & Research, 33*(5), 389–391. https://doi.org/10.1177/0734242X15587641.

Wagner, T. P. (2017). Reducing single-use plastic shopping bags in the USA. *Waste Management, 70*, 3–12. https://doi.org/10.1016/j.wasman.2017.09.003.

Wiesmeth, H., Shavgulidze, N., & Tevzadze, N. (2018). Environmental policies for drinks packaging in Georgia: A mini-review of EPR policies with a focus on incentive compatibility. *Waste Management & Research, 36*(11), 1004–1015. https://doi.org/10.1177/0734242X18792606.

Wilts, H. (2012). National waste prevention programs: Indicators on progress and barriers. *Waste Management and Research, 30*(9_suppl), 29–35. https://doi.org/10.1177/0734242X12453612.

Yong, R. (2007). The circular economy in China. *Journal of Material Cycles and Waste Management, 9*(2), 121–129. https://doi.org/10.1007/s10163-007-0183-z.

Yuan, Z., Bi, J., & Moriguichi, Y. (2006). The circular economy: A new development strategy in China. *Journal of Industrial Ecology, 10*(1–2), 4–8. https://doi.org/10.1162/108819806775545321.

Stephenson, W. (2018). Why plastic recycling is so confusing. (Original work published 2018). Retrieved April 22, 2020, from https://www.bbc.com/news/science-environment-45496884.

24

Textiles in a circular economy

Already in the 1950s, with textile plants mainly located in the industrialised countries, production of textiles raised concern regarding the many sources of waste. The focus at that time was clearly on pollution of rivers through biological treatment of textiles, with starch constituting "one of the most important single factors in textile wastes", although chemicals, including sometimes toxic substances, became relevant, in particular the use of large amounts of synthetic detergents (Ingols, 1958). Synthetic fibres were still in their early stages at that time.

In the 1970s, textile waste research "aimed at better control and prevention of chemical spills" and "more effective treatment" (Alspaugh, 1972) with numerous articles on textile waste problems and their possible solution. Clearly, chemicals continued to gain attention in textile production, including effluents of synthetic fibre production. However, probably not many people talked about potential problems with waste textiles and not just with textile wastes from production.

This changed gradually in the years thereafter. Groff (1992), referring to issues involving the manufacture of textiles, and addressing "the environmental impact of natural fibres production versus man-made fibre production",

mentions also "the use of natural resources, including land, energy and water", and points to methods of waste minimisation and reuse of process water. Moore and Ausley (2004) characterise this development with a focus on the United States (U.S.), and also "green manufacturing" had a significant role in textile industries (Alay, Duran, & Korlu, 2016). Textile production had therefore arrived in the year of the "Earth Summit" in Rio de Janeiro in 1992.

Globalisation of trade brought about significant changes for the textile and clothing industry. EEA (2014) points out that clothing consumption has increased significantly in recent decades, mainly due to comparatively low prices, but also to lifestyle changes ("fast fashion"). These developments can be attributed to changes in the value chain, also the removal of trade barriers, "making clothing one of the world's most traded manufactured products" (p. 107f). This significant increase in demand and production of textiles has led to similarly significant increases in waste textiles. Xu, Cheng, Liao, and Hu (2019) are reviewing the development in China, pointing to the obvious priority of recycling waste textiles.

Fig. 24.1 shows the situation with textile waste or, more precisely, waste textiles, in the

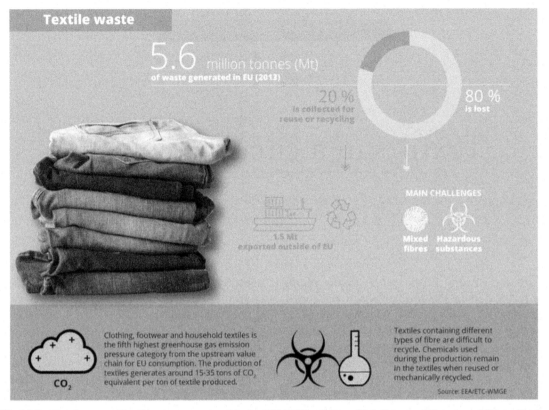

FIG. 24.1   Textile waste. Textiles waste generated in the EU (2013). © *From EEA. Resource Efficiency and Waste. Retrieved April 24, 2020 from https://www.eea.europa.eu/media/infographics/textile-waste/view (Original work published 2019).*

European Union (EU) with only 20% of textiles being collected separately for reuse and recycling, thus losing 80%.

With production of textiles moving mainly to developing countries, a great deal of the environmental impacts of textile manufacturing moved to these countries, too. The implementation of the "Globally Harmonised System of Classification and Labelling of Chemicals" (GHS) was encouraged by COP 8, the Conference of the Parties to the Kyoto Protocol in Johannesburg in 2002, and helped to adapt the safety assessment of chemicals and products in the context of international trade. The Swedish Chemicals Agency has identified in a "nonexhaustive list" around 1900 chemicals used during the production of clothing, of which 165 are classified in the EU as hazardous to health or the environment (see KEMI, 2013, p. 23). The Ellen MacArthur Foundation points to the greenhouse gas emissions from textiles production, which at 1.2 billion tonnes annually are more than those of all international flights and maritime shipping combined.

The greatest impact on health and the environment occurs in the countries where textiles are produced with process chemicals. Although there should be no measurable amounts in finished products, unfortunately, a fraction of these chemicals can sometimes be found – with direct (skin) and indirect (emission) exposure to consumers (see KEMI, 2013, p. 26).

Surprisingly, however, new risks arose with first attempts to implement a circular economy: commodities produced with recycled material, perhaps contaminated with additives or other potentially hazardous chemicals, could contain higher quantities of these chemicals (Friege, 2013). These risks increased with insufficient knowledge of the origin of certain materials in global trade. This issue, already covered in Section 11.2.2, played a role in the project RISKCYCLE (see also Bilitewski, 2012 and the contributions in Bilitewski, Darbra, & Barcel, 2011).

These observations, together with an increasing awareness about what is happening in the countries manufacturing textiles, brought the issue of textiles wastes and waste textiles back to the industrialised countries, complemented by many innovations in the field of technical textiles, which are of relevance in a variety of areas – including automotive and aircraft construction as wind turbines and e-textiles (Elmogahzy, 2020). And, one must not forget that products manufactured by means of technical textiles, will also reach their end of life at some point of time.

Undoubtedly, the production and use of textiles, both for clothing and textiles in commercial applications, can lead to environmental degradation and even endanger human health. This fact is aggravated by a sharp increase in demand for clothing ("fast fashion") on the one hand and important innovations in the field of technical textiles on the other. According to a briefing addressed to the European Parliament (EP, 2019), only a comparatively small share of waste textiles is collected and consigned to recycling with large differences between countries, and worldwide less than 1% of all materials used in clothing is recycled back into clothing.

There is therefore a certain need for environmental policies that better align the production and consumption of textiles with the objectives of a circular economy. The following section sets out the objectives of an "Integrated Environmental Policy" (IEP) for textile wastes and waste textiles.

24.1 Objectives of an IEP for textiles

With a view to the implementation of a circular economy, the waste hierarchy for textile wastes and waste textiles should be used as a benchmark for sustainable development in this context. However, particular attention must be paid to textile wastes, mainly from production, and waste textiles, mainly from consumption, which could play a different role in the objectives of an IEP.

One of the reasons is the significant increase in demand for textiles – both for clothing and for technical textiles. And, unlike the situation with packaging, it is difficult from a social point of view to reduce this "societal" demand. There are initiatives such as Green Strategy, "an innovative and dedicated consultancy specialised in sustainability and circularity issues of the global fashion and apparel industry", and there is, of course, the possibility of reusing garments, and there is the possibility for a design for environment (DfE), especially in terms of technical textiles, but breaking the habit of "fast fashion", to disrupt this societal habit, seems to remain a critical challenge, at least for some time. Moreover, reducing demand for garments could also have a significant impact on the economy of the textile producing countries – mainly developing countries.

Thus the task that has been discussed since the 1950s (Ingols, 1958) remains: the reduction of textile wastes. The task of increasing collection, reuse and recycling, perhaps also remanufacturing or reprocessing of old textiles, is gaining importance (Leal Filho et al., 2019), emphasising technology-guidance (Section 14.1.2).

24.1.1 Regulations for textile wastes

Efforts to reduce waste from textile production must be continued with a view on existing regulations and initiatives that provide an appropriate framework: REACH, GPP, and Green Chemistry.

REACH is a regulation of the EU concerning the "Registration, Evaluation, Authorisation and Restriction of Chemicals", which entered into force in 2007 and was adopted to improve the protection of human health and the environment from the risks that can be posed by chemicals. As "downstream users", textile manufacturers in the EU need to review their obligations when handling chemicals. Textile companies outside the EU are not directly bound by the obligation of REACH, even if they export their garments, etc., to the EU. However, the importers of these textiles have the responsibility to comply with the regulations of REACH. The European Chemicals Agency (ECNA) provides more detailed information on REACH.

Both the Environmental Protection Agency (EPA) of the U.S. and the Green Chemistry Institute of the American Chemical Society (ACS) have significantly promoted research and education to prevent pollution with chemicals, and to find innovative ways to reduce waste in production processes using chemicals and to replace hazardous substances. In contrast to REACH, Green Chemistry is an initiative, not an official regulation, and the first of the 12 Principles of Green Chemistry is prevention of waste, "as it is better to prevent waste than to treat or clean up waste after it has been created". To, Uisan, Ok, Pleissner, and Lin (2019) review recent trends in Green Chemistry and their applicability for textile wastes.

It remains to be seen, how and to what extent these regulations and initiatives can play a role in an IEP for textiles, in particular textile wastes. The general objectives of any environmental policy in this context should then be to control the risks associated with actually or potentially hazardous chemicals and additives in products and recovered material. This includes incentives to reduce the application of these chemicals wherever and whenever possible in the sense of a DfE, to recover used products and recycle chemicals and additives in an environmentally friendly way, and to control trade in products containing hazardous substances to countries without adequate recycling capacity.

24.1.2 Regulations for waste textiles

The EU green public procurement (GPP) criteria for textiles products and services, resulting from a voluntary initiative of the EU, "are designed to make it easier for public authorities to purchase goods, services and works with reduced environmental impacts". Interestingly, the criteria refer also to a take-back system including a collection system with post-collection sorting and segregation based on fibre, colour and condition of garment. Moreover, the technique of "life cycle costing" is considered to estimate the total cost of ownership for textile products including environmental externalities. These aspects could, of course, be part of an IEP for waste textiles and textile wastes, and GPP bridges the gap between production and consumption of textiles.

In the case of waste textiles, the focus of an IEP has to be on collecting old garments, but also old technical textiles – with a certain emphasis, of course, on reuse. As indicated above, changing societal habits such as "fast fashion" with its economic consequences especially for developing countries, seems to be difficult. This leads then to a focus on reprocessing and recycling of waste textiles, depending on the availability of suitable recycling technologies. If recycling certain waste textiles is profitable, a market-oriented approach could help to address this issue: potential profits could attract businesses and startups. Note that, in contrast to the situation with plastics recycling, companies' vested interests are in this case limited: besides diverting old textiles from reuse there are not many possibilities to increase the volume of waste textiles for recycling.

So far some thoughts on objectives of an IEP for textiles. The following subsection addresses collection and segregation of waste textiles which need to be strengthened for a successful IEP.

24.2 Collection of waste textiles

It is clear from the previous proposals of IEPs that a somewhat complete collection of waste commodities, in this case waste textiles, is important in order to create incentives for a DfE which helps to minimise the environmental impact of the products in a life cycle context. In addition, the collection of waste textiles is also important to reduce landfilling and incineration with other ecological challenges – such as additives that pollute the air, soil and water (Darbra et al., 2012). A study on the environmental improvement potential of textiles shows that increasing collection rates for reuse and recycling are among the most efficient options for reducing environmental impact of waste textiles (JRC, 2014).

24.2.1 Some empirical results

According to a study by the European Clothing Action Plan (ECAP), which focused on six members of the EU, there seem to be large differences: estimates point to 11% of used clothing and household textiles collected in Italy in 2015, but 75% in Germany in 2013 (see Watson, Aare, Trzepacz, & Petersen, 2018, Table 2). The authors of this study admit that there is some uncertainty in these figures, and that various factors, cultural differences, activities of charities and others, cause these differences. Nørup, Pihl, Damgaard, and Scheutz (2018) point first to the necessity to have a "uniform definition of textile waste and a stringent sorting procedure" as a precondition of investigating waste textiles (see also Watson et al., 2018, p. 17), and Nørup, Phil, Damgaard, and Scheutz (2019) assess residual household waste throughout Denmark for the quantity and quality of waste textiles – in order to get a clearer picture on textiles disposed of in the waste.

There are also major differences regarding physical collection methods and how collection

of used and waste textiles is organised, and the role municipalities have taken in this context. One important aspect seems to be transparency: what will happen with the collected garments. This may influence the quality of used textiles to be delivered to the containers.

All these different experiences gathered in different countries with their own collection system for used and waste textiles should be respected when setting up a more targeted collection system. This corresponds to Constitutive Element I of an IEP – respecting local conditions. Moreover, an IEP for textile wastes and waste textiles should take into account that manufacturing of clothing and household textiles is mainly located in developing countries and emerging economies, whereas technical textiles are the domaine of industrialised countries.

24.2.2 A take-back system for used and waste textiles

The collection of used and waste textiles is important, on the one hand, for the prevention of environmental impacts from landfilling, etc., and, on the other hand, for the implementation of the EPR principle. Therefore, a take-back system for used and waste textiles seems to be an appropriate proposal as a policy tool for an IEP. This is, for example, in accordance with the criteria of the EU GPP, the green public procurement recommendations of the EU. In France, there is Eco TLC, a not-for profit organisation, in fact the only PRO for clothing, household textiles and footwear accredited by the public authorities in France to cover this sector. Any company introducing these commodities on the French market under its own brand must either set up its own internal collection and recycling program, or pay Eco TLC to provide for collection and recycling. Some other multinational retailers of garments, H&M, for example, organise their own voluntary collection schemes.

The collection of old textiles naturally requires a certain technical and organisational infrastructure. There have to be official collection points, perhaps depending on the local situation, with some possibilities to separate waste textiles. In view of the EPR principle, manufacturers and importers of textiles should pay for collection. This will be discussed more carefully in the context of implementing the EPR principle. And, of course, waste technical textiles need as well be collected and recycled.

In order to additionally motivate consumers to returning used and old textiles, there could be a refund, if local conditions allow such a system. In order to prevent fraud, this necessitates labelling the textiles included in such a system – similar to a refund system for waste electrical and electronic equipment (WEEE) or a mandatory deposit system for drinks packaging. Again, some more details regarding a refund system will be provided in the context of implementing the EPR principle (see also Section 21.1.2).

The important task is now to implement the EPR principle and, in particular, to achieve the objectives of an IEP for textiles.

24.3 Implementing the EPR principle for textiles

As indicated, a holistic approach to the handling of chemicals and additives in the manufacture and use of textiles in general, and an IEP for textile waste and waste textiles in particular, should take into account that manufacturers and consumers are at least to some extent separate: production activities for clothing and household textiles, also environmentally sensitive ones, are often located in developing countries and emerging economies. Moreover, current policy approaches to textiles are most often limited to command-and-control policies based on regulations, with REACH the most prominent one. Authorised EPR schemes such as Eco TLC in France, are the exemption.

Voluntary initiatives, with the ACS Green Chemistry Institute for the industry and H&M for retailers as prominent examples among them, complete the picture.

In order to establish an IEP for textiles, it is necessary to briefly consider the constitutive elements of such a holistic policy.

Constitutive Element I: The local situation plays a role in terms of the technical and organisational details of the collection system, but also, of course, in terms of the targets for collecting and recycling waste textiles, perhaps also the targets referring to textile wastes from production in developing countries. In addition, the general conditions for producing textiles in developing countries must not be forgotten.

Constitutive Element II: Stakeholders in terms of the policy objectives are, of course, all consumers, producers and importers of textiles. The large group of consumers should be addressed through appropriate framework conditions that provide incentives for themselves and for other stakeholders to reduce all types of wastes and health hazards related to the production and consumption of textiles. For example, in addition to refunds for the return of used and waste textiles, consumers could be addressed through a labelling policy. It goes without saying that awareness-raising campaigns can also encourage consumers to act in an environmentally friendly manner when buying and using textiles.

Manufacturers and importers should, as usual, comply with various regulations in the form of command-and-control policies. First of all, there are the legal provisions of REACH for manufacturers and importers in the EU, and perhaps also the recommendations of Green Chemistry, which could be further developed into official environmental regulations. It could also be a good idea to bring REACH and the principles of Green Chemistry closer together.

Then, again as usual, manufacturers and importers of textiles will have to join a compliance scheme in order to licence the various categories of textiles. In return, the compliance scheme takes care of collection of waste textiles in accordance with the collection targets that correspond to their market share. The collected textiles are either reused or consigned to recycling. The compliance schemes are independent from manufacturers and importers and operate in a competitive environment.

Constitutive Element III: The final and decisive step is to link these various policy parts appropriately to the objectives of the IEP. This presupposes that the licence fees for the various types of textiles are specific and include the costs of collection, possible refunds, but perhaps also an ecological component, which therefore leads to fees that reflect the life cycle costs of textiles.

These "ecological" licence fees, when levied, have several important consequences. They are primarily reflected in the consumer prices of the various textiles, which may divert demand to products with lower life cycle costs. They can also motivate local manufacturers for production processes and products that reduce threats to human health and the environment. This motivation applies also to importers and manufacturers in developing countries: with ecological licence fees, importers will look for the cheaper, more environmentally friendly production processes and products. Thus, there is a clear motivation for a DfE to reduce textile wastes.

Manufacturers and importers with a DfE in this sense should be encouraged to affix a label, an eco-label to the garments, showing that they are, for example, "free of hazardous substances" or "produced in an environmentally friendly way" (Clancy, Fröling, & Peters, 2015). The report EEA (2014) highlights that labelling can allow "citizens to differentiate between products on the basis of their environmental and social impacts" (see p. 121). This label should be awarded by an independent research institute recognised at least in the EU to cover a significant part of global markets. It goes without saying that this label should refer to a website explaining what this label means and offering a complete list of additives that are still permitted. This list must be regularly updated in accordance with REACH or other official regulations in order to allow for new insight and scientific progress. Of course, manufacturers have to respond to changes in this list by either changing production or finishing processes or removing the label from their products.

The question arises as to whether this "advertising policy" could be strengthened by an explanation of the substances contained in the garments. However, this additional information could mean that it requires too much from ordinary consumers and therefore could prove to be less effective than the above-mentioned declaration. There is an interesting approach from circular.fashion with its "circularity.ID", which is created for approved garments and which can be attached as label.

The incentives associated with this policy approach are as follows: health-conscious consumers with a presumably high level of environmental awareness will quickly learn about this label and adjust their demand. Since their health could be affected by their purchases, the Tragedy of the Commons, which is prevalent in environmental decisions, is significantly weakened – if it matters at all.

Consumer decision-making is further influenced by relatively higher prices for "polluted" textiles, as these garments are subject to the licence fees, including a possible refund. As described in Section 21.1.2 for the case of WEEE, it might be useful to structure the refund system such that "clean" manufacturers can be compensated for temporarily higher production costs. The resulting demand effect will motivate manufacturers, including in developing countries, to switch to cleaner production technologies, at least gradually. See also Wiesmeth and Häckl (2015) for some more details on this policy proposal.

24.4 The waste hierarchy regarding textiles

The above considerations do not explicitly relate to the various parts of the waste hierarchy. For this reason, this section will address the main steps: prevention, reuse and recycling of waste textiles, respectively textile wastes, including some comments on the remanufacturing of waste textiles.

Prevention of Waste: As explained in the introductory remarks to this chapter, it seems appropriate to distinguish between the prevention of textile wastes, which come mainly from the manufacture of textiles, and the prevention of waste textiles, resulting from the consumption of textiles.

As far as textile wastes are concerned, the above-mentioned policy proposal motivates producers and importers for a DfE with less wastes and less use of hazardous substances, also in developing countries. This implies prevention of textile wastes and an aspect of sustainable development for countries that depend on textile production. The use of "ecological" licence fees will certainly lead to more serious efforts for a DfE. The revenue from the ecological fees could then be used, for example, to promote further innovations and recycling technologies. The compliance schemes that charge these fees have an interest in retaining their licensees and will therefore use such revenue appropriately. In this context, it is, of course, important that there is a competitive system of for-profit compliance schemes.

The situation is different for waste textiles. Despite efforts such as "Circular Fashion" and related measures, it will be difficult to redirect societal paths towards more sustainable consumption of textiles, although there are clear ways to prevent waste by redesigning certain garments. Strengthening tools, such as the EU's GPP, can provide additional incentives.

In this context, the example of Eco TLC in France must be mentioned. This EPR system refers to clothing and household textiles, also points to eco-designs, but mainly deals with collection and recycling. To encourage participating companies to use recycled fibres from waste clothing, Eco TLC has introduced some variability in the financial contributions. However, Eco TLC is only one non-profit compliance scheme with incentives which, as already mentioned (see Section 20.4.3), do not always have to be in line with the objectives of the IEP. In addition, companies can set up and organise their own collection and recycling system, which corresponds to the "individual" schemes discussed previously (see Section 17.4.2), including the incentive issues associated with such schemes. It remains to be seen, to what extent and how this compliance scheme will affect textile production and use in France.

This "fashion aspect" does not apply to technical textiles in the same way. But the demand for technical textiles is rising due to a supply effect associated with a steady stream of innovations that are also related to various environmental issues (Yasin & Sun, 2019). While there are probably many opportunities for a DfE, the rapidly growing volume of these technical textiles is likely to "overcompensate" most of these efforts. For all these reasons, it might therefore be better to focus more, at least for the time being, on the collection and recycling of old textiles, including technical textiles, or on the possibilities of reuse.

Reusing Clothes: Second-hand shops and shops renting out evening attire, charitable organisations such as the Red Cross or the Salvation Army, and other physical market places for used clothes have been around for a long time. With the digital revolution most of them gained and continue to gain even more importance: e-commerce also offers platforms for second-hand clothing, including for luxury clothing (Orlean, 2019). Meanwhile there are even global trade networks for "the large-scale international trade in second-hand clothes which are exported from the Global North to Africa" (Brooks, 2013), which raises some environmental concerns given transport activity. Also Cruz-Cárdenas, Guadalupe-Lanas, and Velín-Fárez (2019) investigate factors that influence the reuse of clothing in Ecuador. In contrast to the situation about 100 years ago, however, the spread of infectious diseases through second-hand clothing (Joseph, 1915) should no longer play a role today, although the Corona crisis seems to raise awareness of this issue again.

The question of interest in this context is whether the implementation of the EPR principle proposed above promotes the reuse of garments and thus helps to prevent additional waste. The EPR system with licensing of the various types of textiles, in particular with ecological licence fees, will lead to higher prices for clothing, especially for problematic garments. In addition, the price of used clothes will fall as motivated consumers return more old clothes to the collection points, which could also be stimulated by a refund. This effect is enhanced by returning more clothing, which is often disposed of before the end of its life, to the collection stations. The task of the collection points is to sort the returned clothes and send the reusable ones to appropriate destinations. In addition, there is likely to be more activity on online platforms for second-hand clothing, which will also increase the supply of used clothing. It is therefore expected that the reuse of clothing will become more important in near future.

However, there is a problem that needs to be taken into account: as mentioned above, according to the report (Ellen MacArthur Foundation, 2017), a large proportion of the reusable clothing collected in Europe and the U.S. is sold to textile retailers, who also resell it in developing countries (see p. 105). Similarly to WEEE, this may lead to saturated markets with further environmental problems. The

above proposed EPR system should be carefully monitored in this regard, as vested interests of the compliance schemes can play a role (see also Section 20.4.2).

The reuse of old textiles is sometimes understood to reuse the various fibres. In their review, Sandin and Peters (2018) provide a clear definition of textile reuse and recycling and analyse the relevant literature in this context. This is obviously comparable to "remanufacturing" addressed already in Section 21.1.4. Fig. 24.2 points to the many applications of remanufactured or recycled textile waste.

Recycling Waste Textiles: A large proportion of old textiles is deposited or incinerated worldwide (see JRC, 2014, p. 57, for the case of the EU). In addition, textiles collected for reuse are often resold to developing countries. Thus, recycling of waste textiles in industrialised countries generally does not seem to be a profitable business, although the literature points to "socioeconomic, environmental and ecological advantages brought about by recycling textiles" (Leal Filho et al., 2019). In their "account of the textile waste policy in China" from 1991 to 2017, Xu et al. (2019) highlight the dominance of recycling regulations.

Nevertheless, as has already been mentioned and discussed, it is currently difficult to attract a larger proportion of consumers to "sustainable fashion", although there are a number of initiatives such as "circular fashion" with their core principles, including a "design for recycling". However, JRC (2014) points to some figures (see p. 121): "across Europe, about 20% of the clothing waste is collected, of which about 40% are reused, 50% are recycled, and 10% are disposed of by incineration or landfilling" (see also Fig. 24.1). Consequently, the recycling of textiles must be given a boost. The IEP introduced in the last section supports such a development.

If textile manufacturers and importers have to licence their materials with a compliance scheme, and if the compliance schemes in

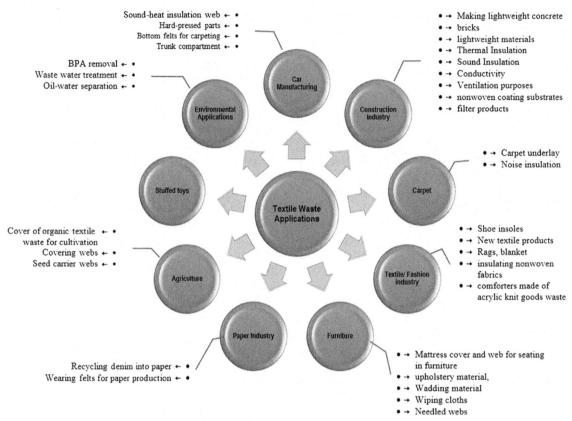

FIG. 24.2 Applications of textile wastes and waste textiles. The graphical abstract shows the large variety of applications of textile wastes and of remanufactured or recycled waste textiles. © *From Shirvanimoghaddam, K., Motamed, B., Ramakrishna, S., & Naebe, M. (2020). Death by waste: Fashion and textile circular economy case. Science of the Total Environment, 718. https://doi.org/10. 1016/j.scitotenv.2020.137317 (web archive link).*

competition face specific collection and recycling targets, then they will have to support the development of textile recycling by commissioning the services of such businesses, thereby encouraging further innovations in this area. Thus, a "profitable" recycling business can develop if compliance schemes are prevented from exporting too much waste textiles to third countries.

In this way, the IEP for textiles creates space for private companies to recycle waste textiles. At the same time, the licence fees, which depend

also on the amount of waste textiles collected and the recycling costs, motivate producers for a DfE, and the resulting higher prices could also drive demand towards more sustainable textiles. Depending on the local situation, a refund for old textiles could help to increase the collection and thus strengthen this link.

This proposal for a holistic approach to textiles illustrates the essential part of such an IEP: appropriate involvement of all stakeholders and linking their decisions to the objectives of the policy.

References

Alay, E., Duran, K., & Korlu, A. (2016). A sample work on green manufacturing in textile industry. *Sustainable Chemistry and Pharmacy, 3,* 39–46. https://doi.org/10.1016/j.scp.2016.03.001.

Alspaugh, T. A. (1972). Textile waste. *Journal of the Water Pollution Control Federation, 44*(6), 1081–1088. Retrieved from http://www.jstor.org/stable/25037501.

Biletewski, B. (2012). The circular economy and its risks. *Waste Management, 32*(1), 1–2. https://doi.org/10.1016/j.wasman.2011.10.004.

Biletewski, B., Darbra, R. M., & Barcel (2011). Global risk-based management of chemical additives I: Production, usage and environmental occurrence. In *The handbook of environmental chemistry.* https://doi.org/10.1007/978-3-642-24876-4.

Brooks, A. (2013). Stretching global production networks: The international second-hand clothing trade. *Global Production Networks, Labour and Development, 44,* 10–22. https://doi.org/10.1016/j.geoforum.2012.06.004.

Clancy, G., Fröling, M., & Peters, G. (2015). Ecolabels as drivers of clothing design. *Journal of Cleaner Production, 99,* 345–353. https://doi.org/10.1016/j.jclepro.2015.02.086.

Cruz-Cárdenas, J., Guadalupe-Lanas, J., & Velín-Fárez, M. (2019). Consumer value creation through clothing reuse: A mixed methods approach to determining influential factors. *Journal of Business Research, 101,* 846–853. https://doi.org/10.1016/j.jbusres.2018.11.043.

Darbra, R. M., Dan, J. R. G., Casal, J., Àgueda, A., Capri, E., Fait, G., … Guillén, D. (2012). *Additives in the textile industry* (pp. 83–107). Berlin, Heidelberg: Springer Berlin Heidelberg. https://doi.org/10.1007/698_2011_101.

EEA (2014). *Environmental Indicator report: Environmental impacts of production-consumption systems in Europe.* Luxembourg: Publications Office of the European Union. https://doi.org/10.2800/22394.

Ellen MacArthur Foundation (2017). *A new textiles economy: Redesigning fashion's future. Retrieved from (2017).* https://www.ellenmacarthurfoundation.org/assets/downloads/publications/A-New-Textiles-Economy_Full-Report_Updated_1-12-17.pdf.

Elmogahzy, Y. E. (2020). 14—Performance characteristics of technical textiles: Part I: E-textiles. In *The Textile Institute Book Series* (pp. 347–364). Woodhead Publishing. https://doi.org/10.1016/B978-0-08-102488-1.00014-9.

EP. (2019). Environmental impact of the textile and clothing industry. Retrieved April 23, 2020, from https://www.europarl.europa.eu/RegData/etudes/BRIE/2019/633143/EPRS_BRI(2019)633143_EN.pdf (Original work published 2019)

Friege, H. (2013). Waste—Valuables—Secondary resources—Contaminants—Waste again? *Environmental Sciences Europe. 25*(1). https://doi.org/10.1186/2190-4715-25-9.

Groff, K. A. (1992). Textile waste. *Water Environment Research, 64*(4), 425–429. Retrieved from http://www.jstor.org/stable/25044177.

Ingols, R. S. (1958). Textile waste problems. *Sewage and Industrial Wastes, 30*(10), 1273–1277. Retrieved from https://www.jstor.org/stable/25033718.

Joseph, G. W. N. (1915). Some remarks on second-hand clothing and the spread of infectious disease. *Public Health, 28*(4), 267–270. https://doi.org/10.1016/S0033-3506(15)80779-3.

JRC (2014). Environmental improvement potential of textiles (IMPRO textiles). In *JRC scientific and policy reports.* Luxembourg: Publications Office of the European Union. https://doi.org/10.2791/52624.

KEMI (2013). *Hazardous chemicals in textiles report of a government assignment.* Stockholm: Swedish Chemicals Agency. Retrieved from (2013). https://www.kemi.se/global/rapporter/2013/rapport-3-13-textiles.pdf.

Leal Filho, W., Ellams, D., Han, S., Tyler, D., Boiten, V. J., Paço, A., … Moora, H. (2019). A review of the socio-economic advantages of textile recycling. *Journal of Cleaner Production, 218,* 10–20. https://doi.org/10.1016/j.jclepro.2019.01.210.

Moore, S. B., & Ausley, L. W. (2004). Systems thinking and green chemistry in the textile industry: Concepts, technologies and benefits. *SME's and Experiences with Environmental Management Systems, 12*(6), 585–601. https://doi.org/10.1016/S0959-6526(03)00058-1.

Nørup, N., Pihl, K., Damgaard, A., & Scheutz, C. (2018). Development and testing of a sorting and quality assessment method for textile waste. *Waste Management, 79,* 8–21. https://doi.org/10.1016/j.wasman.2018.07.008.

Nørup, N., Pihl, K., Damgaard, A., & Scheutz, C. (2019). Quantity and quality of clothing and household textiles in the Danish household waste. *Waste Management, 87,* 454–463. https://doi.org/10.1016/j.wasman.2019.02.020.

Orlean, S. (2019). *The RealReal's online luxury consignment shop*: (pp. 28–32). The New Yorker. (10/21). Retrieved from https://www.newyorker.com/magazine/2019/10/21/therealreals-online-luxury-consignment-shop.

Sandin, G., & Peters, G. M. (2018). Environmental impact of textile reuse and recycling—A review. *Journal of Cleaner Production, 184,* 353–365. https://doi.org/10.1016/j.jclepro.2018.02.266.

To, M. H., Uisan, K., Ok, Y. S., Pleissner, D., & Lin, C. S. K. (2019). Recent trends in green and sustainable chemistry: Rethinking textile waste in a circular economy. *Bioresources, Biomass, Bio-Fuels and Bioenergies, 20,* 1–10. https://doi.org/10.1016/j.cogsc.2019.06.002.

Watson, D., Aare, A. K., Trzepacz, S., & Petersen, C. D. (2018). *Used textile collection in European cities. Study commissioned by Rijkswaterstaat under the European clothing action plan (ECAP)*. Retrieved from http://www.ecap.eu.com/wp-content/uploads/2018/07/ECAP-Textile-collection-in-European-cities_full-report_with-summary.pdf.

Wiesmeth, H., & Häckl, D. (2015). Integrated environmental policy: Chemicals and additives in textiles. *Waste Management*, 46, 1–2. https://doi.org/10.1016/j.wasman.2015.10.028.

Xu, C., Cheng, H., Liao, Z., & Hu, H. (2019). An account of the textile waste policy in China (1991–2017). *Journal of Cleaner Production*, 234, 1459–1470. https://doi.org/10.1016/j.jclepro.2019.06.283.

Yasin, S., & Sun, D. (2019). Propelling textile waste to ascend the ladder of sustainability: EOL study on probing environmental parity in technical textiles. *Journal of Cleaner Production*, 233, 1451–1464. https://doi.org/10.1016/j.jclepro.2019.06.009.

PART VI

Concluding remarks

25

Circular economy – A summary in times of corona

This chapter attempts to briefly review some important aspects regarding the implementation of a circular economy. According to the concept of the circular economy introduced by Pearce and Turner (1989), it is decisive to respect and sustainably preserve the fundamental functions of the environment: supplier of natural resources, receiver of waste, and direct provider of utility.

The first section highlights the role of respecting the waste hierarchy in relation to a sustainable development in general and to the fundamental functions of the environment in particular. The section following thereafter provides a review of the characteristics of the different IEPs drafted in the chapters of Part V. The Corona crisis undoubtedly influences the implementation of a circular economy in a manifold of ways. Some of this aspects are presented in the final section together with a short outlook.

25.1 Sustainability and the implementation of a circular economy

Sustainability has been addressed in various chapters of the book and in various contexts. Similar to a circular economy it is perceived differently depending again on local conditions – the economic, demographic, geographic, and climatic conditions of a country or a region. Some authors, such as Korhonen, Honkasalo, and Seppälä (2018) relate a "successful" circular economy to the social, economic and environmental dimension of sustainable development.

In this context, it is therefore difficult to think about specific policies enhancing sustainable development. Interestingly, Walnum, Aall, and Løkke (2014) investigate whether rebound effects can explain the "manifold negative environmental impacts of transportation", which "are an important contributor to the so-far non-sustainable development in financially rich areas of the world". They arrive at the conclusion that sustainable mobility, to be achieved through gains in efficiency, substitution, and/or volume reduction, is affected by rebound effects that are more generally defined (p. 9512).

Nevertheless, sustainability is of relevance for "smart cities", for "green development", for "circular fashion" and various other concepts, elsewhere discussed in the book – there clearly is a strong technology-guidance. Economy-guidance is, however, necessary for supporting the development of these technological concepts, which are, after all, dependent on the appropriate involvement of the stakeholders and their knowledge.

And sustainability is relevant also for saving natural resources, but also for mining them in a "sustainable way". This aspect needs to be respected in an IEP in this context, perhaps supported through some additional measures – restricting trade in waste products or promoting fair trade concepts.

These remarks point again to close relationships between sustainability and concepts addressed in this book. It seems that a "sustainable" implementation of the waste hierarchy, with greenhouse gases also considered as "waste", contributes significantly to the goal of a sustainable development, at least as far as the environmental and economic dimension is concerned.

25.2 Sustainability and the waste hierarchy

Respecting and maintaining the assimilative capacity of the environment implies in particular to prevent waste and thus save resources, through a design for environment (DfE), for example. Therefore "sustainably" observing the waste hierarchy with the priority goal of waste prevention is a decisive part of implementing a circular economy, in particular for resource-poor countries such as Germany (Friege & Dornack, 2019).

In more details: sustainably respecting the waste hierarchy certainly helps to maintain the assimilative capacity of the environment, it helps to save natural resources. Also the task of the environment as direct provider of utility is to some extent related to the other two: extensive mining activities and landfilling waste can have an impact on an otherwise beautiful landscape or undisturbed view.

For this reason, there is a clear focus in this book on circular economy strategies related to the waste hierarchy. The question arises, how to implement the waste hierarchy? How to allocate the environmental commodities? Especially this last question needs an appropriate answer. Unfortunately, economic theory cannot provide a mechanism in this context, which is comparable to the market mechanism. Due to external effects and public good properties of the environmental commodities, the market mechanism cannot simply be extended to the case of implementing a circular economy. Incomplete and asymmetric information rule out a straightforward "economy-guidance".

Consequently, the possibility of a "technology-guidance", occasionally raised by proponents of industrial ecology, an important root of the circular economy, should be taken into account. However, a detailed analysis reveals a variety of issues, which are also related to informational problems.

What remains is an attempt to combine economy-guidance and technology-guidance through holistic policies which respect the local situation, also the technological environment of a circular economy. In order to profit from outstanding properties of the market mechanism, in particular the aspect of decentralised decision-making, these policies, composed of different tools, are characterised by certain "constitutive elements". Besides the locality principle, stakeholders need to be adequately addressed and involved, and decisions need to be linked, also to the objectives of the policy.

This explains the structural properties of the "Integrated Environmental Policies" (IEP) drafted, or rather "designed" in Part V. Again, due to the non-availability of a general allocation mechanism each environmental area requires a somewhat different approach.

The following review of important specifications of these policies helps to design adequate policies for environmental areas not considered in this book.

25.3 IEPs for the implementation of a circular economy

This section provides a brief review of the features of IEPs designed in Part V. Emphasis is given to the challenges in dealing with "waste" in the various areas, with shortcomings of current policies, with cornerstones of an IEP, including possible additional measures, and critical aspects, which should be monitored. Mechanisms such as the Tragedy of the Commons and the Prisoners' Dilemma, as well as vested interests of stakeholders cannot always be completely excluded with these policies.

25.3.1 Packaging waste

Challenges in dealing with packaging waste in the context of a circular economy:

- Societal path dependencies influence the perception of waste and waste prevention.

Shortcomings of current policies in this area:

- There is a focus on the recycling of packaging waste, less on prevention.
- Vested interests of producers tend to reduce incentives for a DfE and thus for waste prevention.

Proposal for an IEP:

- Separate collection of packaging waste can be based on laissez-faire with technology-guidance, accompanied by awareness raising campaigns.

- The EPR principle should be implemented through independent for-profit compliance schemes in competition.

Possible additional measures:

- The announcement of possible ecological licence fees, for example on plastic packaging, could help to change the mindset and reduce packaging.

Critical aspects:

- Vested interests of compliance schemes must be monitored, as must be trade in packaging waste, particularly with developing countries.

25.3.2 Waste electrical and electronic equipment

Challenges in dealing with waste electrical and electronic equipment (WEEE) in the context of a circular economy:

- Societal path dependencies influence trade in WEEE.
- Societal path dependencies influence the perception of "reuse" of old equipment.

Shortcomings of current policies in this area:

- There is an insufficient consideration of local conditions in the countries, which import WEEE.
- Vested interest of manufacturers tends to reduce incentives for a DfE and thus also for waste prevention.

Proposal for an IEP:

- Separate collection of WEEE can be based on laissez-faire with technology-guidance, preferably in combination with a refund, and accompanied by awareness raising campaigns.
- The EPR principle should be implemented through independent for-profit compliance schemes in competition.

Possible additional measures:

- Trade with WEEE should be further restricted.

Critical aspects:

- Vested interests of compliance schemes must be monitored, as must be trade in WEEE, particularly with developing countries.

25.3.3 End-of-life vehicles

Challenges in dealing with end-of-life vehicles (ELV) in the context of a circular economy:

- Societal path dependencies influence trade in second-hand cars.
- Societal path dependencies influence the perception of the "reuse" of second-hand cars.

Shortcomings of current policies in this area:

- Export of a sizeable number of used cars to developing and emerging countries with likely insufficient possibilities of proper maintenance.
- Vested interest of manufacturers tends to reduce incentives for a DfE and thus also for waste prevention.

Proposal for an IEP:

- The EPR principle should be implemented by means of independent for-profit compliance schemes in competition.

Possible additional measures:

- The export of used cars in developing countries should be monitored.
- Maintenance and recycling centres should be established in import countries with support from car manufacturers.

Critical aspects:

- Societal path dependencies in terms of reuse and trade must be taken into account.
- Remanufacturing that is gaining in importance needs to be observed.

25.3.4 Climate change

Challenges in dealing with climate change in the context of a circular economy:

- Vastly different local conditions have to be considered for this global environmental commodity.
- The Prisoners' Dilemma and an insufficient perceived feedback from actions mitigating climate change influence decisions.

Shortcomings of current policies in this area:

- Current policies are not successful in integrating enough countries with sufficient efforts to mitigate climate change.

Proposal for an IEP:

- The EPR principle should be implemented through a global market for certificates or appropriately linked national markets.
- Campaigns to raise perceived feedback from efforts to mitigate climate change should accompany these measures.

Critical aspects:

- The integration of sufficiently many countries will remain a difficult task for some time, perhaps till the first countries suffer serious damage.
- The "waste hierarchy" regarding greenhouse gases has so far focused mainly on prevention; the "reuse" and "recycling" of CO_2 in chemical and pharmaceutical processes, etc., as well as "landfilling" in the form of carbon capture and storage (CCS), do not yet play a major role in policies.

25.3.5 Plastic waste

Challenges in dealing with plastics and plastic waste in the context of a circular economy:

- Plastics and plastic waste have both a local and a global dimension.

- Plastic products offer many benefits for consumers and producers and will therefore retain or even increase their importance in everyday life.

Shortcomings of current policies in this area:

- Different local conditions have to be considered in dealing with the global dimension of plastic waste.
- Societal path dependencies influence trade in plastic waste.
- Societal path dependencies tend to favour the recycling of plastic waste.

Proposal for an IEP:

- Separate collection of plastic waste can be based on laissez-faire with technology guidance, perhaps refund for larger items, or deposit system for special items such as plastic bottles.
- The EPR principle should be implemented through independent for-profit compliance schemes in competition.

Possible additional measures:

- Trade with plastic waste should be further restricted.

Critical issues:

- Societal path dependencies regarding trade with plastic waste must be taken into account.

25.3.6 Textile waste

Challenges in dealing with textile wastes and waste textiles in the context of a circular economy:

- Production of textiles generating textile wastes is concentrated in developing countries.
- Although there is a tradition of second-hand clothes, the concept of "fast fashion" generates a large quantity of waste textiles.

Shortcomings of current policies in this area:

- There are mainly command-and-control regulations (REACH, Procurement Policy, etc.), but no holistic policy approaches.

Proposal for an IEP:

- Separate collection of waste textiles can be based on laissez-faire with technology guidance, possibly with a refund for larger items.
- The EPR principle should be implemented through independent for-profit compliance schemes in competition.

Possible additional measures:

- The focus should be on the creation of markets for the reuse of second-hand clothing.
- Trade in old and waste textiles should be monitored, perhaps even restricted.

Critical issues:

- The situation with textile waste particularly in developing countries should be monitored.
- Societal trends such as "fast fashion" should be accompanied by awareness-raising campaigns.

So far this short review of the IEPs developed in the chapters of Part V. Clearly, some of these policies do overlap. An IEP for plastic waste, for example, also touches an IEP for textiles, or an IEP for packaging addresses also plastic packaging. These overlaps need to be taken into account when some of these policies are to be implemented at the same time.

25.4 Circular economy in times of Corona

An interesting observation refers to some traditional societal path dependencies, which now appear to be weakened in times of Corona. The constraints on public life brought significant parts of the economies to a standstill – parts which are also of relevance for a circular economy. We have to mention tourism and all kinds of business trips, but also many sectors that depend on logistics and international trade.

Of course, the specific constraints led to this situation supported by an individually perceived feedback: people largely understood the necessity of these restrictions because their own health is at risk. Therefore, there seems to be a close correlation between the constraints, perceived feedback and weaker societal path dependencies. However, it is questionable, whether the initial situation will be restored after Corona in a short period of time.

With respect to international trade, the vulnerability of the global network of trade relations including logistics became visible, especially for car manufacturers. Accordingly, there might be some changes in near future reducing the dependence on international trade. But this remains to be seen.

For some time after the discovery of the threat through the Corona virus environmental issues disappeared from the public agenda. This, in turn, is a direct consequence of a perceived feedback: concern for personal health or employment seemed more important than mitigating a more abstract threat through environmental degradations such as plastic waste or the climate change. Such an observation is well-known: an economic crisis tends to put environmental issues in the background. However, this is problematic for the further implementation of a circular economy.

25.4.1 Developments in favour of a circular economy

What the Corona crisis shows is that changes in societal path dependencies are possible when there are overarching developments which influence the perceived feedback of one's own actions. Unfortunately, this is little more than a confirmation of what has already been experienced: the reaction to the almost complete

deforestation of Saxony some 300 years ago is proof of that (see Section 6.1.1).

The conclusions regarding global efforts to mitigate climate change and also to stop the pollution of rivers and seas with plastic waste are clear: these efforts will then be more successful, when it is possible to establish such a perceived feedback. With respect to plastic waste the issue of micro-plastics showing up in the food chain might turn out to be a tipping point. More and more people start to understand the health risks associated with plastic waste (Peng et al., 2020).

With respect to climate change, the situation seems still to be more complicated. In addition, there is the long period of time between an action and the feedback. Nevertheless, there are some ideas in the European Union (EU) for restructuring the economy with a view to the European Green Deal. It remains to be seen to what extent environmental issues can be successfully combined with the necessary efforts to bring the economies back to full life.

The reduced travel activities for businesses imply more online conferences. The digital transformation could therefore gain from the Corona crisis. Again, it remains to be seen, if this observation can be used for sustainable development. At least, there are efforts not to return to the status quo ante.

To sum up, there now seem to be more initiatives to take the chance for a new orientation of the economy, not to return to the old paths, also to respect the necessities of a circular economy. In this sense, the Corona crisis poses additional challenges, but it also opens up new opportunities for the implementation of a circular economy. However, these opportunities must be seized.

25.5 Concluding remarks

The implementation of a circular economy remains a challenging task. This book attempts to introduce a certain structure into this endeavour, in particular it attempts to explain the relationship between economics and the technological environment, and it attempts to highlight the importance of individual motivation.

The necessary "systems change", often referred to for the implementation of a circular economy, does not seem to be easy to achieve. Our economies are based on the market system, in particular on the profitability of economic activity, on the mutual benefits of international trade. However, the introduction of a circular economy sometimes requires a somewhat more general view. The systems change therefore refers to necessary changes in these societal path dependencies, without, of course, forgetting the positive characteristics of the market mechanism.

References

Friege, H., & Dornack, C. (2019). *Abfall- und Kreislaufwirtschaft: Prioritäten für nachhaltiges Resource management*: (pp. 593–611). Berlin, Heidelberg: Springer Berlin Heidelberg. https://doi.org/10.1007/978-3-662-57693-9_31.

Korhonen, J., Honkasalo, A., & Seppälä, J. (2018). Circular economy: The concept and its limitations. *Ecological Economics*, *143*, 37–46. https://doi.org/10.1016/j.ecolecon.2017.06.041.

Pearce, D. W., & Turner, R. K. (1989). *Economics of natural resources and the environment.* Johns Hopkins University Press.

Peng, L., Fu, D., Qi, H., Lan, C. Q., Yu, H., & Ge, C. (2020). Micro- and nano-plastics in marine environment: Source, distribution and threats—A review. *Science of the Total Environment*, *698*, 134254. https://doi.org/10.1016/j.scitotenv.2019.134254.

Walnum, H. J., Aall, C., & Løkke, S. (2014). Can rebound effects explain why sustainable mobility has not been achieved? *Sustainability (Switzerland)*, *6*(12), 9510–9537. https://doi.org/10.3390/su6129510.

Index

Note: Page numbers followed by *f* indicate figures and *t* indicate tables.